Research in Mathematics Education

Series editors

Jinfa Cai, Newark, DE, USA
James A. Middleton, Tempe, AZ, USA

This series is designed to produce thematic volumes, allowing researchers to access numerous studies on a theme in a single, peer-reviewed source. Our intent for this series is to publish the latest research in the field in a timely fashion. This design is particularly geared toward highlighting the work of promising graduate students and junior faculty working in conjunction with senior scholars. The audience for this monograph series consists of those in the intersection between researchers and mathematics education leaders—people who need the highest quality research, methodological rigor, and potentially transformative implications ready at hand to help them make decisions regarding the improvement of teaching, learning, policy, and practice. With this vision, our mission of this book series is:

1. To support the sharing of critical research findings among members of the mathematics education community;
2. To support graduate students and junior faculty and induct them into the research community by pairing them with senior faculty in the production of the highest quality peer-reviewed research papers; and
3. To support the usefulness and widespread adoption of research-based innovation.

More information about this series at http://www.springer.com/series/13030

Kelly S. S. Mix • Michael T. Battista
Editors

Visualizing Mathematics

The Role of Spatial Reasoning in Mathematical Thought

Springer

Editors

Kelly S. S. Mix
Department of Human Development
and Quantitative Methodology
University of Maryland
College Park, MD, USA

Michael T. Battista
Department of Teaching and Learning
The Ohio State University
Columbus, OH, USA

ISSN 2570-4729 ISSN 2570-4737 (electronic)
Research in Mathematics Education
ISBN 978-3-319-98766-8 ISBN 978-3-319-98767-5 (eBook)
https://doi.org/10.1007/978-3-319-98767-5

Library of Congress Control Number: 2018958712

This Springer imprint is published by the registered company Springer Nature Switzerland AG
The registered company address is: Gewerbestrasse 11, 6330 Cham, Switzerland

Foreword

In 2016, we were approached by series editor, Dr. Jinfa Cai, with a novel idea—invite authors from the fields of developmental psychology and mathematics education to write about their work on spatial visualization and mathematics, and then ask them to write commentaries on one another's chapters. The goal was to provide a unique view of research on this topic that encompassed both disciplines, as well as foster cross-field communication and intellectual synergy. We eagerly took up the challenge and invited scholars whose work we knew to be at the forefront of our respective fields. The chapters and commentaries contained in this volume are the products of this esteemed group. They reflect the state of the art in research on spatial visualization and mathematics from at least two perspectives. They highlight important new contributions, but they also reveal the fault lines between our respective disciplines. The commentaries insightfully point out some of these fault lines, as well as the immense common ground and the potential for deeper collaboration in the future.

The basic question of how spatial skill relates to mathematics has received steady attention over the years. In psychology, most of this work has focused on long-term outcomes in STEM fields for individuals with more advanced spatial skill (e.g., Wai, Lubinski, & Benbow, 2009), the possibility that spatial deficits contribute to poor mathematics outcomes in children (e.g., Geary, Hoard, Byrd-Craven, Nugent & Numtee, 2007), and the use of materials that physically embody (via spatial relations) abstract mathematics concepts (see Mix, 2010, for a review). Running through these disparate research programs is the shared notion that spatial thinking plays a major role in understanding mathematics, but it has not been addressed head on in psychology until recent years.

In mathematics education, Clements and Battista, in their 1992 research review, address just this issue. They wrote that both Hadamard and Einstein (renown mathematicians) claimed that much of the thinking required in higher mathematics is spatial, and they cited positive correlations between spatial ability and mathematics achievement at all grade levels. However, even in that time period, the relations between spatial thinking and learning nongeometric concepts did not seem straightforward, and there were conflicting findings. For some tasks, having high-spatial skill seemed to improve performance, whereas in other tasks, processing mathematical information using verbal-logical reasoning enhanced performance compared to students who processed the information visually. Other mathematics education researchers countered that the understanding of some low-spatial students who did well in mathematics was instrumental, whereas high-spatial students' understanding was more relational, a difference often not captured by classroom or standardized assessments. Clements and Battista concluded that even though there was reason to believe that spatial reasoning is important in students' learning and use of mathematical concepts—including nongeometric concepts—the role that such reasoning plays in this learning remained elusive.

Possibly because of this elusiveness, interest in the topic waned in mathematics education. However, currently there is intense interest in this general topic in both psychology and mathematics education due to its potential educational benefits (Newcombe, 2010) and the insights into the relations found through extensive and detailed student interviews (Bruce et al., 2017; Davis et al., 2015). The chapters contributed to this volume represent various approaches to advancing this work in education or moving the work in both fields toward educational application.

The developmental psychology chapters tended to focus on the underlying mental representations used to understand mathematics, and the extent to which these representations already involve, or could be improved by spatial processing. Cipora, Schroeder, Soltanlou, and Nuerk provide a detailed analysis of the link between spatial and numerical processing purportedly demonstrated by spatial-numerical association (SNA) or mental number line effects. They conclude that spatial skills provide a crucial tool for understanding mathematics, but this relation may not be realized in the form of a fixed mental number line. Congdon, Vasileyva, Mix, and Levine examine a deep psychological structure that may underlie a range of mathematics topics—namely, the structure involved in identifying and enumerating spatial units of measurement. They argue that mastery of this structure has the potential to support mathematics learning throughout the elementary grades and perhaps head off misconceptions related to fractions, proportions, and conventional later on. Similarly, Jirout and Newcombe focus on another spatial relation with strong ties to mathematics—namely, relative magnitude—outline its potential role in improving instruction on whole number ordering, fractions, and proportions. Casey and Fell discuss the difference between general spatial skill and spatial skill instantiated in specific mathematics problems, concluding that the most effective way to leverage spatial training to improve mathematics outcomes is likely the latter. They highlight a number of instructional techniques from existing curricula that successfully use spatial representations. Finally, Young, Levine, and Mix considered the multidimensional nature of spatial processing and mathematics processing and the inherent complexity involved in identifying possible instructional levers. Following a critique of the existing literature, including recent factor analytic approaches, they conclude with a set of recommendations for improving these approaches and applying what is already known in educational settings.

The mathematics education chapters discuss the spatial processes involved in specific topics in mathematics. Sinclair, Moss, Hawes, and Stephenson examine how children can learn "*through* and *from* drawing," focusing on spatial processes and concepts in primary school geometry. They argue that drawing is not innate but can be improved, and they illustrate through fine-grained analysis how the potential benefits of geometric drawing can be realized in classrooms. Gutiérrez, Ramírez, Benedicto, Beltrán-Meneu, and Jaime analyze the spatial reasoning of mathematically gifted secondary school students as they worked on a collaborative, communication-intensive, task in which they were shown orthogonal projections of cube buildings along with related verbal information. The authors related the objectives of students' actions and their visualization processes and students' solution strategies and cognitive demand. Herbst and Boileau argue that high school geometry instruction can do more than provide names for 3D shapes and formulas for finding surface area and volume. They illustrate, and invite reflection on their design of, a 3D geometry modeling activity in which students write and interpret instructions for how to move pieces of furniture up an L staircase. Lowrie and Logan discuss how the frequency of encountering, and interacting with, information in visual/graphic format, including on the web, has increased our need for research on the role of spatial reasoning in students' encoding and decoding of information in mathematics. To this end, they analyze the representational reasoning of students engaged in tasks that permit different types of representations, from diagrams to equations. Battista, Frazee, and Winer describe the spatial processes involved in reasoning about the geometric topics of measurement, shapes, and isometries. They introduce, and use in their analysis, the construct of *spatial-numerical linked structuring* as the coordinated process in which numerical operations on measurement numbers are linked to spatial structuring of, and operation on, the measured objects in a way that is consistent with properties of numbers and measurement.

As the chapters and commentaries illustrate, there are still fundamental differences between how researchers in psychology and mathematics education view and investigate the fundamental relations between spatial and mathematical reasoning. However, these differences provide fertile ground for exciting new investigations as each field respectively has advanced knowledge in some areas while leaving gaps in others. The commentaries are a starting point for identifying these points of contact and complementarity. We encourage readers to reflect on how the research in the two fields might be further integrated and how to build productive collaborations between the two sets of researchers. We thank all of our authors for taking a first step in this direction.

Michael T. Battista
Department of Teaching and Learning
The Ohio State University,
Columbus, OH, USA

Kelly S. S. Mix
Department of Human Development
and Quantitative Methodology
University of Maryland,
College Park, MD, USA

References

Bruce, C., Davis, B., Sinclair, N, McGarvey, L., Hallowell, D., ..., Woolcott, G. (2017). Understanding gaps in research networks: using "spatial reasoning" as a window into the importance of networked educational research. *Educational Studies in Mathematics, 95*(2), 143–161.

Clements, D. H., & Battista, M. T. (1992). Geometry and spatial reasoning. In D. A. Grouws (Ed.), *Handbook of research on mathematics teaching and learning* (pp. 420–464). New York: Macmillan.

Davis, B., & the Spatial Reasoning Study Group. (2015). *Spatial Reasoning in the Early Years: Principles, assertions, and speculations*. New York: Routledge.

Geary, D. C., Hoard, M. K., Byrd-Craven, J., Nugent, L., & Numtee, C. (2007). Cognitive mechanisms underlying achievement deficits in children with mathematical learning disability. *Child Development*, *78*(4), 1343–1359.

Mix, K. S. (2010). Spatial tools for mathematical thought. In K.S. Mix, L. B. Smith, & M. Gasser (Eds.), *The spatial foundations of language and thought* (pp. 41–66). New York: Oxford University Press.

Newcombe, N. S. (2010). Picture this: Increasing math and science learning by improving spatial thinking. *American Educator*, *34*(2), 29–43.

Wai, J., Lubinski, D., & Benbow, C. P. (2009). Spatial ability for STEM domains: Aligning over 50 years of cumulative psychological knowledge solidifies its importance. *Journal of Educational Psychology*, *101*(4), 817–835.

Contents

Part I Psychological Perspectives

1 **How Much as Compared to What: Relative Magnitude as a Key Idea in Mathematics Cognition** 3
Jamie Jirout and Nora S. Newcombe

2 **From Intuitive Spatial Measurement to Understanding of Units**. . . . 25
Eliza L. Congdon, Marina Vasilyeva, Kelly S. S. Mix, and Susan C. Levine

3 **Spatial Reasoning: A Critical Problem-Solving Tool in Children's Mathematics Strategy Tool-Kit** 47
Beth M. Casey and Harriet Fell

4 **More Space, Better Mathematics: Is Space a Powerful Tool or a Cornerstone for Understanding Arithmetic?** 77
Krzysztof Cipora, Philipp Alexander Schroeder, Mojtaba Soltanlou, and Hans-Christoph Nuerk

5 **What Processes Underlie the Relation Between Spatial Skill and Mathematics?**. 117
Christopher Young, Susan C. Levine, and Kelly S. S. Mix

Part II Commentaries

6 **Part I Commentary 1: Deepening the Analysis of Students' Reasoning About Length** 151
Michael T. Battista, Leah M. Frazee, and Michael L. Winer

7 **Part I Commentary 2: Visualization in School Mathematics Analyzed from Two Points of View** 165
A. Gutiérrez

8 **Part I Commentary 3: Proposing a Pedagogical Framework for the Teaching and Learning of Spatial Skills: A Commentary on Three Chapters** 171
Tom Lowrie and Tracy Logan

9 **Part I Commentary 4: Turning to Temporality in Research on Spatial Reasoning** 183
Nathalie Sinclair

Part III Educational Perspectives

10 **Analyzing the Relation Between Spatial and Geometric Reasoning for Elementary and Middle School Students** 195
Michael T. Battista, Leah M. Frazee, and Michael L. Winer

11 **Learning Through and from Drawing in Early Years Geometry** 229
Nathalie Sinclair, Joan Moss, Zachary Hawes, and Carol Stephenson

12 **The Interaction Between Spatial Reasoning Constructs and Mathematics Understandings in Elementary Classrooms** 253
Tom Lowrie and Tracy Logan

13 **Geometric Modeling of Mesospace Objects: A Task, its Didactical Variables, and the Mathematics at Stake** 277
Patricio Herbst and Nicolas Boileau

14 **Visualization Abilities and Complexity of Reasoning in Mathematically Gifted Students' Collaborative Solutions to a Visualization Task: A Networked Analysis** 309
A. Gutiérrez, R. Ramírez, C. Benedicto, M. J. Beltrán-Meneu, and A. Jaime

Part IV Commentaries

15 **Part II Commentary 1: Mathematics Educators' Perspectives on Spatial Visualization and Mathematical Reasoning** 341
Beth M. Casey

16 **Part II Commentary 2: Disparities and Opportunities: Plotting a New Course for Research on Spatial Visualization and Mathematics** 347
Kelly S. S. Mix and Susan C. Levine

17 **Part II Commentary 3: Linking Spatial and Mathematical Thinking: The Search for Mechanism** 355
Nora S. Newcombe

18 On the Multitude of Mathematics Skills: Spatial-Numerical Associations and Geometry Skill? 361
Krzysztof Cipora, Philipp A. Schroeder, and Hans-Christoph Nuerk

Index 371

Contributors

Michael T. Battista Department of Teaching and Learning, The Ohio State University, Columbus, OH, USA

M. J. Beltrán-Meneu Departamento de Educación, Jaume I University, Castellón, Spain

C. Benedicto Departamento de Didáctica de la Matemática, University of Valencia, Valencia, Spain

Nicolas Boileau Educational Studies Program, University of Michigan, Ann Arbor, MI, USA

Beth M. Casey Lynch School of Education, Boston College, Boston, MA, USA

Krzysztof Cipora Department of Psychology, University of Tuebingen, Tuebingen, Germany

LEAD Graduate School and Research Network, University of Tuebingen, Tuebingen, Germany

Eliza L. Congdon Department of Psychology, Bucknell University, Lewisburg, PA, USA

Harriet Fell College of Computer and Information Science, Northeastern University, Boston, MA, USA

Leah M. Frazee Department of Mathematical Sciences, Central Connecticut State University, New Britain, CT, USA

A. Gutiérrez Departamento de Didáctica de la Matemática, University of Valencia, Valencia, Spain

Zachary Hawes Ontario Institute for Studies in Education/University of Toronto, Toronto, ON, Canada

Patricio Herbst Educational Studies Program, University of Michigan, Ann Arbor, MI, USA

A. Jaime Departamento de Didáctica de la Matemática, University of Valencia, Valencia, Spain

Jamie Jirout Curry School of Education, University of Virginia, Charlottesville, VA, USA

Susan C. Levine Departments of Psychology, and Comparative Human Development and Committee on Education, University of Chicago, Chicago, IL, USA

Tracy Logan Faculty of Education, University of Canberra, Bruce, ACT, Australia

Tom Lowrie Faculty of Education, University of Canberra, Bruce, ACT, Australia

Kelly S. S. Mix Department of Human Development and Quantitative Methodology, University of Maryland, College Park, MD, USA

Joan Moss Ontario Institute for Studies in Education/University of Toronto, Toronto, ON, Canada

Nora S. Newcombe Department of Psychology, Temple University, Philadelphia, PA, USA

Hans-Christoph Nuerk Department of Psychology, University of Tuebingen, Tuebingen, Germany

Leibnitz-Institut fur Wissensmedien, Tuebingen, Germany

LEAD Graduate School and Research Network, University of Tuebingen, Tuebingen, Germany

R. Ramírez Departamento de Didáctica de la Matemática, University of Granada, Granada, Spain

Philipp A. Schroeder Department of Psychology and Psychiatry, University of Tuebingen, Tuebingen, Germany

Department of Psychiatry and Psychotherapy, University of Tuebingen, Tuebingen, Germany

Nathalie Sinclair Faculty of Education, Simon Fraser University, Burnaby, BC, Canada

Mojtaba Soltanlou Department of Psychology, University of Tuebingen, Tuebingen, Germany

LEAD Graduate School and Research Network, University of Tuebingen, Tuebingen, Germany

Carol Stephenson Ontario Institute for Studies in Education/University of Toronto, Toronto, ON, Canada

Marina Vasilyeva Lynch School of Education, Boston College, Chestnut Hill, MA, USA

Michael L. Winer Department of Mathematics, Eastern Washington University, Cheney, WA, USA

Christopher Young Consortium on School Research, University of Chicago, Chicago, IL, USA

Part I
Psychological Perspectives

Chapter 1
How Much as Compared to What: Relative Magnitude as a Key Idea in Mathematics Cognition

Jamie Jirout and Nora S. Newcombe

Abstract Most topics beyond basic arithmetic require relative magnitude reasoning. This chapter describes the link between relative magnitude reasoning and spatial scaling, a specific type of spatial thinking. We discuss use of the number line, proportional reasoning, and fractions. Consideration of the relational reasoning involved in mathematics can advance our understanding of its relation to spatial skills, and has implications for mathematics instruction, such as using spatial reasoning interventions in developing effective methods for supporting relative magnitude understanding. We review evidence that interventions can be successful in promoting better relative magnitude understanding and associated spatial-relational reasoning, and suggest that education considers ways of including relative magnitude learning, along with more traditional whole-number operations, in early educational efforts.

Keywords Spatial scaling · Spatial learning · Spatial development · Spatial visualization · Scale · Spatial representations · Representations · Diagrams · Spatial-relational · Relative magnitude · Absolute magnitude · Magnitude reasoning · Number line estimation · Proportional reasoning · Fractions · Symbolic understanding · Benchmark strategy · Manipulatives · Interventions · Spatial play

Is "½" big or small? Reasoning about this question demonstrates two types of numerical reasoning: if you answered that 1/2 is small, because it is less than one, you are reasoning about the number as an absolute value. If your answer is to ask, ½ of what, you are reasoning about relative magnitude. In this chapter, we suggest that number is often interpreted in an absolute sense in mathematics education, though we describe how most topics beyond basic arithmetic require relative

J. Jirout (✉)
Curry School of Education, University of Virginia, Charlottesville, VA, USA
e-mail: jirout@virginia.edu

N. S. Newcombe
Department of Psychology, Temple University, Philadelphia, PA, USA
e-mail: newcombe@temple.edu

K. S. S. Mix, M. T. Battista (eds.), *Visualizing Mathematics*, Research in Mathematics Education, https://doi.org/10.1007/978-3-319-98767-5_1

magnitude reasoning. We discuss how relative magnitude reasoning might involve a specific type of spatial thinking: spatial scaling. Specifically, use of spatial scaling can contribute to precision in relative magnitude reasoning, perhaps by tapping a more generalized magnitude representation. Focusing on relative magnitude reasoning is an important consideration when determining how spatial thinking relates to mathematics learning, and may have implications for approaches to mathematics instruction, such as using spatial reasoning research and interventions in developing effective methods for supporting relative magnitude understanding. More broadly, consideration of relational reasoning involved in mathematics might advance our understanding of its relation to spatial skills, and even support the inclusion of emphasizing spatial learning as a way to prepare for and support mathematics learning.

The National Research Council has outlined goals for mathematics education in which they suggest that numeracy should be the topic most emphasized early on (NRC, 2009). The content standards of the National Council of Teachers of Mathematics concur (NCTM, 2010). But it can be hard to follow these guidelines because it is not always clear what numeracy or "number" means. In early elementary school, number often means integers, using count words to enumerate sets of discrete objects, and to add and subtract from those sets. In a kindergarten mathematics lesson, for example, students might assemble sets of objects to make a specific number or, later in the year, use sets of objects to do addition and subtraction problems. The kindergarten children are encouraged to think about numbers as referring to these collections, such as telling "how many." When children think about this kind of number—say, they imagine "5"—they need to recognize the Arabic symbol and recall the word "five," remember that five comes after four and before six, and, crucially, imagine a set of five objects. They might also know that five can be used to refer to a time or date, or to a five-dollar-bill or a 5-year-old child. But in most cases, they are thinking about the value of a positive integer referring to a collection of discrete objects.

In later grades, lessons would be different. Third graders might be learning how to move a decimal point to convert percentages to proportions. Fifth graders might be learning about remainders in long division problems. These lessons involve a fundamentally different kind of "number" than integers. Older children need to think about numbers as a number system, which can quantify many kinds of referents, including referring to continuous magnitudes that need not, and often do not, denote collections of discrete objects. When comparison of fractions is required, when calculating a proportion or percentage, or when dividing one number into another but with some quantity remaining, relative magnitude is key. How is this shift, from number as discrete objects to the meaning of number, dependent on the specific problem?

To determine how to support children's underlying representation of number as it becomes more complex, it is important to first make the different representations explicit. Recently, Newcombe and colleagues defined the different categories of quantification types across two dimensions (Newcombe, Frick, & Möhring, 2018). A first distinction is that number may refer to collections of discrete objects, as

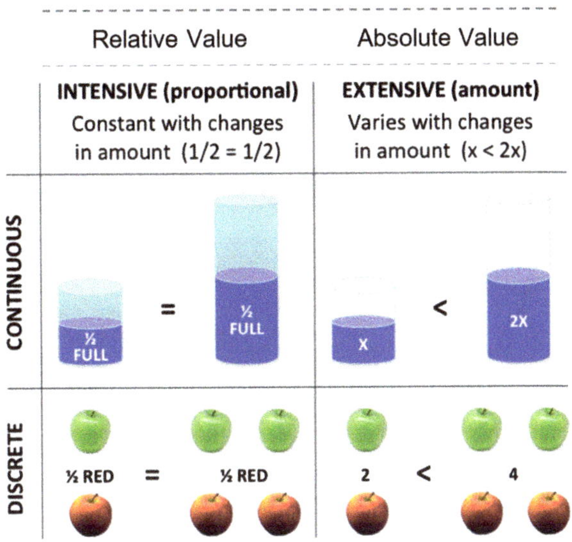

Fig. 1.1 Newcombe et al.'s (2018) categorization of systems of quantification, modified

shown in the bottom row of Fig. 1.1. However, the number system can also be used to quantify continuous magnitudes, as shown in the top row of Fig. 1.1. A second distinction is that numbers may refer to relative (or intensive or proportional) quantities, as shown in the left column of Fig. 1.1, or to absolute (or extensive) quantities, as shown in the right column of Fig. 1.1. Jointly, these two dimensions generate a four-cell classification system. Kindergarten mathematics generally focuses on the bottom right, but eventually, children must learn about the whole system (Common Core State Standards Initiative, 2010). Thus, thinking about numbers in terms of these varied uses and meanings is important in understanding the development of mathematical cognition, and in identifying effective ways of supporting students' representation of number as they continue in their education.

The distinctions of quantification categories are also important in understanding the link between mathematical and spatial thinking (Newcombe et al., 2018; Newcombe, Levine, & Mix, 2015). Spatial reasoning has multiple aspects, including a distinction between the spatial characteristics of an object itself (intensive) and the spatial position of an object in relation to other objects and its surroundings (extensive) (Newcombe & Shipley, 2015). In this chapter, we focus on the overlap between the relational reasoning processes involved in mathematics and spatial reasoning tasks. We focus on relative magnitude, a concept that is fundamental both to spatial scaling and to proportional reasoning of the kind shown in the left column of Fig. 1.1. Spatial relational reasoning skills are seen in young children and even infants, and could help support similar relational reasoning processes in mathematic tasks involving relative magnitude.

This chapter will begin with examples of relative magnitude in mathematics education, examining number line understanding, proportional reasoning, and fraction learning. We then provide a review of research on relative-magnitude reasoning in spatial processes, especially spatial scaling, and how these spatial processes relate to mathematic skills and are utilized when using external representations. We return to number line, proportional reasoning, and fraction tasks to discuss how spatial representations, and thus spatial-relational reasoning, are used in learning. Finally, we conclude with a discussion of interventions shown to improve children's relative magnitude understanding, and possible connections between spatial learning and developing mathematic skills. We end with a discussion of potential implications of the research on spatial thinking and relative magnitude for mathematics education.

Relative Magnitude in Mathematics Learning

The importance of relative magnitude is evidenced by its necessity for many mathematics tasks both in education and in everyday life. Simple questions like whether a number is "big" or "small" must be evaluated relative to some comparison or scale, for example knowing where to put the number nine on a number line depends on the range of the line (i.e., toward the end on a 0–10 line, but toward the beginning on a 0–100 line). Relative magnitude is important in thinking about proportions to determine what things are being compared and how, for example knowing how much sugar to add for a cup of lemonade when using one lemon, if you know you use a whole cup of sugar for a pitcher using four lemons. Early fraction tasks such as dividing something among friends require understanding the meaning of a part-whole relation, for example, that as the number of friends sharing a cake (the denominator) becomes larger, the portion of the cake that each receives becomes smaller. These different tasks share the common cognitive process of relative magnitude reasoning. We explain this idea further for each of three mathematical tasks now, and return to these tasks later in the chapter to discuss relations across relative magnitude tasks and how spatial representations influence relative magnitude reasoning in the tasks.

Number Lines as Relative Magnitude

Although some mathematic tasks in research measure absolute value knowledge, others, including the widely used number line estimation, involve relative magnitude reasoning. In early development, understanding of number is often assessed with tasks asking children to give a specified number of objects, or the "give-N" task (e.g., Wynn, 1990). But many studies of mathematics cognition or number understanding use measures of relative rather than absolute magnitude reasoning (e.g., using estimates of *more* or *less* rather than absolute value). Rather than asking how many, these tasks rely on children considering *how much/many compared to what.* For example, on the widely used number-line estimation task, children are asked to show "how much" of the given line is equivalent to a specific value, but often they

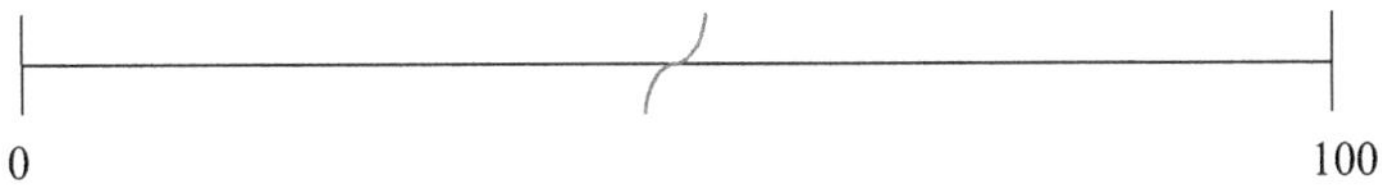

Fig. 1.2 Example of the number line estimation task. In one version, the participant is asked to mark a line to show where a given Arabic numeral would go, relative to the scale provided. In another version, the participant is asked to give the value of the mark provided on the number line, relative to the scale provided. Here, the mark shows a value of 50 relative to the 0–100 scale. If the scale changed to 0–1000, the mark would show 500

are provided beginning and endpoint values, including a scale for comparison to the value (Siegler & Booth, 2004). The child's specific task is to place a mark on the line to represent where a specific number would fall, or to provide an estimate of the number represented by a mark shown on the line (see Fig. 1.2). As opposed to the give N tasks, successful performance on the number line task requires relative magnitude understanding. The magnitude of the space between the provided endpoints is relative to the scale (i.e., 1 in. on a 10-in. number line represents a single unit for a scale of 0–10, but represents 10 units if the scale is 0–100). This requires children to shift from their early mathematics experience in thinking about number as absolute values of discrete objects, to thinking about number as a value relative to the scale. Younger children sometimes use a more familiar discrete counting strategy, ignoring the provided endpoint value. This can result in overestimating lower numbers across the line, and then squeezing the larger numbers toward the end of the scale, resulting in a logarithmic representation of the number scale. As they get older, children begin to use more proportional strategies and thus their representations become more linear, though a shift back to the earlier strategy is observed as the scale increases (e.g., 0–1000 to 0–10,000). That is, children progress from basic concepts such as knowing that the word two means two objects, to having linear representations of numbers on scales of 1–10 (age 3–5), 1–100 (age 5–7), 1–1000 (age 7–11), and eventually understanding fractions in a similar way, beginning around age eight, advancing through adulthood (Siegler & Braithwaite, 2017).

In mathematics education, much emphasis is placed on understanding the absolute magnitude of numbers in early elementary school. Yet, even as early as first grade when addition and subtraction are emphasized, current standards explicitly mention the importance of understanding relative magnitude of numbers as well (Common Core State Standards Initiative, 2010). As children begin to learn more complex mathematics, reasoning about relative magnitude becomes much more central, as the content begins to rely much more on relational reasoning than absolute values of number when fractions, proportions, functions, probabilities, etc. are introduced (Common Core State Standards Initiative, 2010). Beginning in third grade, children whose education thus far focused on the absolute value of numbers are expected to shift to reasoning about relative magnitude, where paying attention to whole numbers would lead to less accurate performance (DeWolf & Vosniadou, 2011). It is important, then, to consider how this shift can be supported in educational practice.

Proportional Reasoning

Proportional reasoning is typically used to solve a problem in order to reach a specific value of interest, yet, like number line estimation, it requires relative magnitude reasoning. Reasoning about proportion is observed in everyday problems, such as when comparing costs of two items that are different prices and amounts or to estimate total cost with sales tax. In mathematics instruction, proportional reasoning is necessary when calculating concentrations of a solution, or in traditional multiplication problems such as using a given proportion—say, the speed a car travels—to determine how long it will take for the car to reach a destination of a specific distance. This task is dependent on the ability to reason about one quantity relative to another, determining a ratio, and often to apply this relational information to another context. Though the algorithmic use of formulas often studied to solve this type of problem does not seem to involve relative magnitude (i.e., plugging numbers in), many direct measures of proportional reasoning require children to use relative magnitude reasoning. For instance, proportional reasoning tasks in research typically use spatial displays of concentrations, such as proportions of water to juice, with children choosing a matching concentration that would taste the same (i.e., have the same concentration) or showing the concentration using a scale from very weak to very strong (Boyer, Levine, & Huttenlocher, 2008; Möhring, Newcombe, Levine, & Frick, 2016a).

Fraction Learning

Like proportions, fractions are part-whole relations in which their meaning is derived from relative magnitude. But unlike proportions, fractions are considered numbers themselves that can be represented on a number line, and they can be greater than one. In fact, many researchers and educators argue that thinking about fractions as numbers by representing them on a number line (relative to other fractions and whole numbers) improves learning. Specifically, the widely used Common Core standards suggest that students learn to place fractions on number lines, implicitly considering them to be absolute magnitudes (Common Core State Standards Initiative, 2010). Yet when fractions are included in mathematics problems, understanding them as relative magnitudes helps conceptual understanding. For example, it is fairly easy to place the fraction one half between zero and one on a number line, but if you are multiplying a number by one half, it is helpful to think of it as signifying one of two equal parts of a quantity, which may of course be much greater than one. The relation between the numerator and denominator is what makes a fraction meaningful, and knowing that this relation can be applied to any quantity just as proportions can be. Understanding fractions is important for success in algebra (NMAP, 2008), with algebra considered to be "the gatekeeper to higher learning in mathematics and science" (Booth & Newton, 2012, p. 247), and it lays the foundation to more advanced mathematics. It has been suggested that the relation between fraction knowledge and later success in algebra may be due to more general underlying knowledge of number systems and magnitude, supporting more abstract mathematical reasoning (Ketterlin-Geller, Gifford, & Perry, 2015).

Relative Magnitude and Spatial Thinking

Similar to the varied conceptualizations of number, spatial thinking is a broad label for many different types of reasoning. For example, planning how to navigate from one point to another requires spatial thinking, but so does determining which size pot will hold the amount of food you plan to cook; children's spatial thinking in play can involve matching up pieces of a puzzle or building with Lego diagrams, or comparing lengths of sticks to choose which should be the *daddy* vs. *baby*. Spatial thinking is also multidimensional, similar to the conceptualizations of number.

Relational thinking can be observed in intensive spatial tasks similar to number tasks. For example, relational position of discrete objects in a model or diagram (i.e., attending to the relative position within an array) is used to solve spatial analogies. In arrays of toys where array 1 is "pig," "dog," "chick" and array 2 is "horse," "mouse," "dog," the relational match to "dog" from array 1 is "mouse" in array 2. This relational matching of relative position can also be done with continuous spaces, such as finding a location in real space on a map. When using a map, the task draws on relative magnitude when determining the position of a target in the continuous space, creating a representation or estimation of relative position that can then be applied to another space. When this relational matching is done across spaces and/or representations of different size, the process of spatial scaling is used. Spatial scaling is one area of focus in research investigating the relation between spatial thinking and mathematics tasks requiring relative magnitude.

Spatial Scaling

Spatial scaling is the ability to reason about spatial relations in one context and to apply this relational information to a different sized spatial area. Just as proportions and fractions often involve identifying relational information and applying it to solve a problem, spatial scaling involves two steps: first, recognizing the relational correspondence between the two areas; second, mentally transforming the spatial-relational information from one space to the other (Möhring, Newcombe, & Frick, 2014). An example of spatial scaling is the process used when looking at the distance between two points on a map and then determining that distance in real space. Just as mathematics-relational reasoning can be assessed with number-line estimation, spatial-relational reasoning can be assessed using a spatial scaling measure (see Fig. 1.3). Instead of a symbolic number, children are shown a scaled map

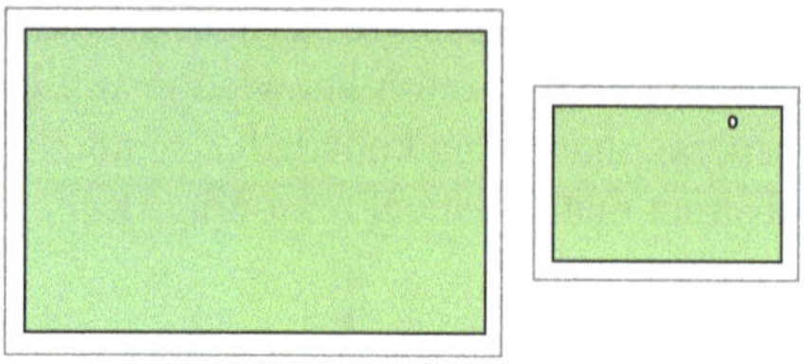

Fig. 1.3 Example of a spatial scaling task with a map (right) and referent space (left)

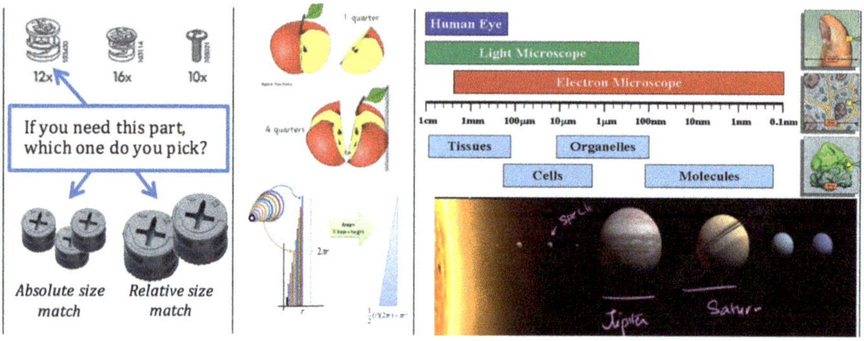

Fig. 1.4 Examples of diagrams requiring relational-reasoning, shown for every-day tasks (left), mathematics learning (middle), and science learning with microscopic (right-top) and telescopic (right-bottom) scales

marking a location, and are asked to match that location on a larger space (Frick & Newcombe, 2012).

By definition, spatial scaling requires a representation and a referent. A key function of representations is to convey various types of relational information about the referent. Examples of everyday representation use include navigating a space (referent) that is larger than one can see from a single viewpoint, such as with scaled maps (representation), or showing things too large or small to be seen with the human eye (referent), like in scaled photographs (representation). Other examples include configuring complex materials into a desired state, like when using building instructions, or for displaying information to show relations visually as charts and graphs do. In all of these examples, relational reasoning is critical, and scaling is often used—both in understanding the scale of the representation and, in several cases, extrapolating the information to apply it to another scale (e.g., navigating through a park, or determining which correct size part to use for a step when building furniture, see Fig. 1.4).

Focus on superficial or perceptual information can lead to ineffective use of the representation, for example not going far enough when driving because the distance on a map looked small, while the scale actually indicated it was quite far. Similarly, focus on perceptual rather than relational information in mathematics learning can cause difficulty. For example, in the original conservation of number tasks, Piaget found that children focused on the perceptually salient spatial information of how far a sequence of objects stretched as indicating that it had more, even after just observing that the sequence was equivalent to another before being rearranged (Piaget, 1952). In more advanced magnitude comparison, children often struggle with ordering fractions by magnitude because although magnitude decreases as the denominator increases, they mistakenly associate this increase with an increase in magnitude (e.g., saying one fourth is larger than one half, because four is greater than two).

Research on performance in proportional reasoning and fractions shows a correlation between these processes and spatial scaling. Despite these tasks being different, this relation supports the common reliance on relative magnitude for successful performance (Möhring et al., 2016a; Möhring, Newcombe, Levine, & Frick, 2016b). More broadly, spatial magnitude information is used when interpreting numerical magnitude, for better or for worse (Newcombe et al., 2015); spatial thinking is necessary when completing tasks like placing numbers on a number line, but can also cause bias or error such as Piaget's number conservation mistakes or when estimations are pulled toward discrete markers or benchmarks. Further, children often learn about proportional reasoning and fractions using diagrams and representations, thus it is important for them to understand how to use them correctly, which often requires spatial scaling.

In fact, the importance of understanding scale of diagrams and using scaled diagrams, as well as an emphasis on scaling skills more generally, is made explicit by its inclusion throughout the common core mathematics standards (Common Core State Standards Initiative, 2010). Although early developmental theory suggested that processes like proportional reasoning are well beyond what young children are capable of (e.g., Piaget, 1952; Piaget & Inhelder, 1956), measures of relational reasoning, especially spatial-relational reasoning and magnitude understanding, suggest that even infants show early aptitude and children become progressively more skilled in early years, with both improved scaling accuracy and more advanced proportional reasoning with age (Lourenco & Longo, 2011).

Spatial Representations

Visual representations like diagrams are common in and important for mathematics education (National Council of Teachers of Mathematics, 2010). These representations rely on spatial thinking, but also have the potential to help develop relational-reasoning skills (Davies & Uttal, 2007; Gentner, 1988) and promote broader learning in mathematics (Woodward et al., 2012). Spatial-relational reasoning might be specifically important for fully understanding and using representations. Representations are not always helpful, for example, if they are perceived as pictures or objects of their own rather than as representing information or a referent (Garcia Garcia & Cox, 2010; Uttal & O'Doherty, 2008). Several varying factors of implementing representations might influence their effectiveness for learning: the structure of representation implementation, type of contact with the representations, the relation of the representation to familiar symbols, and the amount of exposure to the representations (Mix, 2010). The type of representation and the learning context can vary further in the mechanisms of support representations may offer, from reducing fine motor and cognitive demands and to providing metaphors or embodied experiences (Mix, 2010). The implementation methods, support sought and available from representations and needs of different learners, and the specific learning goals more broadly are all important considerations in using representations for mathematics instruction. Importantly, though, children must have the necessary skills for understanding representations in the first place, such as their representational nature (Uttal & O'Doherty, 2008).

As discussed above, interpreting information from spatial representations often requires spatial scaling. Although children rarely receive explicit instruction on scaling, research shows they are capable of spatial scaling from very young ages. Success in using scaled representations relies on matching the *relational structure* of the representation, which they learn to do quite early (Gentner, 1988). Failure in this step, for example, would be in choosing the absolute rather than relative size match in Fig. 1.4 (left). Children develop more accurate scaling ability in the early elementary school years, though even adults perform poorly when reasoning about very small or very large scales, thus there is room for improvement. One potential mechanism for developing spatial scaling is improved relational reasoning, such as learning the related processes of proportions or fractions taught in late elementary and middle school grades. Alternatively, early practice with scaled spatial representations could help prepare children for later mathematics learning by strengthening their relational reasoning.

To understand how spatial representations might help support learning of relative magnitude mathematics tasks, an important question is, how do children learn relational reasoning with spatial representations? Though instances of children's relational thinking can be observed in very young children, and even infants, research shows improvement in representation use, and changes in strategies or processes involved, with age. For example, children demonstrate the ability to understand symbolic correspondence, or that a model represents something else, around the age of 3 years old (DeLoache, 1987). However, understanding spatial representations typically also requires geometric correspondence (Newcombe & Huttenlocher, 2000). Children must be able to encode length or distance within representations and then map that to the referent space, while conducting any necessary scaling if the representation is a different size (Möhring et al., 2014). Young children can successfully use models, typically by using landmark mapping and perceptual coding of the space, and there is some evidence they use categorical or even distance coding (Huttenlocher, Newcombe, & Vasilyeva, 1999). However, coding distances and scaling are not simple tasks. Difficulty can easily be increased by requiring children to code distances along both horizontal and vertical axes or by removing landmark cues. Even adults completing a simple scaling task (i.e., using a map to mark a target on a two-dimensional space) show variation in accuracy as a function of the scale of the representation to the referent space (Möhring et al., 2014).

The adult strategy for using scaled representations is likely determined by the need for accuracy; while adults can quickly and quite accurately interpret scaled information using perceptual methods (on familiar scales), they are also able to use a slower metric coding strategy, which should provide even more precision. This more sophisticated metric strategy is difficult for children, because it relies on proportional reasoning. Though they do not calculate exact proportions, young children have been found to be capable of using basic spatial categories—for example using a midpoint to create categories within the space (Huttenlocher, Newcombe, & Sandberg, 1994). Just as estimation errors on the number line are seen around benchmarking points, suggesting the use of proportional reasoning strategies, spatial estimation responses also show error patterns that suggest they are being slightly biased by these categorical boundaries, where their responses are pulled toward corners or midpoints (Sandberg, Huttenlocher, & Newcombe, 1996; Slusser, Santiago, & Barth, 2013). This more implicit or categorical way of reasoning about proportions might prepare children for later mathematics learning that requires relative magnitude understanding. Further, spatial representations can be used specifically to help children reason about relative magnitude in mathematics learning.

Spatial Representations in Relative Magnitude Mathematics Tasks

We return now to our discussion of the relative magnitude mathematics tasks discussed above, with a focus on spatial representations and how they might support mathematics learning. Common Core standards include using number lines (similar to the research task discussed earlier) to represent whole numbers and fractions (Common Core State Standards Initiative, 2010). Studies show that number line estimation performance relates to spatial thinking, perhaps by emphasizing the spatial importance in number understanding as a linear representation of numbers (Gunderson, Ramirez, Beilock, & Levine, 2012). The question of what specific processes are involved in number line estimation is the subject of much research. Many studies show that there are spatial patterns in developing number line estimation skills by what is described as a representation shift. Performance on the number line estimation task transitions from estimates that are better fit by a logarithmic function (i.e., lower numbers tend to be, with larger numbers being grouped too closely toward the top of the scale) to a linear fit where numbers are placed relatively equally spaced apart and in order. Children's estimations become more linear with age, and the timing of this transition is positively correlated with scale; the older children get, the higher the scale showing a linear representation (Siegler & Opfer, 2003). This body of research emphasizes this representation shift as an important developmental milestone of number understanding, for good reason, but it should also be noted that the shift is referring to children's *spatial* representations.

One reason linear estimates on the number line are important is that they show understanding of equal spacing between whole numbers, and correct knowledge of order in which the numbers follow; this tends to occur when scales are familiar to the participant (Ebersbach, Luwel, Frick, Onghena, & Verschaffel, 2008). Yet even when one knows the order of numbers, the one-to-one correspondence on a number line is not as easy as counting objects—the space provided must be divided into the equal units. Studies investigating *how* people use the given scale to determine where a number goes provide evidence that relational reasoning is involved. Specifically, studies using eye-tracking and differences in error patterns show that both children and adults use proportional reasoning strategies when making estimations (Barth & Paladino, 2011; Schneider et al., 2008; Slusser et al., 2013). These studies indicate the use of landmarks—the endpoints as well as fractional landmarks (e.g., ½, ¼). The level of proportional benchmarking relates to improved performance; the more precise proportions used (i.e., more benchmarks), the more accurate estimations were (Peeters, Verschaffel, & Luwel, 2017). Peeters et al. (2017) found that providing explicit proportional markers can induce finer grained proportional reasoning with landmarks, but the use of implicit landmarks (i.e., the proportional ones) also appears to increase with age (Slusser et al., 2013). Thus, not only is children's developing number line estimation measured as a spatial representation, it also draws on mathematical reasoning related to relative magnitude understanding, and can be improved using spatial-relational interventions with representations.

The use of proportional landmarks is observed in other mathematics tasks involving representations as well. Children at the earliest ages of formal schooling show the ability to use ½ as a benchmark when comparing proportions (Singer-Freeman & Goswamani, 2001; Spinillo & Bryant, 1991). On basic part-whole fraction tasks, emerging understanding appears by 4 years of age (Mix, Levine, & Huttenlocher, 1999; Singer-Freeman & Goswamani, 2001). Some research on early proportional reasoning suggests that the specific features of diagrams or materials used to represent proportions might impact children's attention to the relative vs. absolute magnitude information. Boyer and Levine (2012) asked children to match proportions displayed using a "juice task," where different concentrations of juice to water were displayed using red and blue segments (see Fig. 1.5). Sometimes the diagram included a single red and a single blue bar, differing in size, to show continuous proportion of whole (e.g., 50% each; Fig. 1.5, bottom). Other diagrams showed the

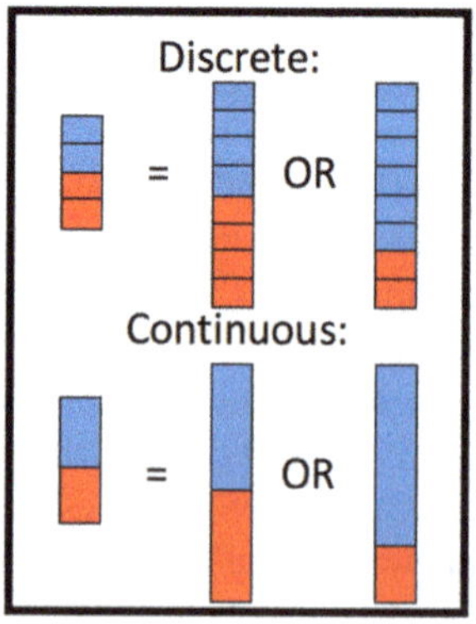

Fig. 1.5 Discrete vs. continuous proportional representations

same information, but the blue and red bars were divided using discrete units (e.g., two red, two blue; Fig. 1.5, top). Children were then asked to choose which of two options was a match to the concentration (i.e., "Which of these two would taste like Wally Bear's Juice?"). Children were better at matching the continuous bars, though their performance decreased as the comparison became less similar by becoming increasingly different in size (for example, choosing 2/8 instead of 4/8 as a match for 2/4). This effect of scale difference is similar to that observed in tasks of spatial scaling, where performance decreases as the scale factor increases (Möhring et al., 2014). When the visualizations included the discrete units, children often defaulted to an inaccurate counting strategy of the units shown rather than using proportional reasoning; when they were asked to produce a matching proportion, the lines of the discrete units actually amplified errors across scale factors, pulling responses further from the correct location compared to the continuous displays (Boyer & Levine, 2012). This difficulty in applying proportional information from one display to another with discrete units seems unfortunate, since real-world use of proportional reasoning often involves discrete units. For example, if you figure out that pancakes come out best when you combined 2-tbs water to 3-tbs boxed mix and want to now make more than a single pancake, you can determine the proportional composition of 2:3 or 40% water to 60% boxed mix, and then apply this proportion to a unit of measurement (rather than estimating), for example 1-cup water to 1.5-cups mix. The key is to translate the initial discrete amounts into a proportion, rather than trying to apply the raw values from the first amounts. It may not be that children are unable to apply proportional information to quantities of discrete units, but that they are being distracted from determining the proportional information in the first place.

Boyer and Levine (2015) tested this by having children select a proportional match with the comparison proportion presented in spatial representations using either discrete or continuous amounts. As expected, when children are presented first with proportions shown as continuous quantities, they then do better on proportions shown as discrete quantities, though only the older children benefited from the different displays (Boyer & Levine, 2015). The authors suggest that providing the continuous representation first encourages more spatial proportional reasoning strategies rather than a "count-and-match" method. More generally, these finding suggest that spatial visualization using representations might help direct attention to relative magnitude (i.e., the continuous presentation) rather than whole number counting (i.e., the discrete presentation).

Spatial representations of proportions like those in Fig. 1.5 can also be used to show fractions concepts, e.g., 1:2 or 50% is the same as ½, and fractions can be similarly displayed using part-whole diagrams. However, fractions can be larger than one to include any rational numbers, sometimes making diagrams of part-whole relations confusing. Reasoning about fractions can also become difficult when the fraction is to be applied to another number, such as multiplying or dividing by a fraction. Even a simple magnitude comparison task can be challenging if the numbers in the larger fraction have smaller components, (e.g., 3/4 vs. 8/20; Lortie-Forgues, Tian, & Siegler, 2015). Siegler and Lortie-Forgues (2017) outline the reasons why fractions and rational numbers more generally are difficult to understand, some of which relate to the differences between proportions and fractions. First, the same value can be symbolized in more than one way, such as 1/2 and 2/4 or 1½, 3/2 and 9/6. Second, the presentation of fractions as a numerator over denominator can cause difficulty in correctly understanding the fraction based on inaccurate application of whole-number knowledge. If much of your experience is with whole numbers, it is hard to switch to thinking of the numerical symbols not as individual whole numbers—one half is not "one" and "two." In this case, spatial representations of fractions could be helpful in demonstrating the meaning of the fraction. Even more complicated is that when performing fraction problems sometimes you *do* treat numerators as whole numbers—but not always. Consider subtraction of fractions that have the same vs. different denominators (e.g., 6/8–4/8 is simply solved as 6–4, to give the answer 2/8, vs. 6/8–1/2 in which 6–1 is incorrect; 1/2 must first be converted to 4/8). And, of course, when multiplying fractions, you treat the numerators and denominators as whole numbers to multiply across whether or not the denominators are the same, however with division of fractions you also multiply across, but using the reciprocal of the devisor. This formulaic thinking involves considering fractions independently as absolute values rather than using relative magnitude reasoning, which perhaps contributes to the difficulty in fraction learning.

Often these "rules" or procedures are memorized by students, but without the conceptual understanding of why they work, it can be difficult to have a sense of the general ballpark the answer should fall, resulting in higher risk of error and mistakes. Using conceptual understanding of relative magnitude could help students catch themselves when they make these mistakes. For example, if they are multiplying by a fraction that is less than one, or finding a proportion of some amount, conceptual understanding would help them catch a mistake where the answer should not be larger than the original value. Similarly, when comparing magnitudes, having an idea of where a fraction would fall on a number line can help students to more quickly compare without having to do calculations. For example, when comparing 112/250 vs. 167/310: realizing that one number is less than half and one is more can allow you to quickly respond with the larger fraction. Understanding why these different procedures work can also help you remember them accurately in the first place, and using relative magnitude is necessary for this understanding.

Some studies have shown that using spatial representations of fractions, such as manipulatives, can help to build stronger foundational knowledge about fractions and support later problem solving (Carbonneau, Marley, & Selig, 2013), though Mix (2010) suggests that there are important considerations that may influence learning. Concrete manipulatives can be used to direct attention from absolute to relative magnitude. These materials are similar to the 2-dimensional diagrams in their representational nature, but they also typically allow physical manipulation, and can provide scaffolding for moving from concrete objects to abstract meaning. Spatial features of the manipulatives like size can be used to help children observe the relational information when making comparisons. For example, in Montessori classrooms children can manipulate bars or beads made up of 10 connected units, comparing its relation to the 100-board made up 10 of the same bars, and then relating it to the 1000-unit cube—supporting understanding of the base-10 system. Cuisenaire and fraction bars provide concrete representations of fixed values (e.g., labeled 2 or ½), which can then be compared to practice part-whole relations, moving from absolute to relative magnitude reasoning. Mix (2010) identifies several possible reasons that this type of interaction with concrete manipulatives has potential for learning. First, it provides embodiment, or movement through space. For example, it might be important for children to be able to feel that the different bars take up different amounts of space as they are aligning them. Having the physical materials in front of them can also help direct attention to the physical features (size or length) rather than focusing attention on writing out problems or knowing what part of a picture diagram to attend to; in this way, the physical objects can also reduce memory demands, similar to taking notes. Finally, the materials can serve as models or metaphors, helping to ground abstract ideas and provide connections across knowledge and experiences, though this may take some explicit support and instruction for success (Mix, 2010). These abstract ideas involve relative magnitude understanding.

Improving Children's Relative Magnitude Understanding

Many interventions to improve children's mathematics learning have been successful, and some of these have included training on relative magnitude specifically or have key features related to spatial-relational reasoning. For example, playing a number board game has shown to be effective in improving children's performance on the number line estimation task (e.g., "Great Race", Ramani & Siegler, 2014; Ramani & Siegler, 2008). Children played on a linearly designed game board with equal rectangular spaces, with the goal of moving from *start* to *finish*. After playing the game in four 15–20 min sessions, children showed significant learning when the board had numbered spaces compared to only colors. These benefits were observed *only* when the game was presented linearly with equal spaces (Siegler & Ramani, 2009), and included counting numbers continuously so that their relative position on the board is noticed (e.g., instead of counting "one, two" when moving, counting on from the space, "four, five"; Laski & Siegler, 2014). Importantly, in multiple different interventions, similar exposure to numbers without the linear spatial-relational information in the board game did not have the same positive effect on children's learning (Laski & Siegler, 2014; Siegler & Ramani, 2009). Thus, spatial features of the external representations facilitated encoding of magnitudes, and the ways in which these games were played also influenced the effectiveness for learning (Laski & Siegler, 2014).

Studies have also shown some success in training approximate number system acuity (ANS). In this task, arrays of dots are displayed and participants respond with which array is larger. There are too many dots to count in the short time they are displayed, and so estimation is used. Determining which set of dots is larger relative to the comparison requires relational reasoning, and there is some evidence that

Weber's Law is seen in comparisons made. That is, it is easier for people to detect differences of equal absolute value in lower numbers than higher numbers (e.g., recognizing 12 is larger than 10 compared to 48 is larger than 46); it is not the absolute difference that matters, but the difference relative to the amounts being compared. In a study on improving ANS, children played a game in which they were trained on approximate arithmetic (Park, Bermudez, Roberts, & Brannon, 2016). They saw arrays of objects that then had additional objects added or subtracted, and compared the approximated magnitude to one or two other sets of objects to determine which had the same amount or more, depending on the trial type. Preschool participants completed about 40 items in each of 10 sessions, and were compared to peers who practiced a memory game. Children who received the ANS training significantly outperformed those who completed the memory game control on the Test of Early Mathematics Achievement (TEMA-3, Ginsburg & Baroody, 2003), a measure including items to test both number literacy (e.g., counting) and relative magnitude (e.g., comparing magnitudes of different numbers). Similar training was also found to show a benefit for college-age students, who improved on exact symbolic arithmetic (Park & Brannon, 2014). A more general ANS task without the approximate arithmetic, however, did not show improved arithmetic skill, and only showed trends of improvement on a non-symbolic numerical comparison task (Park & Brannon, 2014). Further, there is some caution about interpreting ANS training studies thus far, as many include small samples among other challenges to methodologies used (Szűcs & Myers, 2017).

Interventions with a focus on using scaled spatial representations have also been effective for more domain specific skills. For example, interventions using number lines help children to improve on fraction understanding, and using supported self-explanation can add additional benefit to conceptual understanding of fractions (Fuchs et al., 2016). Fuchs and colleagues compared a number-line focused curriculum to the standard mathematics instruction students received, which switched the content focus from part-whole relations (76% of emphasis in control) to number-line instruction (70% of emphasis in intervention). The researchers further found that a supported self-explanation component helped to compensate for low working memory in the at-risk students performance. In another study, hearing a 3-min conceptual explanation of fractions and then playing a fraction video game with feedback led to greater improvements in fraction magnitude understanding and fraction number line estimation than a group who played the game with no instruction or feedback (Fazio, Kennedy, & Siegler, 2016).

Understanding of relative magnitudes in broader domains can also be improved by spatial representations. Resnick and colleagues found that by using scaled external representations scaled to time magnitudes, adults significantly improved on their understanding of the geological scale—importantly, the representations that supported relational reasoning showed greatest improvement (Resnick, Davatzes, Newcombe, & Shipley, 2017). Students used analogical reasoning to compare familiar time frames on linear paper representation from one scale to the next, with scales becoming less familiar (i.e., personal timeline to recorded history to Hadeon Eon), and saw each time scale on a physical linear representation relative to the others as they increased in magnitude. The activity led to improved estimation of extreme abstract magnitudes and reduced errors in understanding both geologic and astronomic scales.

Interventions that promote relational reasoning and use spatial representations seem to have positive impacts on relative magnitude understanding. It is reasonable to expect that more general spatial learning interventions that promote stronger relational reasoning skills may transfer to improved relative magnitude understanding, as well as transferring to other related domains. Unfortunately, although many intervention studies have shown the malleability in spatial skills (Uttal et al., 2013), interventions on spatial-relational thinking specifically are sparse. Yet studies using similar methods of providing benchmarks to support hierarchical coding (e.g., using midpoints), could be one method that works. There is also reason to believe that early spatial-relational skills could be developed through children's playful use of scaled representations.

Like many areas of development, there is reason to believe that spatial play, such as playing with puzzles, blocks, and board games, might promote spatial learning and related mathematics skills. As discussed above, play with a linear board game related to children's number line estimation, and the game can be used more informally as a classroom activity and still lead to learning (Ramani, Siegler, & Hitti, 2012). Less structured toys like blocks, puzzles, and mazes are found to relate to spatial performance (Fisher, Hirsh-Pasek, Newcombe, & Golinkoff, 2013; Jirout & Newcombe, 2014; Levine, Ratliff, Huttenlocher, & Cannon, 2012), and a block play intervention led to improved spatial thinking (Casey et al., 2008) and geometry performance when paired with storytelling (Casey, Erkut, Ceder, & Young, 2008). Children who play with spatial toys frequently tend to have higher spatial performance, even when controlling for several other cognitive abilities (e.g., working memory, processing speed, verbal ability), and when this relation was compared with other types of play, *only* spatial play mattered for spatial performance (Jirout & Newcombe, 2015). Toy and activity characteristics are important considerations to promote learning. For example, children's play with card games was associated with better number line estimation (Ramani & Siegler, 2008), but in a more recent study, the type of cards and activity mattered. Children did best in a playful intervention similar to the game War, where the magnitude of cards was compared as the main objective (i.e., to have the higher card), compared to games with the same cards played as Memory (i.e., matching rather than comparing magnitudes) or playing an Uno-type game with colors and shapes (Scalise, Daubert, & Ramani, 2017). Play can have lasting effects—the benefits of playing the magnitude comparison card game were still observed 2 months later. To take advantage of spatial play in promoting relational reasoning, activities can include physical representations of another object, space, perspective, or goal state (e.g., a diagram showing a finished block design, or a map to locate a hidden object). Representational use during play might facilitate children's learning to form more accurate, useful mental representations—helping them better understand spatial scale and reason about spatial relations (Uttal, 2000), perhaps through improved encoding of spatial relations based on feedback during representation use.

A final important consideration to determine effective means of promoting relative magnitude understanding is to explore directionality of these related processes. Perhaps spatial skills have a mediating effect on mathematics interventions, and could be targeted. Or it is also reasonable to ask whether improving relative magnitude understanding could help children learn to better attend to relational information, and if that could transfer to spatial-relational reasoning. While it is likely that the relation between spatial skills and mathematics skills is bidirectional, this is a question that can and should be explored in further research, as it has obvious implications for designing educational and play interventions to promote improved mathematics reasoning and spatial learning.

Conclusions

Researchers state that "precise representations of numerical magnitudes are foundational for learning mathematics" (Fazio, Bailey, Thompson, & Siegler, 2014, p 53), and in this chapter, we attempted to demonstrate the importance of relative magnitude understanding as a part of this foundation needed for mathematics learning. Relative magnitude reasoning may be more reliant on spatial-relational reasoning and more general spatial visualization skills than understanding exact number magnitude, and thus might involve different mental processing and representations. This idea is consistent with recent studies suggesting the relation between children's early exact-number knowledge and their approximate number system acuity is weak or nonexistent (Negan & Sarnecka, 2014). Relative magnitude tasks perhaps require a more flexible and abstract representation that can be applied to the information in a given task and that use spatial thinking processes, such as spatial-relational reasoning.

In some areas of mathematics learning found to be especially difficult, such as fraction learning, too much emphasis might be placed on symbolic numbers as exact values rather than thinking about numbers in more relative terms. In fact, this is what is often seen in errors made in learning topics like fractions, where students incorrectly choose the fraction with the denominator that has a larger absolute value than the distractor on a magnitude-comparison task, based solely on comparing absolute values of the denominator (e.g., one half vs. one fourth). Longitudinal data show that early magnitude reasoning skills relate to later conceptual understanding of fractions (Ye et al., 2016). Similarly, general cognitive abilities are important for mathematics learning, but specific early number skills relate to later fraction ability (Vukovic et al., 2014). Ability in kindergarten to recognizing number sets that—together—can produce a specific quantity (i.e., require consideration of multiple numbers rather than a single), as well as second grade arithmetic and number line estimation, were all predictive of better fraction understanding in fourth grade. The study authors suggest, "proportional reasoning might have a different learning trajectory separate from whole-number concepts and procedures" (Vukovic et al., 2014). For this reason, and because there is convincing evidence that interventions are successful in promoting better relative magnitude understanding and associated spatial-relational reasoning, education should consider ways of explicitly prioritizing relative magnitude learning along with more traditional whole-number knowledge and arithmetic processes. Further, because even children just beginning elementary school have relatively developed spatial-relational reasoning skills, this focus should begin early to better prepare students for later mathematics learning.

References

Barth, H. C., & Paladino, A. M. (2011). The development of numerical estimation: Evidence against a representational shift. *Developmental Science, 14*(1), 125–135.

Booth, J. L., & Newton, K. J. (2012). Fractions: Could they really be the gatekeeper's doorman? *Contemporary Educational Psychology, 37*(4), 247–253.

Boyer, T. W., & Levine, S. C. (2012). Child proportional scaling: Is 1/3= 2/6= 3/9= 4/12? *Journal of Experimental Child Psychology, 111*, 516–533.

Boyer, T. W., & Levine, S. C. (2015). Prompting children to reason proportionally: Processing discrete units as continuous amounts. *Developmental Psychology, 51*(5), 615.

Boyer, T. W., Levine, S. C., & Huttenlocher, J. (2008). Development of proportional reasoning: where young children go wrong. *Developmental Psychology, 44*(5), 1478.

Carbonneau, K. J., Marley, S. C., & Selig, J. P. (2013). A meta-analysis of the efficacy of teaching mathematics with concrete manipulatives. *Journal of Educational Psychology, 105*(2), 380.

Casey, B., Erkut, S., Ceder, I., & Young, J. M. (2008). Use of a storytelling context to improve girls' and boys' geometry skills in kindergarten. *Journal of Applied Developmental Psychology, 29*(1), 29–48.

Casey, B. M., Andrews, N., Schindler, H., Kersh, J. E., Samper, A., & Copley, J. (2008). The development of spatial skills through interventions involving block building activities. *Cognition and Instruction, 26*, 269–309.

Common Core State Standards Initiative. (2010). *Common core state standards for mathematics*. Washington, DC: National Governors Association Center for Best Practices and the Council of Chief State School Officers. Retrieved from http://www.corestandards.org/the-standards/mathematics.

Davies, C., & Uttal, D. H. (2007). Map use and the development of spatial cognition. In J. Plumert & J. Spencer (Eds.), *The emerging Spatial Mind* (pp. 219–247). New York, NY: Oxford University Press.

DeLoache, J. S. (1987). Rapid change in the symbolic functioning of very young children. *Science, 238*(4833), 1556–1557.

DeWolf, M., & Vosniadou, S. (2011, January). The whole number bias in fraction magnitude comparisons with adults. In Proceedings of the annual meeting of the cognitive science society (Vol. 33, No. 33).

Ebersbach, M., Luwel, K., Frick, A., Onghena, P., & Verschaffel, L. (2008). The relationship between the shape of the mental number line and familiarity with numbers in 5-to 9-year old children: Evidence for a segmented linear model. *Journal of Experimental Child Psychology, 99*(1), 1–17.

Fazio, L. K., Bailey, D. H., Thompson, C. A., & Siegler, R. S. (2014). Relations of different types of numerical magnitude representations to each other and to mathematics achievement. *Journal of Experimental Child Psychology, 123*, 53–72. https://doi.org/10.1016/j.jecp.2014.01.013

Fazio, L. K., Kennedy, C. A., & Siegler, R. S. (2016). Improving children's knowledge of fraction magnitudes. *PLoS One, 11*(10), e0165243. https://doi.org/10.1371/journal.pone.0165243

Fisher, K. R., Hirsh-Pasek, K., Newcombe, N., & Golinkoff, R. M. (2013). Taking shape: Supporting preschoolers' acquisition of geometric knowledge through guided play. *Child Development, 84*, 1872–1878.

Frick, A., & Newcombe, N. S. (2012). Getting the big picture: Development of spatial scaling abilities. *Cognitive Development, 27*, 270–282.

Fuchs, L. S., Malone, A. S., Schumacher, R. F., Namkung, J., Hamlett, C. L., Jordan, N. C., ... Changas, P. (2016). Supported self-explaining during fraction intervention. *Journal of Educational Psychology, 108*(4), 493.

Garcia Garcia, G., & Cox, R. (2010). "Graph-as-picture" in Misconceptions in young students. In A. K. Goel, M. Jamnik, & N. H. Narayanan (Eds.), *Diagrammatic representation and inference. Diagrams 2010. Lecture notes in computer science* (pp. 310–312). Berlin: Springer.

Gentner, D. (1988). Metaphor as structure mapping: The relational shift. *Child Development, 59*, 47–59.

Ginsburg, H. P., & Baroody, A. J. (2003). *Test of early mathematics ability*. Austin, TX: PRO-ED.

Gunderson, E. A., Ramirez, G., Beilock, S. L., & Levine, S. C. (2012). The relation between spatial skill and early number knowledge: The role of the linear number line. *Developmental Psychology, 48*(5), 1229.

Huttenlocher, J., Newcombe, N., & Sandberg, E. H. (1994). The coding of spatial location in young children. *Cognitive Psychology, 27*(2), 115–147.

Huttenlocher, J., Newcombe, N., & Vasilyeva, M. (1999). Spatial scaling in young children. *Psychological Science, 10*(5), 393–398.

Jirout, J., & Newcombe, N. S. (2014). Mazes and maps: Can young children find their way? *Mind, Brain, and Education, 8*(2), 89–96.

Jirout, J. J., & Newcombe, N. S. (2015). Building blocks for developing spatial skills evidence from a large, representative US sample. *Psychological Science, 26*(3), 302–310.

Ketterlin-Geller, L. R., Gifford, D. B., & Perry, L. (2015). Measuring middle school students' algebra readiness: Examining validity evidence for three experimental measures. *Assessment for Effective Intervention, 41*(1), 28–40.

Laski, E. V., & Siegler, R. S. (2014). Learning from number board games: You learn what you encode. *Developmental Psychology, 50*(3), 853–864.

Levine, S. C., Ratliff, K. R., Huttenlocher, J., & Cannon, J. (2012). Early puzzle play: A predictor of preschoolers' spatial transformation skill. *Developmental Psychology, 48*, 530–542.

Lortie-Forgues, H., Tian, J., & Siegler, R. S. (2015). Why is learning fraction and decimal arithmetic so difficult? *Developmental Review, 38*, 201–221. https://doi.org/10.1016/j.dr.2015.07.008

Lourenco, S. F., & Longo, M. R. (2011). Origins and development of generalized magnitude representation. In S. Dehaene & E. Brannon (Eds.), *Space, time, and number in the brain: Searching for the foundations of mathematical thought* (pp. 225–244). Waltham, MA: Academic Press.

Mix, K. S. (2010). Spatial tools for mathematical thought. Space and language, 41-66. In K. S. Mix, L. B. Smith, & M. Gasser (Eds.), *The spatial foundations of cognition and language: Thinking through space (no. 4)*. Oxford: Oxford University Press.

Mix, K. S., Levine, S. C., & Huttenlocher, J. (1999). Early fraction calculation ability. *Developmental Psychology, 35*, 164–174.

Möhring, W., Newcombe, N. S., & Frick, A. (2014). Zooming in on spatial scaling: Preschool children and adults use mental transformations to scale spaces. *Developmental Psychology, 50*(5), 1614–1619.

Möhring, W., Newcombe, N. S., Levine, S. C., & Frick, A. (2016a). The relation between spatial thinking and proportional reasoning in preschoolers. *Journal of Experimental Child Psychology, 132*, 213–220.

Möhring, W., Newcombe, N. S., Levine, S. C., & Frick, A. (2016b). Spatial proportional reasoning is associated with formal knowledge about fractions. *Journal of Cognition and Development, 17*(1), 67–84.

National Council of Teachers of Mathematics. (2010). *Principles and standards for school mathematics*. Reston, VA: Author.

National Mathematics Advisory Panel (NMAP). (2008). *Foundations for success: The final report of the National Mathematics Advisory Panel*. Washington, DC: US Department of Education.

National Research Council. (2009). *Mathematics Learning in Early Childhood*. Washington, DC: The National Academic Press.

Negan, J., & Sarnecka, B. W. (2014). Is there really a link between exact-number knowledge and approximate number system acuity in young children? *British Journal of Developmental Psychology, 33*(1), 92–105.

Newcombe, N. S., Frick, A., & Möhring, W. (2018). How big is many? Development of spatial and numerical magnitude understanding. In A. Henik & W. Fias (Eds.), *Heterogeneity of function in numerical cognition*. Academic.

Newcombe, N. S., & Huttenlocher, J. (2000). *Making space: The development of spatial representation and spatial reasoning*. Cambridge, MA: MIT Press.

Newcombe, N. S., Levine, S. C., & Mix, K. S. (2015). Thinking about quantity: The intertwined development of spatial and numerical cognition. *WIREs in Cognitive Science, 6*(6), 491–505.

Newcombe, N. S., & Shipley, T. F. (2015). Thinking about spatial thinking: New typology, new assessments. In J. Gero (Ed.), *Studying visual and spatial reasoning for design creativity* (pp. 179–192). Dordrecht: Springer.

Park, J., Bermudez, V., Roberts, R. C., & Brannon, E. M. (2016). Non-symbolic approximate arithmetic training improves mathematics performance in preschoolers. *Journal of Experimental Child Psychology, 152*, 278–293.

Park, J., & Brannon, E. M. (2014). Improving arithmetic performance with number sense training: An investigation of underlying mechanism. *Cognition, 133*, 188–200.

Peeters, D., Verschaffel, L., & Luwel, K. (2017). Benchmark-based strategies in whole number line estimation. *British Journal of Psychology, 108*, 668–686. https://doi.org/10.1111/bjop.12233

Piaget, J. (1952). *The child's conception of number*. New York: Norton.

Piaget, J., & Inhelder, B. (1956). *The child's conception of space*. New York: Humanities Press.

Ramani, G. B., & Siegler, R. S. (2008). Promoting broad and stable improvements in low-income children's numerical knowledge through playing number board games. *Child Development, 79*, 375–394.

Ramani, G. B., & Siegler, R. S. (2014). How informal learning activities can promote children's numerical knowledge. In R. C. Kadosh & A. Dowker (Eds.), *Oxford handbook of mathematical cognition*. New York: Oxford University Press. https://doi.org/10.1093/oxfordhb/9780199642342.013.012

Ramani, G. B., Siegler, R. S., & Hitti, A. (2012). Taking it to the classroom: Number board games as a small group learning activity. *Journal of Educational Psychology, 104*, 661–672.

Resnick, I., Davatzes, A., Newcombe, N. S., & Shipley, T. F. (2017). Using relational reasoning to learn about scientific phenomena at unfamiliar scales. *Educational Psychology Review, 29*(1), 11–25.

Sandberg, E., Huttenlocher, J., & Newcombe, N. (1996). The development of hierarchical representation of two-dimensional space. *Child Development, 67*, 721–739.

Scalise, N. R., Daubert, E. N., & Ramani, G. B. (2017, April). *Short- and Long-term effects of playing card games on low-income children's early mathematics skills*. Paper presented at the biennial meeting of the Society for Research in Child Development, Austin, TX.

Schneider, M., Heine, A., Thaler, V., Torbeyns, J., de Smedt, B., Verschaffel, L., ... Stern, E. (2008). A validation of eye movements as a measure of elementary school children's developing number sense. *Cognitive Development, 23*(3), 409–422.

Siegler, R. S., & Booth, J. L. (2004). Development of numerical estimation in young children. *Child Development, 75*(2), 428–444.

Siegler, R. S., & Braithwaite, D. W. (2017). Numerical development. *Annual Reviews in Psychology, 68*, 187–213. https://doi.org/10.1146/annurev-psych-010416-044101

Siegler, R. S., & Lortie-Forgues, H. (2017). Hard lessons: Why rational number arithmetic is so difficult for so many people. *Current Directions in Psychological Science, 26*(4), 346–351.

Siegler, R. S., & Opfer, J. (2003). The development of numerical estimation: Evidence for multiple representations of numerical quantity. *Psychological Science, 14*, 237–243.

Siegler, R. S., & Ramani, G. B. (2009). Playing linear number board games—But not circular ones—Improves low-income preschoolers' numerical understanding. *Journal of Educational Psychology, 101*, 545–560.

Singer-Freeman, K. E., & Goswamani, U. (2001). Does half a pizza equal half a box of chocolates? Proportional matching in an analogy task. *Cognitive Development, 16*, 811–829.

Slusser, E. B., Santiago, R. T., & Barth, H. C. (2013). Developmental change in numerical estimation. *Journal of Experimental Psychology: General, 142*(1), 193–208.

Spinillo, A. G., & Bryant, P. (1991). Children's proportional judgments: The importance of "half". *Child Development, 62*(3), 427–440.

Szűcs, D., & Myers, T. (2017). A critical analysis of design, facts, bias, and inference in the approximate number system training literature: A systematic review. *Trends in Neuroscience and Education, 6*, 187–203.

Uttal, D. H. (2000). Seeing the big picture: Map use and the development of spatial cognition. *Developmental Science, 3*, 247–286.

Uttal, D. H., Meadow, N. G., Tipton, E., Hand, L. L., Alden, A. R., Warren, C., & Newcombe, N. S. (2013). The malleability of spatial skills: A meta-analysis of training studies. *Psychological Bulletin, 139*, 352–402.

Uttal, D. H., & O'Doherty, K. (2008). Comprehending and learning from 'visualizations': A developmental perspective. In *Visualization: Theory and practice in science education* (pp. 53–72). Dordrecht: Springer.

Vukovic, R. K., Fuchs, L. S., Geary, D. C., Jordan, N. C., Gersten, R., & Siegler, R. S. (2014). Sources of individual differences in children's understanding of fractions. *Child Development, 85*(4), 1461–1476.

Woodward, J., Beckmann, S., Driscoll, M., Franke, M., Herzig, P., Jitendra, A., … & Ogbuehi, P. (2012). Improving Mathematical Problem Solving in Grades 4 through 8. IES Practice Guide. NCEE 2012-4055. *What Works Clearinghouse*.

Wynn, K. (1990). Children's Understanding of Counting. *Cognition, 36*(2), 155–193.

Ye, A., Resnick, I., Hansen, N., Rodriguez, J., Rinne, L., & Jordan, N. C. (2016). Pathways to fraction learning: Numerical abilities mediate the relation between early cognitive competencies and later fraction knowledge. *Journal of Experimental Child Psychology, 152*, 242–263.

Chapter 2
From Intuitive Spatial Measurement to Understanding of Units

Eliza L. Congdon, Marina Vasilyeva, Kelly S. S. Mix, and Susan C. Levine

Abstract The current chapter outlines children's transition from an intuitive understanding of spatial extent in infancy and toddlerhood to a more formal understanding of measurement units in school settings. In doing so, the chapter reveals that children's early competence in intuitive spatial thinking does not translate directly into success with standardized measurement units without appropriate scaffolding and support. Findings from cognitive science and education research are integrated to identify (a) the nature of children's difficulties with measurement units, (b) some effective instructional techniques involving spatial visualization, and (c) suggestions for how instruction could be further modified to address children's specific conceptual difficulties with standardized measurement units. The chapter ends by suggesting that the most effective instruction may be that which directly harnesses the power of children's early intuitive reasoning as those children navigate the transition into a deeper conceptual understanding of standardized units of measure.

Keywords Mathematical development · Spatial thinking · Spatial visualization · Units · Linear units of measure · Ruler measurement understanding · Spatial extent · Area · Angle · Misconceptions · Manipulatives · Gestures · Children · Infants · Cognitive development · Education · Instruction · Mathematics learning · Procedural understanding of measurement · Conceptual understanding of measurement

---

E. L. Congdon (✉)
Department of Psychology, Bucknell University, Lewisburg, PA, USA
e-mail: eliza.congdon@bucknell.edu

M. Vasilyeva
Lynch School of Education, Boston College, Chestnut Hill, MA, USA

K. S. S. Mix
Department of Human Development and Quantitative Methodology, University of Maryland, College Park, MD, USA

S. C. Levine
Departments of Psychology, and Comparative Human Development and Committee on Education, University of Chicago, Chicago, IL, USA

K. S. S. Mix, M. T. Battista (eds.), *Visualizing Mathematics*, Research in Mathematics Education, https://doi.org/10.1007/978-3-319-98767-5_2

The current chapter focuses on children's ability to understand and visualize spatial units of measurement, a foundational concept in mathematics. As stated by Gal'perin and Georgiev (1969), "Mastery of the initial concept of the unit is the most important step in the formation of elementary arithmetic concepts (they are all built on the unit or presuppose it)." In reviewing existing findings, we consider evidence of early measurement competence and evidence of later measurement struggles, and work to integrate and reconcile these seemingly disparate findings. We also outline some successful instructional techniques that have come out of basic cognitive science research. In doing so, we suggest that units of measure, an inherently spatial concept, are a fitting case study for understanding how children's learning outcomes are improved when spatial visualization techniques are employed during instruction.

Units are powerful because they allow us to meaningfully discretize continuous quantities, thereby allowing for extremely accurate comparisons across space and time. But the concept of "units" has important implications beyond this—it is also foundational to humans' understanding of quantity and numeracy more generally (e.g., Davydov, 1975; Gal'perin & Georgiev, 1969; Sophian, 2007). For example, when counting a set of shoes, one could count each shoe, or one could count each pair of shoes as one unit. Just as when measuring length, the numerosity obtained depends on the unit one adopts. Children under the age of 4 struggle with this idea. They tend to count any discrete physical object as a unit, even if the object is actually part of a larger unit (Brooks, Pogue, & Barner, 2011; Shipley & Shepperson, 1990). For example, when asked, "How many forks?", they might count one whole fork and one fork broken into two pieces as two forks, and respond that there are three forks altogether. At around age four, they instead respond that there are two forks, which is consistent with the way adults tend to answer this question. This suggests that with age, learners become increasingly sensitive to the unit-based information represented by nouns. In fact, when parts of objects have readily accessible names (e.g., wheels), children are able to focus on these part-of-object units at an earlier age than if the parts do not have labels (Shipley & Shepperson, 1990). These biases to attend to and count discrete physical entities that are readily labeled ultimately help children count different kinds of units (Shipley & Shepperson, 1990). This is critical since later in development, having an understanding that units are flexible and depend on the question one is addressing, becomes the backbone of children's understanding of topics such as place value, measurement, geometry, part-whole relations, and fractions (Piaget, Inhelder, & Szeminska, 1960).

Despite the importance of units in the ontogenetic development of mathematical thinking, there are well-documented challenges children face in understanding units of measure and how they are applied in problem-solving scenarios. Jean Piaget, a master observer of children's behavior, claimed that children were not capable of reasoning accurately about distance, length, or angle measure until middle childhood (Piaget et al., 1960). For example, children up until about 7 years of age were likely to fail a conservation of length test, stating that if one of two equal sticks was shifted with respect to the other, it had become "longer." In another classic experiment, Piaget showed children ages 3–7 years a tower of blocks and then asked them to construct a tower of equal height with smaller blocks on the other side of the room. Children did not create a metric of conversion (e.g., "two large blocks are equal to 4 smaller blocks") nor did they spontaneously use available resources to aid in transitive inference (wooden sticks and strips of paper) until at least 7 years of age. Similarly, children who were asked to replicate a drawing of an angle figure tended to approximate the drawing and did not spontaneously measure with available paper, string, or compasses until mid-to-late elementary school. These findings suggest that children may have a fundamental misunderstanding of the form and function of formal systems of measurement.

Even when children do receive instruction about the proper usage of measurement tools, they continue to demonstrate conceptual difficulties. Recent international assessments of children's mathematics performance indicate that children perform particularly poorly on measurement test items as compared to other mathematics assessment items at least through fourth grade (TIMSS, 2011), echoing similar patterns of findings reported decades ago (Carpenter et al., 1988; Lindquist & Kouba, 1989). Children also struggle with test items about angle measures through elementary school and even middle school (e.g., Clements & Battista, 1992; Mitchelmore & White, 2000).

In stark contrast to these dire assessments of children's understanding of formal spatial units, there is ample evidence that young children and infants are able to reason intuitively about continuous extent, length, and angle (e.g., Baillargeon, 1987; Lourenco & Huttenlocher, 2008; Slater, Mattock, Brown, & Bremner, 1991; Spelke, Lee, & Izard, 2010). For example, 2- to 4-month-old infants who are habituated to an angle figure will dishabituate to a change in angle measure (Cohen & Younger, 1984), and 5- to 7-month-old infants can encode an object's height and make subsequent predictions about its behavior even when the object is not visible (Baillargeon & Graber, 1987). What, then, can explain how children's intuitive understanding of spatial extent gets "lost in translation" when encountering similar concepts in formal schooling contexts? In the current chapter, we propose that a formal understanding of units requires children to overcome two challenges. First, they must integrate their intuitive understanding of continuous spatial extent with discrete, countable entities. In other words, one challenge of mastering units of measure is that they lie squarely at the intersection of intuitive spatial understanding and learned numerical representations. Secondly, children must connect intuitive, non-verbal understandings with the corresponding formal concepts that are referenced by newly acquired spatial language terms (e.g., units, angle, length, area, volume).

The present chapter reviews the literature related to these developmental achievements. In Part I, we review evidence that young infants have the perceptual capabilities to process and compare various dimensions of continuous extent. In Part II, we discuss how these perceptual abilities of infancy fail to directly translate to success with formal units of measure in school settings. In Part III, we end with some optimistic evidence from successful training interventions that help school-aged children to bridge the gap between intuitive understanding of extent and formal units of measure.

Part I: Intuitive Understanding of Extent

Before they are introduced to formal measurement and numerical systems, there is evidence to suggest that even infants can make judgments that reflect their sensitivity to continuous spatial extent, a developmental precursor of measurement skills. In a series of violation of expectation paradigms investigating infants' ability to reason about an occluded object, 5- to 7-month-old infants can encode an object's height and make subsequent predictions about its behavior when the object is not visible (Baillargeon, 1987; Baillargeon & Graber, 1987). For example, 7-month olds expected a rotating screen to stop sooner when a taller object was placed behind the screen than when a shorter object was placed behind the screen. In a separate study, 5-month olds were surprised when a taller rabbit's path of movement behind a barrier did not show the rabbit's head poking above the barrier (Baillargeon & Graber, 1987). Extensions of these findings show that infants as young as 5.5 months can simultaneously track the width of one object—in this case, a cylinder—and the displacement distance of a second object, a small bug toy, to reason appropriately about collision events (Kotovsky & Baillargeon, 1998). Further, 6.5-month-old infants can use proportional information about objects that are partially resting on a surface to predict when the object has sufficient support and when it will fall (Baillargeon, Needham, & DeVos, 1992).

In addition to this research evidencing infants' qualitative judgments about height, width, and distance, there has been research suggesting that young children can reason quantitatively about extent. That is, some researchers have proposed that infants may be able to encode and reason about the absolute size of objects. In one study, 6-month olds were habituated to a glass cylinder with a certain amount of red liquid (Gao, Levine, & Huttenlocher, 2000). At test, infants dishabituated to the same size cylinder with a novel amount of liquid, but not to the same size cylinder with the same amount of liquid. In an experiment where objects were hidden in a long, narrow rectangular sandbox, children as young as 5-months old were surprised when the object was revealed in a location 6 in. from where it was initially hidden (Newcombe, Huttenlocher, & Leamonth, 2000; Newcombe, Sluzenski, & Huttenlocher, 2005). These findings could indicate that infants are capable of encoding approximate absolute extent without the explicit presence of a measurement standard or comparison object.

However, subsequent evidence has called this conclusion into question and has shown that this early reasoning about height and length may be based on intuitive proportional reasoning rather than a true understanding of absolute extent. In all of the work described above, target stimuli were presented within some sort of container (e.g., sandbox, glass cylinder), next to another comparison object, or in relation to the salient frame of a computer screen. Because of this, the absolute height of stimuli (e.g., more liquid in a cylinder) was conflated with the relative proportion the stimulus occupied within a container or relative to a frame (e.g., the liquid fills a larger proportion of the cylinder).

Evidence for encoding of relative extent. To disentangle the question of whether infants encode absolute or relative spatial extent, several experiments were conducted. In one study, infants were habituated to a wooden dowel in one of three conditions: alone, within a glass cylinder, or next to a wooden stick (Huttenlocher, Duffy, & Levine, 2002). Infants only dishabituated to a novel dowel when the dowels during habituation and test were presented either inside the glass container or next to the wooden stick. The most parsimonious explanation of these data is that infants were using the container or the stick to encode relative height. Yet it remained a possibility that the mere presence of a second object or container heightened infants' awareness of absolute extent of the original object. To directly address this possibility, a second study directly compared infants' sensitivity to absolute versus relative extent in the presence of a container (Duffy, Huttenlocher, Levine, & Duffy, 2005). In this work, 6.5-month-old infants were habituated to a wooden dowel that was a specific height, say three inches, and filled a certain proportion, say three-fourths, of the clear cylinder in which it was placed. In the key test conditions, infants were shown a larger cylinder with a wooden dowel that either filled the same proportion (e.g., three-fourths), or was the exact same absolute height as the original dowel (e.g., three inches). Infants dishabituated to the latter display—the same size object as in the original display with a different proportional relation to the container. These findings indicate that infants were encoding the height of the dowel relative to its container, and not its absolute height.

Using a different experimental technique where 2- to 4-year-old children were asked to remember the height of a target object and then select the matching object in a two-option test trial, these investigators found that it was not until 4 years of age that children were able to accurately encode the height of the target objects. Even then, they were only able to make the correct selection at test in the presence of a salient comparison standard and a distractor that was substantially different in size from the target (Huttenlocher et al., 2002). By 8 years of age, the ability to focus on absolute extent was more refined and children could differentiate lengths that were closer in size, perhaps by imposing a mental unit, such as a mental inch (Vasilyeva, Duffy, & Huttenlocher, 2007).

Continued development of proportional reasoning. If young children are indeed encoding relative and not absolute extent as a kind of proportional reasoning, to do so still requires an impressive set of reasoning skills and an emerging ability to unitize, with resulting improvements in precision. In the sandbox search paradigm mentioned above, work with toddlers has shown that by the age of 24 months, most children can remember the location of a hidden object long enough to go retrieve it from the sandbox (Huttenlocher, Newcombe, & Sandberg, 1994). The patterns of children's errors were biased toward the center of the sandbox, suggesting a rudimentary unitizing of the continuous space into two equal parts (Huttenlocher et al., 1994). With increasing age, children's errors cluster around smaller division points (e.g., by dividing the space into quarters). This intuitive unitization of continuous extent sharpens over developmental time and has been hypothesized to represent a Bayesian combination of categorical and continuous information, and may be a necessary precursor to children's understanding of later unit-based concepts (Mix, Levine, & Newcombe, 2016; Newcombe, Levine, & Mix, 2015). Similar patterns have been identified in number line estimation tasks, where children's improvement over developmental time can be explained by improvements in proportional reasoning, rather than the previously proposed qualitative shift from logarithmic to linear representations of the number line (e.g., Barth & Paladino, 2011).

As children get older, their increasing ability to represent and reason about proportional relations of continuous quantities predicts success on more demanding spatial reasoning tasks, such as map reading (Huttenlocher, Newcombe, & Vasilyeva, 1999) and symbolic fraction tasks (Möhring, Frick, Newcombe, & Levine, 2015). Though toddlers struggle with map reading tasks and tend to rely on matching of object features, by the age of 4, they are able to read a simple map that indicates the location of a hidden object. Researchers have argued that success on this task relies, in part, on the same skills young infants use when coding relative extent on simpler object comparison tasks (Duffy, Huttenlocher, & Levine, 2005; Huttenlocher et al., 2002; Vasilyeva & Lourenco, 2012). In this case, the scaling of distance from the map to the object search space is akin to encoding the relative extent between, say, a dowel and its container, and being able to identify this same proportional relation in a test trial. This emerging map reading ability in children, while impressive, remains quite fragile through many more years of developmental change. Children struggle with reading maps when the referent space is misaligned or shifted in orientation (Liben & Downs, 1993), when the scale of the referent space becomes too large (Davis & Uttal, 2007), or when the space becomes too complex or includes distracting but salient landmark features (Liben & Yekel, 1996).

Understanding of angles. Just as with judgments of length, young infants show an intuitive understanding of angle well before they learn about angles in formal school settings. For example, 2- to 4-month-old infants who are habituated to an angle figure composed of two line segments will dishabituate to a change in angle, or a change in the relative position of the lines composing the angle figure, but do not dishabituate to a change in orientation of the entire figure (Cohen & Younger, 1984). In addition, infants differentiate between acute and obtuse angles (Cohen & Younger, 1984; Lourenco & Huttenlocher, 2008), and even newborn infants are capable of tracking the relation between two components or features of an angle figure (Slater, Mattock, Brown, Burnham, & Young, 1991). This work suggests that the perceptual skills needed to encode angles are present very early in development, and indeed, may be innate (Izard, O'Donnell, & Spelke, 2014).

Children approaching kindergarten age begin to make explicit decisions and judgments based on these early percepts of angle. For example, 4-year-old children can accurately identify which of six figures drawn on a card looks different from the others when the key dimension of difference is angular measure (Izard & Spelke, 2009). And though performance is more variable, some 4-year-olds can match fragments of geometric figures from two-dimensional to three-dimensional space when the only informative dimension available in the fragments is angle measure (Izard et al., 2014).

Navigation tasks, which require children to use angle on a 3-D scale, follow a more protracted developmental trajectory than 2-D tasks. For example, 4-year-old children are much more likely to spontaneously use distance cues to succeed on a map task than they are to use angle or orientation (Shusterman, Ah Lee, & Spelke, 2008). However, by the age of 5 or 6, children can successfully use angular relations in map reading and navigation tasks (Spelke, Gilmore, & McCarthy, 2001).

Similar to what has been found for other forms of measurement, the early sensitivity to angle does not confer immediate success in understanding more formal systems of angular measure, which children struggle with until much later in development. Indeed, as we will discuss, they are often confused by irrelevant information—such as the absolute the lengths of the lines composing the angle figure—in making judgments about which of two angles is larger.

Interim conclusions. These and other findings show that before children are exposed to any explicit training in formal systems for linear measurement or angle measure, they are sensitive to continuous spatial extent. This sensitivity is largely confined to reasoning about relative or proportional rather than absolute quantities, but there is some indication that successive divisions of continuous space—which can be regarded as nascent measurement units—might help young children gain greater precision through the first 5 years of development. Taken together, this work suggests that the foundations for reasoning about spatial units emerge quite early.

Part II: Transition to Understanding of Conventional Measurement

Infants' and young children's intuitive reasoning and perceptual sensitivity to differences in length, distance, and rotational measure are necessary but not sufficient for success on unit-based tasks in formal school settings. In this section, we begin by reiterating the key concepts children must learn to navigate the transition from intuitive reasoning about continuous extent to more formal reasoning that involves the application of discrete units to gain precision about these extents. We then briefly discuss how educators assist children in this transition by looking at common teaching techniques for unit-based measurement topics in mathematics. Finally, we identify some potential shortcomings in current instructional practices, and in doing so, strive to characterize some of the common misconceptions children develop regarding unit-based tasks.

Key concepts for children to master. One reason that measurement may prove difficult for young children is that it requires them to integrate their preexisting imprecise intuitions about quantity and continuous extent with conventional, number-based measurement tools such as rulers. When using a simple tool like the ruler, children must understand a set of not-so-simple conventional rules such as what to count, where to start (and stop) counting, that the beginning of the ruler is the zero-point, and the significance of the hatch marks and numbers on the ruler (Solomon, Vasilyeva, Huttenlocher, & Levine, 2015). Simultaneously, children must master several key ideas about spatial units more generally—they are consistent in size within a given measurement instance; a single unit can be iterated to determine length; and units follow an inverse relation rule. That is, as the unit size increases, the number of units needed decreases.

Beyond learning the conventions of measurement tools and units, a true conceptual understanding of measurement requires that children can make transitive inferences (e.g., if the length of A = B and B = C then A = C). More concretely, children must understand that to compare the length of two objects, one can measure the first object and then the second, providing a way to compare the sizes of two objects even when the measurements are separated by time, physical distance, or both (Sophian, 2007). This kind of transitive inference is not intuitive, and there is evidence that children are not capable of this kind of thinking until at least 4 or 5 years of age (e.g., Bryant & Kopytynska, 1976; Miller, 1989; Piaget et al., 1960).

Cognitive biases can inhibit learning. There are several cognitive biases that may inhibit children's ability to master these important conventional rules about measurement. First, children have a tendency to attend to and count bounded objects (e.g., Sophian, 2007) or "countable entities" (Shipley & Shepperson, 1990). Yet the units on conventional measurement tools are spatial intervals, which are, in a sense, "non-objects." The numbers and hatch-marks on rulers serve as countable distractors, obscuring the link between ruler markings and the spatial interval units they represent. Indeed, children who fail on unaligned ruler measurement problems (see Fig. 2.1a) succeed on measurement problems where the to-be-measured object is unaligned with respect to a set of discrete, adjacent circles, which are more readily countable objects than spatial interval units on a ruler (see Fig. 2.1b) (Solomon et al., 2015). On unaligned measurement problems with rulers, young children tend to make one of two kinds of errors (Lehrer, Jenkins, & Osana, 1998; Solomon et al., 2015). They either read off the number on the ruler that aligns with the rightmost edge of the object (i.e., read-off error) irrespective of where the object begins, or count the hatch marks rather than the intervals of space that fall between an object's left-most and right-most edges (i.e., hatch-mark counting error). This later strategy likely reflects an object counting bias because children are drawn to count the object (i.e., lines or nubers) rather than the spaces.

In addition to their documented bias to count objects, children also show a bias to estimate continuous quantities based on perceptual spatial cues alone even when a salient, helpful discrete cue is present. For example, in one experiment by Huntley-Fenner (2001), preschool children were presented with two boxes. One box had three clear glasses full of sand, and one had two. When asked which box had more glasses, children could easily say that the box with three glasses full of sand had more glasses than one with two glasses. In this first case, participants were asked to make a judgment about discrete quantities using discrete units. But when asked which of the two boxes had more *sand*, children's performance dropped significantly and was no better than when they were asked to compare piles of sand consisting of

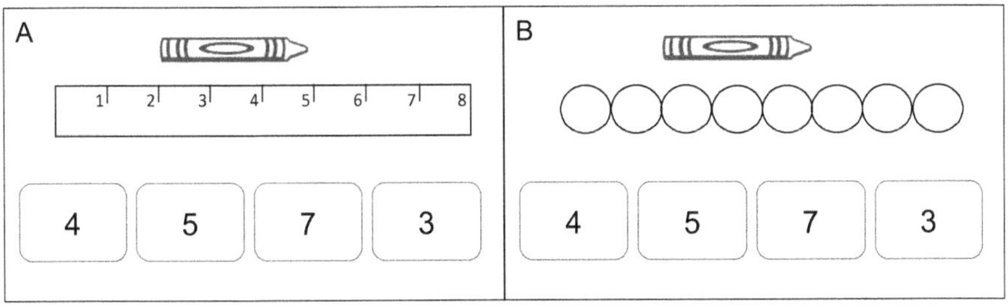

Fig. 2.1 Sample measurement items: (**a**) unaligned ruler with inch units and (**b**) circles representing units. Question: "How many units long is the crayon?" The correct answer is 3. Common answers for the ruler item are 4 and 5, for the circles item the most common answer is 3

these same amounts. In this second case, children were asked to make a judgment about a continuous dimension using a discrete unit—the glasses. Thus, although a discrete unit was readily available, children did not spontaneously use this option, and performance reflected the noisy guesses one would expect from a task asking children to compare continuous quantities based on approximate perceptual estimation. Together, these cognitive biases likely interfere with children's ability to grasp the function of spatial units, how they can be used, and how they are incorporated into conventional tools such as the ruler.

Traditional classroom instruction. Children in American schools are typically given two different types of measurement instruction to help them understand and visualize spatial units. In the first, they are provided with unconventional units (e.g., paperclips, shoes, coins) and asked to measure an object or distance by lining up the units, end to end. While the goal of such an exercise is ostensibly to teach children about the importance of utilizing same-size spatial interval units—a key measurement concept—there is research suggesting that children do not spontaneously make the link between objects and spatial intervals. Children may see such an activity as a game in which the goal is to count objects, not to measure. Indeed, children often leave gaps between objects, overlap objects during an iterative procedure (Bragg & Outhred, 2004; Lehrer, 2003) or select units of differing, nonstandard sizes to line up along an object's edge (Lehrer et al., 1998). Such errors indicate that children do not understand a fundamental aspect of measurement—that it requires the use of adjacent equal-size units. Moreover, even if they execute the measurement correctly, they may not grasp that the objects used represent underlying spatial extents and may instead view the exercise as an object counting task (Solomon et al., 2015).

A second common classroom activity is to ask children to measure objects with a ruler by aligning the object with the leftmost edge (zero-point) of the ruler and reading off the number at the rightmost edge of the object (Smith, Males, Dietiker, Lee, & Mosier, 2013). Such a procedure is effective when children perform it properly. Yet there is evidence that this type of instruction leaves children with a relatively shallow, procedural understanding of measurement. The measurement performance of children in early elementary school is particularly poor on test items where objects to be measured are not aligned with the "0" point on the ruler, such as the problem depicted in Fig. 2.1a above (Clements, 2003; Lehrer et al., 1998; Solomon et al., 2015; Wilson & Rowland, 1993) or when they are a given a "broken" ruler that does not begin at 0, but rather at some other non-0 starting point (Nunes & Bryant, 1996). Importantly, both the read-off and hatch-mark counting strategies, described above, consistently result in correct answers when the object to-be-measured is properly aligned at the 0-point of the ruler. In other words, classroom instruction that does not challenge children with difficult, shifted-object problems may allow misconceptions to go unnoticed by educators.

Different instructional needs. Performance on these difficult shifted ruler problems raises questions about whether children may require different types of instruction depending on their specific misconceptions. Children who use the read-off strategy tend to be younger or from lower socio-economic status backgrounds (Kwon, Levine, Ratliff, & Snyder, 2011; Solomon et al., 2015) and therefore may have less experience with measurement problems. They also tend to have lower spatial working memory than their peers who make hatch-mark counting errors, despite equal scores on a verbal working memory task (Congdon & Levine, 2017). Irrespective of the specific cause of their misconception, children who use the read-off strategy have no trouble perceiving that an object does not begin at the start of the ruler, but do not know how to adapt their strategy to account for this unusual arrangement when asked how long an object is. This suggests that there may be something about how these questions are typically asked (i.e., "How long is the X?") that leads children to assume they are being asked to determine the end-point of an object no matter where that object begins. Children of this age, around 5 years old, are also likely to say that two walking paths, one straight and one with a large bend in it are the same length if they have the same starting and ending points (Clements, 1999).

By contrast, children who count hatch marks are aware that determining length involves counting something, but appear to be distracted by an object-counting bias that draws their attention to the lines rather than the spaces. These children may have a firmer understanding of the pragmatics of the problem, but have not yet mastered the ways in which rulers represent discrete spatial units. In other words, these children do not understand the relation between a single unit and the whole object that is being measured, and how that relation is represented by the ruler (Lehrer, 2003). Overall, neither the read-off nor the hatch-mark counting strategy indicates an understanding of the concept of a unit as a measure of that involves uniform spatial intervals (Kamii, 2006; Martin & Strutchens, 2000).

International performance. Alarmingly, children in the United States score lower on test items assessing measurement skills than on items assessing most other mathematical topics (Carpenter et al., 1988; Clements & Bright, 2003; Lindquist & Kouba, 1989; Mullis, Martin, Gonzalez, & Chrostowski, 2004; National Center for Educational Statistics, 2009). Specifically, when given a multiple choice test item akin to the one shown in Fig. 2.2, only 20% of US fourth grade students answered correctly, a rate that is lower than chance and that was significantly lower than the international average (TIMSS, 2011). These struggles could potentially be due to limitations in current classroom instructional practices in the United States. Though there is little published research comparing specific instructional methods in measurement across countries, there is some work suggesting that in the countries where children are generally more successful on measurement test items, like Japan (where 52% of fourth grade children answered that same problem correctly), children are given more opportunities to engage in creative problem-solving and critical thinking than they are in the US (Kawanaka, Stigler, & Hiebert, 1999). These children spend less time practicing memorized procedures, and more time discussing and exploring ideas with the teacher. Though a causal link cannot be drawn between general cultural differences in teaching practices and differences in understanding of units, the parallels are suggestive of the idea that children in the US would benefit from deeper engagement in exploring the conceptual underpinnings of measurement.

Higher order measurement skills. Even if children master the basic procedures of linear measurement, they continue to struggle with unit-related concepts later into childhood. For example, many children find it difficult to understand the inverse relation of units—that you will need more units to measure something if the units are smaller, and fewer units if they are larger (Hiebert, 1984). First grade children overwhelmingly rely on the number of units in a task, and will attempt to keep that number constant when re-measuring an object, even if the experimenter has changed the unit size. Hiebert argued these inverse relations are difficult for young learners because understanding them requires both (1) an understanding of conservation (i.e., the idea that an object does not change length even when moved in space or measured a second time), and (2) an understanding of transitivity (i.e., the idea that two objects can be compared with a standardized measurement tool).

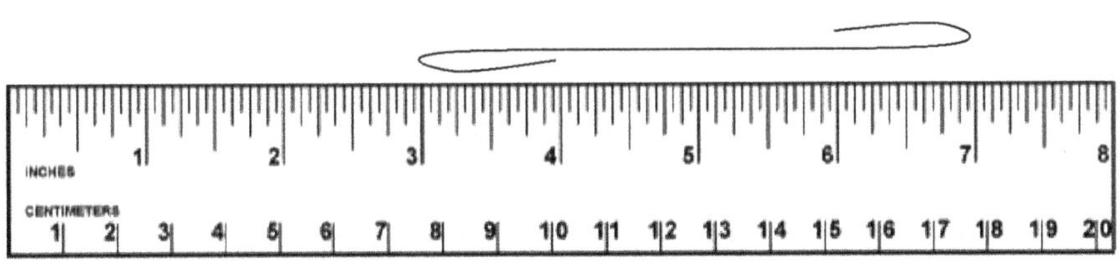

If the string above was pulled straight, which of the following would be closest to its length?

a) 4 inches
b) 6 inches
c) 7 inches
d) 8 inches

Fig. 2.2 A test item similar to one included in the 2011 version of the Trends in International Mathematics Fourth Grade Assessment. The correct answer is b

Hiebert claimed that children who have not yet mastered these two ideas tend to fall back on counting strategies to compare object lengths, irrespective of changes in unit size. Some research suggests that when task complexity is decreased, children as young as 7 years can succeed with unit conversion (Sophian, Garyantes, & Chang, 1997), but other work suggests that children continue to struggle as late as fourth grade, or 9–10 years of age (Vasilyeva, Casey, Dearing, & Ganley, 2009). For example, when given the following scenario, "It took Marc 8 steps to cross the room and it took Peter 5 steps. Who has the longer step?," Fourth grade students tended to respond incorrectly that Marc had larger steps.

Another complex idea involved in a mature understanding of measurement is unit conversion. Even when given a short lesson to demonstrate that 5 centimeters is about the same as 2 in., children up through 8 years old make mistakes in judging the relative lengths of two objects that have been measured in different standard units (Nunes & Bryant, 1996). For example, they may state that a 3-cm stick is longer than a 2-in. stick, suggesting that they rely primarily on the number of units than on the number *and* size of units. Lastly, even after children understand the importance of standardized units, they continue to struggle with selecting the appropriate units for certain measuring tasks (Tipps, Johnson, & Kennedy, 2011). For example, first grade children might not know whether it is more appropriate to measure a computer screen with inches or feet. Currently, the Common Core State Standards do not explicitly suggest introducing these more advanced unit-related concepts until second or third grade, perhaps explaining why children show protracted understanding of these mathematical ideas (e.g., Clements, 1999; Tipps et al., 2011).

Area measurement. While children's understanding of linear measurement is important on its own terms, understanding units and how they represent equal parts of space also lays the groundwork for later understanding of mathematical concepts such as perimeter, area, and fractions. In linear measurement, there is only one dimension on which to compare two objects. In area measurement, there are two relevant dimensions, making this a more difficult concept. Research shows that children consistently struggle on area measurement problems through at least fifth grade, frequently confusing perimeter and area, for example (Lin & Tsai, 2003; Strutchens, Martin, & Kenney, 2003). Without a thorough understanding of units, children tend to fall back on visual comparison techniques that may have worked when comparing the linear extent of two objects, but are more difficult to apply successfully when comparing the area of two differently shaped objects (e.g., Yuzawa, Bart, & Yuzawa, 2000). Additionally, there is evidence that children rely heavily on memorized formulas to calculate the area of shapes without developing a conceptual understanding of why this procedure works (Barrantes & Blanco, 2006; Strutchens, Harris, & Martin, 2001).

Proportions and fractions. Despite the fact that infants and young children have an intuitive understanding of proportion, understanding conventional fractions is notoriously difficult. Fractions, unlike whole numbers, require children to keep track of the relative magnitude of two different sets of units—the denominator representing the number of partitions of a whole unit and the numerator, the number of these units. It also requires that children understand the non-intuitive inverse relation between the size of the denominator and the magnitude of the fraction, a skill that is fundamentally linked to the idea of units of measure and is parallel to understanding the inverse relation between the size of the unit and the number of units in measurement (Sophian, 2007; Sophian et al., 1997).

Even with simple fractions, such as one-half, 4-to-7-year-old children who show an intuitive understanding of one half of a continuous quantity have extreme difficulty when the quantity is discretized into units (Hunting & Sharpley, 1988). For example, they may be able to bisect a cookie roughly in half, but struggle to decide what constitutes half of a set of 12 blocks. Similarly, 6- to 10-year-olds succeed on proportional equivalence tasks when continuous quantities are used but not when these same quantities are divided into countable units. In this latter condition, instead of reasoning proportionally, children tend to rely on counting the number of "shaded" units, akin to attending only to the numerator of a fraction, or they count the total number of units (e.g., Boyer, Levine, & Huttenlocher, 2008).

These findings suggest that access to approximate proportional magnitudes is not sufficient to learn how to map the number words of fractions to their proper unit referents. With more complex fractions, older children will commonly apply a label like "three-fifths" to an image with, say, three shaded parts and five unshaded parts, rather than three shaded parts and two unshaded parts (e.g., Mix & Paik, 2008; Newcombe et al., 2015). Even middle school and high school students will try to add two fractions by simply adding both numerators and both denominators (Kerslake, 1986). These errors indicate that children have a fundamental misunderstanding about how the denominator of a fraction delineates unit size while the numerator indicates the number of units.

Angular measurement. Children's difficulties with angle measurement share some parallels with their misconceptions about linear measurement (Clements & Battista, 1992; Mitchelmore & White, 2000). For example, children must master the ideas of equal partitioning of space, and must understand unit iteration (Clements & Stephan, 2004). There is also evidence that young learners have difficulty understanding the proper referent of the word *angle* (Gibson, Congdon, & Levine, 2015). Because of a quirk of the English language, the word *angle* can actually be used to refer to both the figure of an angle, composed of two rays extending from the same point, and to the measure of rotation between the rays (Clements & Stephan, 2004). This linguistic ambiguity likely contributes to longstanding misconceptions for children in elementary and even middle school who will focus on irrelevant properties such as the length of an angle's sides in a figure, the area contained within the sides, or the absolute distance between the sides when making judgments about the size of angles (see Fig. 2.3; Clements & Battista, 1989; Lindquist & Kouba, 1989).

In school settings, angles are not typically introduced until second or third grade. Before that point, curricula tend to avoid proper spatial labels, instead calling angles "corners." In addition, there is some evidence from case study observations that protractors (i.e., tools for measuring angular rotation), are challenging for children to understand and may be imbuing them with a sense of angle as a static measure rather than allowing them to imagine angles as a dynamic measure of rotation

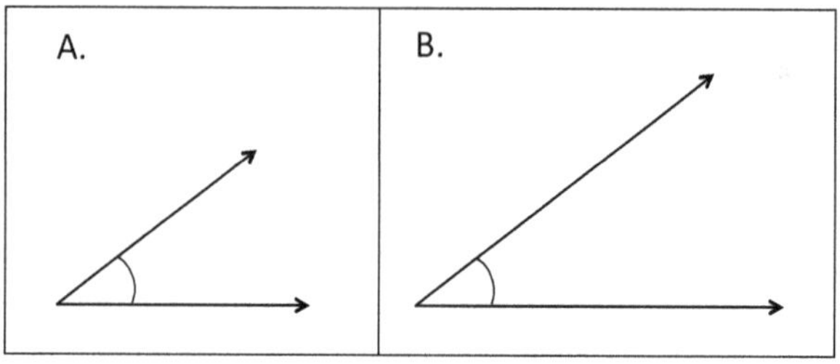

Fig. 2.3 These two angle measures are equivalent, but as late as middle school, children will assume B is a larger angle due to overall size, line length, or distance between the rays

(Clements & Burns, 2000). Given a paucity of research on the topic, it is currently unclear whether confusing use of the word "angle," late exposure, conventional tools, or a combination of the three are to blame for children's long-term misconceptions about angles. What is clear is that similar to other kinds of measurement, children struggle to make the transition from intuitive, perceptual reasoning about angles to a more formal understanding of angular rotation and angle size.

Interim conclusions. Taken together, this rich literature on children's difficulties with measurement reveals a few consistent patterns. First, the transition from reasoning non-verbally about continuous spatial extent to understanding and visualizing discrete units of spatial extent is challenging for children across many subdomains of measurement including linear measurement, angle, and higher-level skills like area and fraction understanding. The specific challenges include learning the proper referents of newly acquired spatial language (e.g., "length/long" does not always mean end-point; "units" on a ruler are not hatch marks or numbers; and "angle size" refers to a measure of rotation rather than the length of the lines that comprise the angle). Second, children must learn to use conventional unit-based tools and understand how those tools allow for transitive inference. Third, children must understand units themselves, which are a way to integrate intuitive understanding of continuous properties with exact numerical representations. Lastly, some of this work suggests that current instructional practices may be overemphasizing rote procedures or improper or ambiguous use of spatial language that could be leaving learners with poor conceptual understanding and thus, a shaky foundation for later mathematics success.

Part III: Training Interventions

It is clear that children face many challenges when making the transition from an intuitive understanding of continuous extent to a formal understanding of unit measures. In the final section of the chapter, we review interventions designed by researchers to scaffold children's learning in the domains of linear measurement, area measurement, and angle understanding. Our aim is to showcase proven instructional techniques, while further clarifying the nature of children's difficulties.

Improving spatial visualization of linear measurement units. To date, the majority of research on linear measurement has documented the nature of children's difficulties and misconceptions. Only a small number of studies have focused on correcting those misconceptions and helping children visualize how discrete units can comprise continuous lengths. One research group did an in-depth case study with eight students who were given a number of different measurement activities and who were continuously assessed across nearly a full year (Barrett et al., 2012). Based on their findings, the authors proposed several instructional tasks that could move children from one conceptual stage of measurement understanding to the next. For example, having children draw their own rulers, having an instructor overlap units to get children to think about why that is problematic, or having an instructor explicitly teach about how to deal with fractions of a unit. At the highest level, the authors argue that learning about intervals as countable units was not sufficient to promote a full conceptual understanding of measurement, so they proposed a lesson to link the ruler, hatch marks, spaces, and numbers all at once. This work, while certainly valuable, used many instructional strategies at once, and did so over a long period of time, making it difficult to ascertain which specific features of the instruction might have driven children's improvement.

In a more recent study, researchers tested whether exposure to and training on measurement test items with objects shifted away from the start of the ruler (unaligned problems) might be beneficial to learning (Kwon, Ping, Congdon, & Levine, under revision). The children completed a brief training lesson with either unaligned ruler problems or more traditional aligned ruler problems, with the object starting at the 0-point on the ruler. The results showed that exposure to unaligned ruler problems during training was crucial for learning. The authors argued that the unaligned ruler problem training was powerful because it provided children with self-discovered evidence that disconfirmed their previous strategies, a technique that can lead to better learning outcomes (e.g., Ramscar, Dye, Popick, & O'Donnell-McCarthy, 2011; Rescorla & Wagner, 1972). For example, if a child who used the hatch-mark counting strategy initially believed an object to be five units long, they would generate a guess of 5, then count the spaces and quickly discover that they were only at the number 4 when reaching the end of the object (Kwon et al., under revision).

A second study used a similar procedure to test the relative efficacy of different ways of drawing attention to a spatial interval as a unit of measure. One group of children was given practice on shifted-object measurement problems with discrete plastic unit chips, and a second group was given the same instruction but was taught to use a thumb-and-forefinger "pinching" gesture instead of the unit chips (Congdon, Kwon, & Levine, 2018). Results showed that children who started the session by counting hatch-marks improved markedly after either type of instruction, whereas children who began the session with the read-off strategy improved much more after unit-chip instruction than gesture-based instruction. These findings suggest that even within a single age group and single domain, children at a lower level of conceptual understanding may need more concrete scaffolding to promote learning. Notably, children who used the read-off strategy and received unit chip training occasionally switched their strategy to a hatch-mark counting strategy after training.

This strategy shift suggests that the training helps by causing children to reevaluate their understanding of the referent of "unit"—a process that occasionally goes awry due to an object counting bias.

Teaching area measurement. A solid understanding of linear measurement can help children when they encounter more difficult problems, such as measuring the area of a two-dimensional figure. As discussed in Part II, a true understanding of area measurement requires children to coordinate multiple dimensions and to understand, conceptually, how formulas for calculating area represent two-dimensional space. In one training study, researchers tested what type of instruction best promoted this understanding (Huang & Witz, 2011). They taught three groups of fourth grade students. One group received practice with applying formulas (i.e., procedural instruction). Another group focused on the properties and features of 2-D geometric shapes and how those features conceptually related to surface area (i.e., conceptual instruction). The third group received both types of instruction simultaneously in an integrated lesson. The results revealed that children who received both types of instruction made better decisions about and more accurately explained challenging area calculation problems than children who received either procedural and conceptual instruction in isolation. The findings echo those of linear measurement training studies, and suggest that optimal interventions for unit-based tasks should target both procedural and conceptual understanding (e.g., Congdon, Kwon, & Levine, 2018; Kwon et al., under revision).

Spatial visualization of angular measurement. Another unit-based concept that is not typically introduced until later in school is that of angle measurement comparison. In one recent study, researchers tested whether children's word-learning biases might explain children's well documented misconceptions about angles (Gibson et al., 2015). The study focused on preschool aged participants who had not yet been introduced to angles in formal school settings. All children were taught about angles, but half of the children were given a second nonsense word to represent the angle figures (i.e., "toma"), while the control group heard the word "angle" used as it is in traditional instruction, ambiguously referring to both the angle figure and the measure of the angle. Children in the experimental condition improved significantly more than the control group after training. The finding was driven by improvement on trials in which the larger overall angle figure was not the figure with the larger angle measure (Fig. 2.4, panel c). These results suggested that children's early misconceptions about angle may stem, in part, from their propensity to apply novel labels to an entire object rather than a feature of that object (e.g., Hollich, Golinkoff, & Hirsh-Pasek, 2007; Landau, Smith, & Jones, 1988; Markman & Hutchinson, 1984). Only when given a label for the angle figure did children then search for another referent of their newly acquired spatial vocabulary. This study also offered some convincing evidence that children as young as 4 years old are capable of successfully learning about angles—a much younger age than is traditionally targeted for this type of lesson.

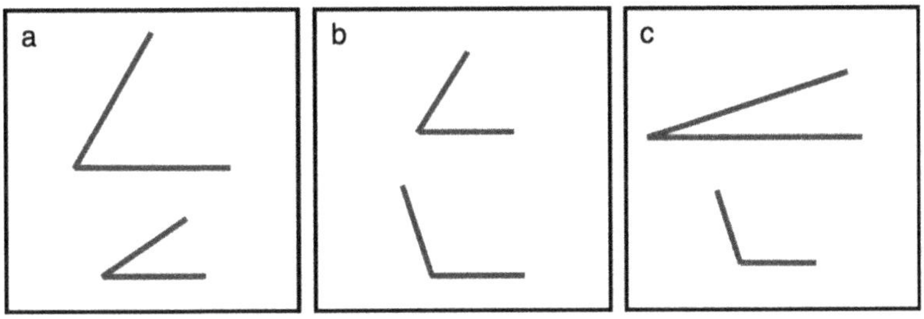

Fig. 2.4 Sample test trials from Gibson et al. (2015). Panel c was the type of trial that was most difficult for all children at pre-test, with children incorrectly selecting the larger figure significantly more than chance. After training, the experimental group selected the correct answer (i.e., the larger angle) at rates significantly above chance

In the 1990s, there was some promising research with older children using a computer programming language, Logo, which was adapted to help children to learn simultaneously about angle and linear measurement. In this platform, learners could direct a small, computerized turtle to turn a certain number of degrees left or right and move certain distances forward to accomplish simple goals (e.g., "go around the pond to get to the house" or "draw a rectangle"). Researchers argued that such a game accomplished two goals. First, it required children to apply numerical values to their perceptual intuitions, and second, it revealed the dynamic nature of mathematics, by, for example, emphasizing that degrees of an angle are really about rotation and the length of a side is about the distance it transverses (Clements & McMillen, 1996). Indeed, after playing with a game like Logo, children in middle school and high school age had more accurate, precise ideas about mathematical concepts like shapes, length and angle than those who followed more traditional instructional methods (Clements & Battista, 1989; Clements & Battista, 1992).

Conclusion

Despite young children's initial successes perceiving and processing continuous spatial properties, understanding how units represent those properties is a difficult transition, rife with the misconceptions. In this chapter, we have argued that to successfully make this leap, children must integrate continuous spatial properties with discrete representations of exact number, and they must identify the proper referents of newly acquired key spatial terms, including unit, angle, length, area, perimeter. Only in doing so can they begin to master higher-order unit-based concepts like transitivity, conservation of length/area/rotation, and the inverse relations between number of units and unit size. It may be helpful for researchers and educators interested in improving children's learning outcomes to be aware of the potential pitfalls children face as a result of their cognitive biases.

Studies aimed at teaching children the role and function of units have revealed several effective techniques. First, by exposing children to difficult exemplars of unit-based problems, such as shifted-object or "broken ruler" linear measurement problems, we can help children avoid applying a memorized procedure, and challenge them instead to reevaluate their preexisting strategies through self-discovered disconfirming evidence about their intuitive strategies. Second, we can make the referents of ambiguous spatial language more transparent by becoming aware of children's misconceptions and then explicitly pointing out through words and actions what spatial language does and does not mean. Third, hands-on, dynamic practice counting units, particularly in the presence of a conventional measuring tool, may help children use their intuitive reasoning about continuous properties to visualize and interpret discrete units in a structured way. Finally, a more radical suggestion is to augment existing mathematics curricula in a way that helps children establish a stronger foundation in proportional reasoning and relative comparisons of magnitude well before transitioning to numerical unit-based instruction. Such a modification could take advantage of children's natural propensity to reason about the intensive (e.g., proportional, comparative) properties of measurement problems before they are asked to master formal systems of extensive measurement (e.g., absolute extent, units). It is an open question as to whether this approach could ease the ultimate transition to formal systems of measurement. In the meantime, it seems that activities that link the intensive and extensive properties of measurement by using representations of units to help to concretize abstract labels and spatial properties of extent are maximally beneficial for improving student learning outcomes.

The lessons learned in this domain of mathematics, measurement, can likely be applied to many other areas. In this chapter we reiterate that the goal of a modern education is not for children to memorize tricks and procedures, but rather to develop a deep conceptual understanding of general principles, irrespective of the specific domain. We use units of measure as an example to outline some of the ways in which findings from cognitive science and psychology may assist in this goal by exploring the cognitive underpinnings of mathematical understanding in infants and young children, explaining the mechanisms that underlie some of the errors and misconceptions children face in formal schooling, and helping to promote crucial development beyond procedural knowledge to deeper conceptual understanding.

References

Baillargeon, R. (1987). Young infants' reasoning about the physical and spatial properties of a hidden object. *Cognitive Development, 2*, 179–200.

Baillargeon, R., & Graber, M. (1987). Where's the rabbit? 5.5-month-old infants' representation of the height of a hidden object. *Cognitive Development, 2*, 375–392.

Baillargeon, R., Needham, A., & DeVos, J. (1992). The development of young infants' intuitions about support. *Early Development and Parenting, 1*, 69–78.

Barrantes, M., & Blanco, L. J. (2006). A study of prospective teachers' conceptions of teaching and learning school geometry. *Journal of Mathematics Teacher Education, 9*, 411–436.

Barrett, J., Clements, D., Sarama, J., Cullen, C., McCool, J., Witkowski, C., & Klanderman, D. (2012). Evaluating and improving a learning trajectory for linear measurement in elementary grades 2 and 3: A longitudinal study. *Mathematical Thinking and Learning, 14*(1), 28–54.

Barth, H., & Paladino, A. M. (2011). The development of numerical estimation: Evidence against a representational shift. *Developmental Science, 14*, 125–135.

Bragg, P., & Outhred, L. (2004). A measure of rulers—The importance of units in a measure. *Proceedings of the 28th Conference of the International Group for the Psychology of Mathematics Education* (pp. 159–166). Bergen, Norway.

Brooks, N., Pogue, A., & Barner, D. (2011). Piecing together numerical language: Children's use of default units in early counting and quantification. *Developmental Science, 14*, 44–57. https://doi.org/10.1111/j.1467-7687.2010.00954.x

Boyer, T. W., Levine, S. C., & Huttenlocher, J. (2008). Development of proportional reasoning: Where young children go wrong. *Developmental Psychology, 44*(5), 1478.

Bryant, P. E., & Kopytynska, H. (1976). Spontaneous measurement by young children. *Nature, 260*(5554), 773–773.

Carpenter, T., Lindquist, M., Brown, C., Kouba, V., Edward, A., & Swafford, J. (1988). Results of the fourth NAEP assessment of mathematics: Trends and conclusions. *Arithmetic Teacher, 36*, 38–41.

Clements, D. H. (1999). Teaching length measurement: Research challenges. *School Science and Mathematics, 99*(1), 5–11.

Clements, D. H. (2003). *Learning and teaching measurement (2003 yearbook)*. Reston, VA: National Council of Teachers of Mathematics.

Clements, D. H., & Battista, M. T. (1989). Learning of geometric concepts in a Logo environment. *Journal for Research in Mathematics Education, 20*(5), 450–467.

Clements, D. H., & Battista, M. T. (1992). Geometry and spatial reasoning. In D. Grouws (Ed.), *Handbook of research on mathematics teaching and learning* (pp. 420–464). New York, NY: Macmillan.

Clements, D. H., & Bright, G. (2003). *Learning and teaching measurement: 2003 yearbook*. Reston, VA: National Council of Teachers of Mathematics.

Clements, D. H., & Burns, B. A. (2000). Students' development of strategies for turn and angle measure. *Educational Studies in Mathematics, 41*(1), 31–45.

Clements, D. H., & McMillen, S. (1996). Rethinking "concrete" manipulatives. *Teaching Children Mathematics, 2*(5), 270–279.

Clements, D. H., & Stephan, M. (2004). Measurement in pre-K to grade 2 mathematics. *Engaging young children in mathematics: Standards for early childhood mathematics education* (pp. 299–317).

Cohen, L. B., & Younger, B. A. (1984). Infant perception of angular relations. *Infant Behavior and Development, 7*(1), 37–47.

Congdon, E.L. & Levine, S.C. (2017, April). Making measurement mistakes: How actions and gestures can rectify common student misconceptions. In B. Kolvoord (Chair), *Supporting Spatial Thinking to Enhance STEM Learning. Symposium conducted at the Annual Meeting of the American Educational Research Association*, San Antonio, TX.

Congdon, E. L., Kwon, M. K., & Levine, S. C. (2018). Learning to measure through action and gesture: Children's prior knowledge matters. *Cognition, 180*, 182–190.

Davis, C., & Uttal, D. H. (2007). Map use and development of spatial cognition. In J. Plumert & J. Spencer (Eds.), *The emerging spatial mind* (pp. 219–247). New York: Oxford University Press.

Davydov, V. V. (1975). The psychological characteristics of the prenumerical period of mathematics instruction. In L. P. Steffe (Ed.), *Soviet studies in the psychology of learning and teaching mathematics* (Vol. 7, pp. 109–206). Chicago: The University of Chicago Press.

Duffy, S., Huttenlocher, J., & Levine, S. (2005). It is all relative: How young children encode extent. *Journal of Cognition and Development, 6*(1), 51–63.

Duffy, S., Huttenlocher, J., Levine, S. C., & Duffy, R. (2005). How infants encode spatial extent. *Infancy, 8*, 81–90.

Gal'perin, P. Y., & Georgiev, L. S. (1969). The formation of elementary mathematical notions. In J. Kilpatrick & I. Wirszup (Eds.), *Soviet studies in the psychology of learning and teaching mathematics*. Chicago: The University of Chicago.

Gao, F., Levine, S., & Huttenlocher, J. (2000). What do infants know about continuous quantity? *Journal of Experimental Child Psychology, 77*, 20–29.

Gibson, D. J., Congdon, E. L., & Levine, S. C. (2015). The effects of word-learning biases on children's concept of angle. *Child Development, 86*(1), 319–326.

Hiebert, J. (1984). Why do some children have trouble learning measurement concepts? *Arithmetic Teacher, 3*(7), 19–24.

Hollich, G., Golinkoff, R. M., & Hirsh-Pasek, K. (2007). Young children associate novel words with complex objects rather than salient parts. *Developmental Psychology, 43*, 1051–1061. https://doi.org/10.1037/0012-1649.43.5.1051

Huang, H. M. E., & Witz, K. G. (2011). Developing children's conceptual understanding of area measurement: A curriculum and teaching experiment. *Learning and Instruction, 21*(1), 1–13.

Hunting, R. P., & Sharpley, C. F. (1988). Fraction knowledge in preschool children. *Journal for Research in Mathematics Education, 19*(2), 175–180.

Huntley-Fenner, G. (2001). Why count stuff? Young preschoolers do not use number for measurement in continuous dimensions. *Developmental Science, 4*(4), 456–462.

Huttenlocher, J., Duffy, S., & Levine, S. (2002). Infants and toddlers discriminate amount: Are they measuring? *Psychological Science, 13*, 244–249.

Huttenlocher, J., Newcombe, N., & Sandberg, E. H. (1994). The coding of spatial location in young children. *Cognitive Psychology, 27*, 115–148.

Huttenlocher, J., Newcombe, N., & Vasilyeva, M. (1999). Spatial scaling in young children. *Psychological Science, 10*(5), 393–398.

Izard, V., O'Donnell, E., & Spelke, E. S. (2014). Reading angles in maps. *Child Development, 85*(1), 237–249.

Izard, V., & Spelke, E. S. (2009). Development of sensitivity to geometry in visual forms. *Human Evolution, 23*(3), 213.

Kamii, C. (2006). Measurement of length: How can we teach it better? *Teaching Children Mathematics, 13*(3), 154–158.

Kawanaka, T., Stigler, J. W., & Hiebert, J. (1999). Studying mathematics classrooms in Germany, Japan and the United States: Lessons from the TIMSS videotape study. In G. Kaiser, E. Luna, & I. Huntley (Eds.), *International comparisons in mathematics education* (Vol. 11, p. 86). London, UK: Falmer.

Kerslake, D. (1986). *Fractions: Children's strategies and errors*. A Report of the Strategies and Errors in Secondary Mathematics Project. NFER-NELSON Publishing Company, Ltd., Windsor, England.

Kotovsky, L., & Baillargeon, R. (1998). The development of calibration-based reasoning about collision events in young infants. *Cognition, 67*, 311–351.

Kwon, M. K., Levine, S. C., Ratliff, K., & Snyder, C. (2011, January). The importance of alignable differences in teaching linear measurement. In: *Proceedings of the Cognitive Science Society* (Vol. 33, p. 1156).

Kwon, M. K., Ping, R., Congdon, E. L., & Levine, S. C. (under revision). Overturning children's misconceptions about ruler measurement units: The power of disconfirming evidence.

Landau, B., Smith, L. B., & Jones, S. S. (1988). The importance of shape in early lexical learning. *Cognitive Development, 3*, 299–321. https://doi.org/10.1016/0885-2014(88)90014-7

Lehrer, R. (2003). Developing understanding of measurement. In J. Kilpatrick, W. G. Martin, & D. Schifter (Eds.), *A research companion to principles and standards for school mathematics* (pp. 179–192). Reston, VA: National Council of Teachers of Mathematics.

Lehrer, R., Jenkins, M., & Osana, H. (1998). Longitudinal study of children's reasoning about space and geometry. In R. Lehrer & D. Chazan (Eds.), *Designing learning environments for developing understanding of geometry and space* (Vol. 1, pp. 137–167). Mahwah, NJ: Lawrence Erlbaum Associates Publishers.

Liben, L. S., & Downs, R. M. (1993). Understanding person-space-map relations: Cartographic and developmental perspectives. *Developmental Psychology, 29*(4), 739.

Liben, L. S., & Yekel, C. A. (1996). Preschoolers' understanding of plan and oblique maps: The role of geometric and representational correspondence. *Child Development, 67*(6), 2780–2796.

Lin, P.-J., & Tsai, W.-H. (2003). Fourth graders' achievement of mathematics in TIMSS 2003 field test. (in Chinese). *Science Education Monthly, 258*, 2e20.

Lindquist, M., & Kouba, V. (1989). Measurement. In M. Linduist (Ed.), *Results from the fourth mathematics assessment of the National Assessment of Educational Progress* (pp. 35–43). Reston, VA: National Council of Teachers of Mathematics.

Lourenco, S. F., & Huttenlocher, J. (2008). The representation of geometric cues in infancy. *Infancy, 13*(2), 103–127.

Markman, E. M., & Hutchinson, J. E. (1984). Children's sensitivity to constraints on word meaning: Taxonomic versus thematic relations. *Cognitive Psychology, 16*, 1–27. https://doi.org/10.1016/0010-0285(84)90002-1

Martin, W. G., & Strutchens, M. E. (2000). Geometry and measurement. *Results from the seventh mathematics assessment of the National Assessment of Educational Progress* (pp. 193–234).

Miller, K. F. (1989). Measurement as a tool for thought: The role of measuring procedures in children's understanding of quantitative invariance. *Developmental Psychology, 25*, 589–600.

Mitchelmore, M. C., & White, P. (2000). Development of angle concepts by progressive abstraction and generalisation. *Educational Studies in Mathematics, 41*, 209–238. https://doi.org/10.1023/A:1003927811079

Mix, K. S., Levine, S. C., & Newcombe, N. S. (2016). Development of quantitative thinking across correlated dimensions. In A. Henik (Ed.), *Continuous issues in numerical cognition: How many or how much* (pp. 1–33). San Diego: Academic.

Mix, K. S., & Paik, J. H. (2008). Do Korean fraction names promote part-whole reasoning? *Journal of Cognitive Development, 9*, 145–170.

Möhring, W., Frick, A., Newcombe, N., & Levine, S. C. (2015). Spatial proportional reasoning is associated with formal knowledge about fractions. *Journal of Cognition and Development, 17*(1), 67–84.

Mullis, I., Martin, M., Gonzalez, E., & Chrostowski, S. (2004). *TIMSS 2003 international mathematics report: Findings from IEA's trends in international mathematics and science study at the fourth and eighth grades*. Chestnut Hill, MA: Boston College. National Center for Educational Statistics.

National Center for Educational Statistics. (2009). NAEP Questions. Retrieved on September 3, 2017, from http://nces.ed.gov/nationsreportcard/itmrls/startsearch.asp

Newcombe, N., Huttenlocher, J., & Leamonth, A. (2000). Infants' coding of location in continuous space. *Infant Behavior and Development, 22*, 483–510.

Newcombe, N. S., Levine, S. C., & Mix, K. S. (2015). Thinking about quantity: The intertwined development of spatial and numerical cognition. *Wiley Interdisciplinary Reviews: Cognitive Science, 6*(6), 491–505.

Newcombe, N. S., Sluzenski, J., & Huttenlocher, J. (2005). Pre-existing knowledge versus on-line learning: What do infants really know about spatial location? *Psychological Science, 16*, 222–227. https://doi.org/10.1111/j.0956-7976.2005.00807.x

Nunes, T., & Bryant, P. (1996). *Children doing mathematics*. Wiley-Blackwell.

Piaget, J., Inhelder, B., & Szeminska, A. (1960). *The child's conception of geometry*. New York: Basic Books.

Ramscar, M., Dye, M., Popick, H. M., & O'Donnell-McCarthy, F. (2011). The enigma of number: Why children find the meanings of even small number words hard to learn and how we can help them do better. *PLoS One, 6*(7), e22501.

Rescorla, R. A., & Wagner, A. R. (1972). A theory of Pavlovian conditioning: Variations in the effectiveness of reinforcement and nonreinforcement. In A. H. Black & W. F. Prokasy (Eds.), *Classical conditioning II: Current research and theory* (Vol. 2, pp. 64–99). New York: Appleton Century Crofts.

Shipley, E. F., & Shepperson, B. (1990). Countable entities: Developmental changes. *Cognition, 34*, 109–136.

Shusterman, A., Ah Lee, S., & Spelke, E. S. (2008). Young children's spontaneous use of geometry in maps. *Developmental Science, 11*(2), F1–F7.
Slater, A., Mattock, A., Brown, E., & Bremner, J. G. (1991). Form perception at birth: Cohen and Younger (1984) revisited. *Journal of Experimental Child Psychology, 51*, 395–406.
Slater, A., Mattock, A., Brown, E., Burnham, D., & Young, A. (1991). Visual processing of stimulus compounds in newborn infants. *Perception, 20*(1), 29–33.
Smith, J. P., Males, L. M., Dietiker, L. C., Lee, K., & Mosier, A. (2013). Curricular treatments of length measurement in the United States: Do they address known learning challenges? *Cognition and Instruction, 31*(4), 388–433.
Solomon, T. L., Vasilyeva, M., Huttenlocher, J., & Levine, S. C. (2015). Minding the gap: Children's difficulty conceptualizing spatial intervals as linear measurement units. *Developmental Psychology, 51*(11), 1564.
Sophian, C. (2007). *The origins of mathematical knowledge in childhood.* Lawrence Erlbaum Associates.
Sophian, C., Garyantes, D., & Chang, C. (1997). When three is less than two: Early developments in children's understanding of fractional quantities. *Developmental Psychology, 33*(5), 731.
Spelke, E. S., Gilmore, C. K., & McCarthy, S. (2001). Kindergarten children's sensitivity to geometry in maps. *Developmental Science, 14*, 809–821.
Spelke, E., Lee, S. A., & Izard, V. (2010). Beyond core knowledge: Natural geometry. *Cognitive Science, 34*, 863–884. https://doi.org/10.1111/j.1551-6709.2010.01110.x
Strutchens, M. E., Harris, K. A., & Martin, W. G. (2001). Assessing geometry and measurement. Understanding using manipulatives. *Mathematics Teaching in the Middle School, 6*, 402–405.
Strutchens, M. E., Martin, W. G., & Kenney, P. A. (2003). What students know about measurement: perspectives from the national assessments of educational progress. In D. H. Clements & G. Bright (Eds.), *Learning and teaching measurement. 2003 Year book* (pp. 195–207). Reston, VA: NCTM.
Tipps, S., Johnson, A., & Kennedy, L. M. (2011). *Guiding children's learning of mathematics.* Belmont, CA: Cengage Learning.
TIMSS 2011 Assessment. Copyright © 2012 International Association for the Evaluation of Educational Achievement (IEA). Publisher: TIMSS & PIRLS International Study Center, Lynch School of Education, Boston College, Chestnut Hill, MA and International Association for the Evaluation of Educational Achievement (IEA), IEA Secretariat, Amsterdam, the Netherlands.
Vasilyeva, M., Casey, B., Dearing, E., & Ganley, C. (2009). Measurement skills in low-income elementary school students: Exploring the nature of gender differences. *Cognition and Instruction, 27*, 401–428.
Vasilyeva, M., Duffy, S., & Huttenlocher, J. (2007). Developmental changes in the use of absolute and relative information: The case of spatial extent. *Journal of Cognition and Development, 8*, 455–471.
Vasilyeva, M., & Lourenco, S. F. (2012). Development of spatial cognition. *Wiley Interdisciplinary Reviews: Cognitive Science, 3*(3), 349–362.
Wilson, P. S., & Rowland, R. (1993). Teaching measurement. In R. J. Jensen (Ed.), *Research ideas for the classroom: Early childhood mathematics* (Vol. 30, pp. 171–194). Old Tappan, NJ: Macmillan.
Yuzawa, M., Bart, W. M., & Yuzawa, M. (2000). Development of the ability to judge relative areas: Role of the procedure of placing one object on another. *Cognitive Development, 15*(2), 135–152.

Chapter 3
Spatial Reasoning: A Critical Problem-Solving Tool in Children's Mathematics Strategy Tool-Kit

Beth M. Casey and Harriet Fell

Abstract This chapter reviews the spatial literature from the perspective of potential mechanisms for widening the range of spatially-based strategies available when solving math problems. We propose that teaching generalized spatial skills, disconnected from specific math content, may not be the best direction to go in future spatial interventions. Students who do not start out with strong spatial skills may need to learn to develop different types of "spatial sense" specific to each content area. Thus, acquiring and applying spatial strategies may depend in part on developing spatial sense within these specific math domains. In this chapter, we present an overview of evidence for different types of spatial sense that may serve as a prerequisite for effectively applying spatial strategies within these math content areas. The chapter also provides examples of math activities designed to help children acquire spatial sense and apply spatial strategies when solving diverse types of math problems.

Keywords Spatial skills · Math achievement · Math strategies · Decomposition · Visualization · Elementary school · Fractions · Geometry · Arithmetic · Mapping · Word problems · Gender differences · 2- and 3-dimensional representations · Mental images · Spatial sense · Image generation · Surfaces · Solids · Common Core Math Standards · Numerical concepts

The strong association between spatial skills and mathematics achievement has been demonstrated across a wide range of mathematics content areas and ages (Mix & Cheng, 2012; Wai, Lubinski, & Benbow, 2009). One dynamic in this association

---

This material is based on work supported by the National Science Foundation under NSF #HRD-1231623.

B. M. Casey (✉)
Lynch School of Education, Boston College, Boston, MA, USA
e-mail: caseyb@bc.edu

H. Fell
College of Computer and Information Science, Northeastern University, Boston, MA, USA
e-mail: fell@ccs.neu.edu

K. S. S. Mix, M. T. Battista (eds.), *Visualizing Mathematics*, Research in Mathematics Education, https://doi.org/10.1007/978-3-319-98767-5_3

may be that students with good spatial skills have access to a unique subset of strategies utilizing spatial sense and spatial reasoning, which enables them to draw upon critical tools in their problem-solving tool set that are not available to students with poorer spatial skills. We would argue that this advantage is not just due to higher fluid reasoning ability, i.e., the ability to draw on novel and effective problem solving strategies in general, but instead is based on the unique association between spatial and mathematical processing (Li & Geary, 2017). We further propose that spatial reasoning can be improved through the development of mathematics activities that facilitate the acquisition of spatial sense and the use of spatially based strategies within different content areas. This approach should eventually result in greater potential by students to draw on spatially based strategies when approaching difficult mathematics problems across a range of mathematics content areas.

In the first part of the chapter, we start by examining the literature on early arithmetic strategy use in relation to spatial processing. We will present an overview of research on the relation between early spatial processing and the early use of higher-level mental arithmetic strategies as predictors of mathematics performance. We briefly address the literature on early spatial skills and mathematics, and then focus mainly on the role of spatial skills in acquiring the use of advanced mental strategies, and how use of these spatially based mental strategies is beneficial to greater strategy-choice flexibility and mathematics achievement at later points in time.

In the second part of the chapter, we address the question of how students may develop the ability to apply spatially based strategies when solving diverse types of mathematics problems. We consider the importance of developing “spatial sense” within particular mathematics content areas as a prerequisite to applying spatially based strategies in these areas. In particular, we briefly review the literature on the ability to *generate images* as a critical component of developing spatial sense within the following mathematics content areas: (1) visual representations of magnitude with respect to fractions, (2) the ability to translate verbal descriptions into visual representations, and (3) the ability to generate and maintain images of 2-d and 3-d visualizations in relation to one another when solving mathematics problems. Each content area is followed by instructional examples for ways of teaching these different types of image generation/visualization activities.

Part 1: Spatial Processing and Arithmetic Strategy Choices

This section focuses on spatial sense and strategy-choice related to early arithmetic skills. The goal is to discuss the link between spatial skills and the use of high-level mental arithmetic strategies as a possible mechanism for encouraging young learners to apply spatially based strategies when approaching arithmetic problems.

Relation Between Early Spatial Skills and Mathematics Achievement

Early spatial skills are related to the development of mathematical skills via multiple pathways, involving both early number line and calculation skills (Gunderson, Ramirez, Beilock, & Levine, 2012; LeFevre et al., 2013). Moss and her associates recently implemented an extensive spatial intervention for 4-to-8 year olds, consisting of approximately 46 h of in-class time over the academic year. The geometry intervention involved carrying out lessons and activities designed to primarily target spatial visualization skills (i.e., forming, maintaining, and manipulating visual-spatial information) (Hawes, Moss, Caswell, Naqvi, & MacKinnon, 2017; Moss, Bruce, Caswell, Flynn, & Hawes, 2016). Results revealed that compared to an active control group, children in the spatial intervention demonstrated gains on three separate measures of spatial thinking; spatial language, visual-spatial geometrical reasoning, and 2-d mental rotation. Interestingly, while there were no group differences on a non-symbolic magnitude comparison task or a number knowledge test, children in the intervention group demonstrated significant gains relative to the control group on a symbolic magnitude comparison task with a substantial effect size. More of this type of intensive spatial intervention research needs to be conducted in the future to experimentally examine the effects of early spatial skills on mathematics achievement.

Spatial reasoning depends in part on good visuospatial memory resources (Miyake, Friedman, Rettinger, Shah, & Hegarty, 2001), and visuospatial memory pathways have been shown to be critical predictors of numerical skills (Geary, 2011; Li & Geary, 2013, 2017; Nath & Szücs, 2014). After controlling for non-verbal IQ and verbal memory measures, Nath and Szücs (2014) found that the association between Lego block-building skills and numerical achievement in 7-year olds was mediated by visuospatial memory. Further, controlling for the central executive, IQ, and phonological memory, Li and Geary (2013, 2017) found that growth in visuospatial memory skills from first-to-fifth grade was predictive of later numerical operations, but not reading achievement. This visuospatial memory/mathematics association extends into high school with visuospatial memory becoming even more important to numerical operations across successive grades (Li & Geary, 2017).

In kindergarten and first grade, the association between visuospatial memory and mathematics achievement is not as strong as in later grades, with phonological and linguistic processing showing a greater relation with numerical development at this age (LeFevre et al., 2010; Li & Geary, 2013). This may be in part because the type of mathematics assessed at this early age (e.g., number words and numerals) may require less spatial processing (Li & Geary, 2013). In fact, Krajewski and Schneider (2009) found that in kindergarten, phonological awareness had a stronger impact on *lower* numeric competencies in first grade (i.e., when number words were isolated from quantities) than for *higher* numerical competencies (i.e., when number words needed to be linked with quantities). The reverse was true for kindergarten visuospatial memory skills, which predicted the higher level numeric competencies at first grade.

Stronger early spatial-mathematics associations are found when more complex spatial reasoning processes (such as forming, maintaining, and manipulating visual-spatial information to solve mathematics problems) are examined in the literature. Mix and her colleagues (Mix et al., 2016) found that at kindergarten, mental rotation performance was a better predictor of mathematical performance than visuospatial memory performance. Studies have shown that as early as preschool, kindergarten, and first grade, these more complex spatial reasoning skills (e.g., mental rotation, spatial visualization, and block building) relate to numeracy and addition and subtraction skills (Casey, Dearing, Dulaney, Heyman, & Springer, 2014; Gunderson et al., 2012; Mix et al., 2016; Nath & Szücs, 2014; Verdine et al., 2014).

In summary, there is critical evidence for an association between early use of spatial skills and numeracy and addition and subtraction skills. The next step is to consider the pathways through which early spatial skills might impact these mathematics skills. One such pathway is the application of spatial skills when solving addition and subtraction problems by using mental arithmetic strategies that draw upon visuospatial memory processes and spatial reasoning.

Spatial Processing and Use of Higher Level Arithmetic Strategies

What are the mechanisms by which spatial reasoning skills might impact early addition and subtraction? One connection may occur through developing ability to visualize quantity along the mental number line. Researchers found that children's spatial skill (i.e., mental transformation ability) at the beginning of first and second grades predicted improvement in linear number line knowledge over the course of the school year. Second, in a separate sample, children's spatial skill at age 5 predicted their performance on an approximate symbolic calculation task at age 8 and this relation was mediated by children's linear number line knowledge at age 6 (Gunderson et al., 2012).

Arithmetic strategy choices. Another mechanism connecting spatial skills to early mathematics may be through the strategies that children use to solve addition and subtraction problems (Foley, Vasilyeva, & Laski, 2017; Laski et al., 2013; Siegler & Shrager, 1984). There are a variety of strategies that children use to solve arithmetic problems. When it comes to solving basic addition and subtraction problems, children generally choose from among four different strategies: count-all, count-on, decomposition, and retrieval (Laski et al., 2013). The count-all strategy involves counting out the first number, then counting out the second number, and then finally counting the total of the two numbers (e.g., to solve 7 + 5, one would first count to 7, then count to 5, and then finally count from 1 to 12). A slightly more sophisticated strategy is count-on, which involves counting up from one number the value of the other number (e.g., to solve 7 + 5, one would start from 7 and count 8, 9, 10, 11, 12).

Higher-level mental strategies involve retrieval and decomposition. The retrieval strategy involves recalling the solution purely from memory (Laski et al., 2013). The decomposition strategy involves breaking the numbers down into simpler mathematics facts that the child knows, and then adding or subtracting the value that remains. For example, to solve 7 + 5, one might first add 5 + 5 to get 10 and then add the remaining 2 to arrive at 12. A retrieval strategy is typically applied to single digit addition and subtraction. When a mathematics fact is not known, use of a decomposition strategy is often applied. Decomposition strategies are typically applied to addition and subtraction problems involving mixed digit (one single digit and one double digit number) or two double digit arithmetic problems.

Decomposition and retrieval are generally considered more advanced strategies for a number of reasons. Decomposition and retrieval are more efficient because they require less time to arrive at an answer than do concrete counting strategies (Ashcraft & Fierman, 1982). In addition, decomposition and retrieval are more sophisticated strategies because they avoid use of concrete counting strategies with fingers and other manipulatives, and instead depend on mental processes, drawing on memory-based mental representations of numerical information that depend on prior knowledge of mathematics facts (Geary, 2011).

Furthermore, frequency of use of decomposition and retrieval in solving arithmetic problems tends to be linked with higher mathematics performance. For example, Carr, Hettinger Steiner, Kyser, and Biddlecomb (2008) examined the association between strategy use and mathematics competence as measured by standardized test scores in a group of second grade students. They found that uses of higher-level cognitive strategies (retrieval and/or decomposition) were stronger predictors of mathematics competence above and beyond fluency and accuracy of solving basic mathematics facts. Both attempted and accurate uses of higher-level cognitive strategies were the strongest predictors of mathematics competency.

A number of longitudinal investigations of the relation between arithmetic strategy use in early grades and later mathematics achievement suggest that mental arithmetic strategies used in first and second grades, such as decomposition and retrieval, positively predicted mathematics performance in third, fourth, and fifth grades (Carr & Alexeev, 2011; Fennema, Carpenter, Jacobs, Franke, & Levi, 1998; Geary, 2011). Carr and Alexeev (2011) followed a group of children longitudinally from second grade through fourth grade. They found better mathematics outcomes in fourth grade for students who had attempted to use higher-level mental strategies previously when solving the basic arithmetic problems in second grade. As measured by standardized test scores in fourth grade, those children who attempted to use and those who correctly used mental strategies in second grade had an increased probability that they would meet or exceed the standards set for the mathematics competency test in fourth grade. These findings suggest that early preferences for these types of mental strategies may have long-term influences on mathematics competency. Studies also show that both retrieval and decomposition strategy-use increase with age (Bjorklund & Rosenblum, 2001; Siegler, 1987).

Spatial skills and arithmetic strategy choice. Both decomposition and retrieval depend on retrieving mathematics facts, and may draw on visuospatial memory processes and spatial reasoning. In a study of the association between spatial reasoning skills (mental rotation and spatial visualization) and arithmetic strategy preferences, Laski and Casey and associates (Laski et al., 2013) found that among first grade girls, these early spatial reasoning skills were positively related to use of both retrieval and decomposition. Furthermore, spatial skills were negatively associated with the count-all strategy and unrelated to the count-on strategy. Verbal skills were related to decomposition, but not retrieval. In a longitudinal study, M. Carr (personal communication, May 19, 2017) also found evidence of an association between 2-d spatial visualization skills and use of the decomposition strategy from first through fourth grades.

When Geary used measures from the Working Memory Test Battery for Children as predictors of frequency of decomposition strategy choice in first grade, he found a visuospatial measure (the Mazes Memory task) to be the strongest predictor at that age (D. C. Geary, personal communication, May 10, 2017). Foley et al. (2017) propose that visuospatial memory can serve as a mental sketchpad for storing problem information. They suggest that this may be particularly important for decomposition as, "a child's capacity for holding information in short-term memory—the storage component of working memory (Baddeley & Hitch, 1974; Gathercole, Pickering, Ambridge, & Wearing, 2004)—may contribute to his or her selection of decomposition because it requires maintenance of intermediate solutions and procedures" (p. 4). They found that short-term visuospatial memory was positively related to the frequency of children's decomposition use in second and fourth graders, while verbal memory was not. Most importantly, frequency of use of decomposition mediated the relation between visuospatial memory and arithmetic accuracy.

Use of decomposition. Decomposition differs from other arithmetic strategies because it does not just involve implementation of rote procedures, such as counting with fingers or recalling mathematics facts, but also requires active problem solving and a more complex series of reasoning processes. Thus, using decomposition for solving arithmetic problems in the early grades may provide a foundation for later mathematics problem solving.

Mix and her associates in a detailed analysis of kindergarten, third, and sixth grade spatial-mathematics associations proposed that across grade levels, the spatial-mathematics association is stronger when students are encountering novel problems and decrease as skills become more automatic or procedural (Mix et al., 2016). Therefore, at the start, as children begin to learn to apply complex decomposition strategies when trying to solve addition and subtraction problems, the ability to draw on spatial processes may be particularly beneficial.

Other longitudinal research has shown that the early use of decomposition arithmetic strategies (specifically in first grade) is a strong predictor of later numerical mathematics performance when controlling for intelligence, working memory, and processing speed (Geary, 2011). In a recent longitudinal study on first grade girls (Casey, Pezaris, Fineman, Pollock, Demers, & Dearing, 2015), we compared early spatial, verbal, and arithmetic skills as predictors of two types of mathematics reasoning skills 4 years later in fifth grade: (1) geometry and measurement problems specifically selected and designed to tap spatial mathematics reasoning skills, and (2) numerical and algebraic problems specifically selected and designed to tap analytical logical deductive mathematics reasoning (Casey et al., 2015). As expected, we found that early spatial skills predicted later geometry and measurement, but were more surprised that first grade spatial skills were also the strongest predictors of performance on later numerical and algebra problems in fifth grade. Furthermore, the pathway between early spatial skills and later numerical/algebraic mathematics reasoning was both a direct and an indirect pathway. Of importance here, is that the indirect pathway led from first grade spatial skills to greater frequency of use of first grade decomposition strategies and then to mathematics reasoning in fifth grade. Thus, it was the early relation between spatial skills and decomposition (not the early relation between spatial skills and retrieval) that predicted later numerical and algebraic mathematics reasoning in fifth grade (Casey, Lombardi, Pollock, Fineman, Pezaris, & Dearing, 2016).

There are a number of studies indicating that early use of decomposition is a particularly strong predictor of mathematics performance (Casey et al., 2016; Foley et al., 2017; Geary, 2011). In a longitudinal study that followed children from first grade through third grade, Fennema et al. (1998) investigated the relation between children's use of the invented algorithm (a form of decomposition) in first and second grade and performance on "extension" problems, or more advanced mathematics problems that involved money and three-digit numbers, in third grade. They found that by third grade, the second graders who had preferred the invented algorithm showed better performance than children who had used the standard algorithm. A recent study found that the frequency with which first graders use a decomposition strategy predicted their accuracy on complex addition problems and mediated cross-national differences in accuracy on these complex arithmetic problems (Vasilyeva, Laski, & Shen, 2015).

Gender differences. There is an interesting parallel between the early development of spatial skills and the early development of decomposition use in solving arithmetic problems—both show evidence of early gender differences. In a recent review of the literature on early gender differences in spatial skills, Levine and her colleagues (Levine, Foley, Lourenco, Ehrlich, & Ratliff, 2016) found that there is evidence for gender differences favoring boys in young children's mental transformation and mental rotation skills, but the gender effects are not obtained as consistently when compared to older ages, where strong support for gender differences on mental rotation and transformation tasks has been found (Casey, 2013; Wai et al., 2009). Early evidence for spatial gender differences can vary depending on age and type of task, with most of the evidence for gender differences occurring at age 5 or above and involving mental rotation of abstract shapes (Casey et al., 2008; Cronin, 1967;

Ehrlich, Levine, & Goldin-Meadow, 2006; Frick, Ferrara, & Newcombe, 2013; Levine, Huttenlocher, Taylor, & Langrock, 1999; Tzuriel & Egozi, 2010). Two studies with kindergarten and first grade students (Casey, Erkut, Ceder, & Mercer Young, 2008; Tzuriel & Egozi, 2010) have used extended spatial interventions to improve spatial skills. Both studies found that the boys in the control groups improved in their spatial skills without the intervention, while the girls improved only with the intervention.

Starting as early as kindergarten and first grade, evidence for gender differences in arithmetic strategy use has been found as well, such that boys are more likely than girls to use the more advanced strategies of retrieval and/or decomposition, while girls are more likely than boys to use concrete manipulatives such as counters or fingers to solve arithmetic problems (Carr & Davis, 2002; Carr & Jessup, 1997; Fennema et al., 1998; Shen, Vasilyeva, & Laski, 2016). Carr et al. (2008) reported similar findings among a group of second-grade students. They found that boys were more likely to use cognitive strategies (e.g., mental count-on, decomposition) and girls were more likely to use manipulative strategies. Carr and Davis (2002) found that in a free-choice condition, girls were more likely than boys to correctly use and attempt to use counting strategies, whereas boys were more likely than girls to correctly use retrieval. When children were forced to use retrieval strategies, boys were more likely than girls to use them correctly.

The well-documented link between higher-level mental strategies and later mathematics performance suggests that girls' early preference for counting strategies may put them at risk for poorer mathematics achievement in later grades, and it may possibly have socio-emotional effects as well. Research suggests that a male mathematics gender stereotype is acquired quite early, and that it influences emerging mathematics self-concepts prior to ages at which there are actual gender differences in mathematics achievement (Cvencek, Metzoff, & Greenwald, 2011). One obvious and observable difference among students in classrooms is whether they are still counting on their fingers or doing the addition mentally. Even if boys and girls are equally accurate on arithmetic problems, gender differences in use of higher level mental strategies may well impact gender-based early mathematics self-concepts.

A recent international study found evidence for gender differences in arithmetic strategy use in Russian and US first graders, but not in Taiwanese first graders (Shen et al., 2016). Among the Taiwanese students, there were no gender differences in accuracy, and girls used decomposition more than boys, while both genders outperformed students from the other two countries. In both Russia and the US, boys were more likely to use decomposition on complex arithmetic problems and have higher accuracy scores on the arithmetic problems (Shen et al., 2016). Most importantly, the researchers found that it was the preferred use of a decomposition strategy in boys that mediated the gender difference in accuracy for the US and Russian students. Thus, early gender differences in strategy choice may have long-term impacts for later gender differences.

In conclusion, evidence of early gender differences in both spatial skills and use of decomposition for solving arithmetic problems suggest that use of spatially based instruction tools (such as the number line) to teach and represent decomposition strategy procedures may be a fruitful approach for girls, in particular. This can begin the process of introducing spatial visualization and problem solving at the outset of arithmetic instruction.

Developmental changes in strategy choice. A major theoretical analysis of developmental changes in arithmetic strategy choice was proposed by Siegler (Lemaire & Siegler, 1995). According to Siegler's Adaptive Strategy Choice Model, as children develop over time in strategy use, they: (1) acquire a wider range of strategies, (2) make more adaptive choices among strategies, (3) increase frequencies of more efficient strategies, and (4) make more efficient use of pre-existing strategies. Siegler argues that strategy *preference* is as important as whether the children are *accurate* in using a particular strategy. Even low frequency use of higher level strategies is considered beneficial, and may be a positive early indicator of later effective flexible strategy-use. According to Siegler (2005) "...children usually choose adaptively among strategies; that is, they choose strategies that fit the demands of problems and circumstances and that yield desirable combinations of speed and accuracy, given the strategies and available knowledge that children possess." (p. 771).

A number of researchers have proposed that a major role of educators is to nurture children's adaptive strategy choices, i.e., the ability to solve mathematical tasks flexibly by being able to draw on a range of strategies when approaching mathematics problems (Siegler, 2007; Torbeyns, Verschaffel, & Ghesquière, 2005). Children with a wider range of strategies available to them may have a later advantage in terms of greater flexibility in strategy choice when task demands favor some strategies over others (Siegler, 1987). Thus, at first grade, even those children who have low frequency of use of decomposition may eventually show greater adaptive choices relating to this strategy later on as they encounter more complex problems.

Summary of Part 1. We propose here that having strong spatial skills at an early age may result in more arithmetic strategy-choice flexibility later on. Children with higher spatial skills have greater ability to draw on spatial as well as analytical strategies when solving mathematical problems. Researchers studying use of mental imagery in solving word problems have argued that use of advanced spatial skills should enable children to more easily generate and manipulate mental representations (Boonen, van der Schoot, van Wesel, DeVries, & Jolles, 2013; Hegarty & Kozhevnikov, 1999; Krawec, 2012). A relation between spatial skills and decomposition is likely to occur because students may use spatial representation as an element of their decomposition strategy when solving mathematics problems.

These findings suggest the possibility that the use of a decomposition strategy may depend in part upon the application of effective spatial skills and memory processes, and that interventions involving use of decomposition strategies may be one way of teaching students how to reason mathematically by drawing on their spatial reasoning and memory abilities. Clearly, this needs to be tested empirically through intervention research, but the present review suggests a possible direction for future studies to examine the benefits of using spatially based mathematics strategies at an early age.

Part 2: Development of Visualization Strategies Across Mathematics Content Areas

Before students can apply spatially based strategies to mathematics problems they need to first acquire sufficient spatial sense in relation to a wide range of mathematics content areas in order to be able to draw on their spatial problem-solving skills. Image generation/visualization is an important component of spatial sense. The concept of spatial sense is linked most frequently to geometry problem solving (NCTM, 2000) involving the ability to: spatially visualize and represent geometrical relations, hold images in spatial working memory, and mentally transform geometric shapes. Measurement sense means that students have a conceptual understanding of the processes underlying measurement procedures (Joram, 2003; Shaw & Pucket-Cliatt, 1989). For example, in terms of measurement sense, Battista (2003) describes the underlying processes in gaining competence in measuring area and volume as understanding how to enumerate arrays of squares and cubes. He identifies two mental processes essential to meaningful structuring of arrays: (1) forming and using mental models and (2) spatial structuring. Thus, he proposes that effective measurement performance involves an understanding of the underlying spatial nature of measurement, as well as the numerical and procedural competence to use measuring tools and apply formulas. Number sense also requires an understanding of magnitude through the generation of mental visual arrays displaying the relation of numbers to one another in terms of relative magnitude along a continuum. Recent research has documented the importance of generating a mental number line as a mechanism for spatially representing this relation (Gunderson et al., 2012; LeFevre et al., 2013). Thus, there is substantial literature documenting the key role of image generation/visualization skills as an important underpinning for conceptual understanding within these mathematics content areas.

In the last half of this chapter, we give a brief overview of the literature on visualization skills within three mathematics content areas. These mathematics areas include: (1) visual representations of magnitude with respect to fractions, (2) the ability to translate verbal descriptions into visual representations with respect to mathematics word problems and geometry, and (3) the ability to translate images from one type of representation into another when solving geometry problems, i.e., moving from 3-d representations to 2-d images, or vice versa.

In the present chapter, we have made the argument that rather than spatial skills being taught in isolation as abstract concepts, they should be taught within the framework of specific mathematics content areas. Although individuals with strong spatial skills may be able to apply spatial strategies across mathematics content areas, this may well not be true for those students who are gradually acquiring spatial reasoning skills. Within each mathematics area, the focus should be on the type of spatial sense that is key for that content area, drawing on this spatial reasoning to teach effective spatial strategies in approaching these mathematics problems.

Within each of these mathematics content areas, we will provide examples of mathematics activities that can be used to develop image generation skills likely to develop both spatial sense and increase use of spatially based mathematics strategies. Many of these examples are similar to ideas introduced already into mathematics curricula and into the Common Core State Standards for Mathematics (National Governors Association Center for Best Practices, Council of Chief State School Officers, 2011). Other examples are less common.

Many of the examples here, connecting spatial skills to mathematics within the different mathematics content areas, depend in part on physical instantiations/concrete models or manipulatives, which are arguably also spatial models (Mix, 2010). Mix proposed that rather than depending on abstract number and language-based symbolic representations and known algorithms, these objects construe mathematics as spatial relations and provide spatial tools with which to reason about mathematics. There is disagreement in the literature as to how effective manipulatives are, and Mix (2010) points out that even those who advocate the use of manipulatives caution that it depends on the way they are implemented. In particular, it depends on whether the children can move beyond a dependence on manipulatives. Researchers have not yet identified the critical elements that make manipulative effective. Nevertheless, in the present chapter, we continue to use a number of examples that employ concrete manipulatives, in order to teach children about spatial representations and spatial strategic approaches to mathematics problem solving.

Representing Fraction Magnitudes by Generating Mental Visual Arrays

Siegler and his colleagues have conducted recent research showing that fractions have magnitudes that can be ordered and assigned specific locations on a mental number line just as whole numbers can (Siegler, Thompson, & Schneider, 2011). Recently, Hurst and Cordes (2016) found that fractions, decimals, and whole numbers can be used to represent the same rational-number values, and that adults conceive of these rational-number magnitudes as lying along the same ordered spatial mental continuum. Thus, development of successful number sense and fraction sense and decimal sense in students depends in part on the ability to successfully represent magnitude spatially in terms of generating mental visual arrays.

It is proposed here that once students acquire more experience generating images involving spatial arrays across different types of magnitude estimations, they will be able to more effectively draw on spatially based strategies for solving a wider range of mathematics problems. Understanding of fractions is particularly important because elementary students' fraction knowledge has been found to uniquely predict their knowledge of algebra and overall mathematics achievement 5–6 years later (even after controlling for other key variables) (Siegler et al., 2012).

A brief overview of the theory and research related to *fraction magnitude representations*, along with suggested mathematics activities to further develop fraction magnitude sense will be presented next. Fraction sense involves conceptualizing fractions as a unit rather than perceiving them as separate numerators and denominators (Schneider & Siegler, 2010). A recent study by Hamdan and Gunderson (2017) found that training using a number line estimation of unit fractions showed significant transfer to an untrained fraction magnitude task, whereas equivalent training using an area model estimation task (unit fractions within a pie chart) did not. Research on fraction magnitude representations has shown that performance on these types of linear spatial representations correlates strongly with fraction arithmetic proficiency and general mathematics achievement scores (Siegler et al., 2011). This is somewhat surprising, as fraction magnitude knowledge is reported to be assessed little if at all on typical school mathematics assessments (Siegler et al., 2011).

In addition, fraction magnitude representations account for substantial variance in achievement scores above and beyond that explained by fraction arithmetic proficiency (Siegler et al., 2011). Furthermore, recent intervention research on at-risk fourth graders (Fuchs et al., 2013; Fuchs et al., 2014) found that fraction magnitude interventions, involving placing fractions on number lines as well as comparing and ordering magnitudes, resulted in both greater improvement in fraction arithmetic proficiency and conceptual understanding when compared to control group taught fractions through a school-based mathematics textbook. It was found that the gap between at-risk and low-risk students narrowed for the intervention group, but not the control group.

Examples of linear mental fraction line fraction games and activities. In this next section, a series of examples of mathematics games and activities are described that can be used to teach students how to generate images of fractions in order to develop fraction sense. The idea is for students to learn to represent fractions as a unit, and to compare them to other fractions along a mental number line. The curriculum, *Everyday Mathematics* (University of Chicago School Mathematics Project, 2007) includes lots of activities with fraction cards—including using them along a number line. The goal for presenting these particular games in the present chapter is to provide examples of how to apply this type of spatial sense as a basis for generating effective spatial strategies when solving fraction arithmetic problems.

Use of fraction card games. These games are based on the traditional card game "War." They encourage children to recognize common fractions, where they fit on the number line, and how they compare with each other. The materials are shown in Fig. 3.1.

The Deck

Each card shows one proper fraction. The decks will vary dependent on the age, grade, or abilities of the children involved. The deck might be suitable for children in K through 2.

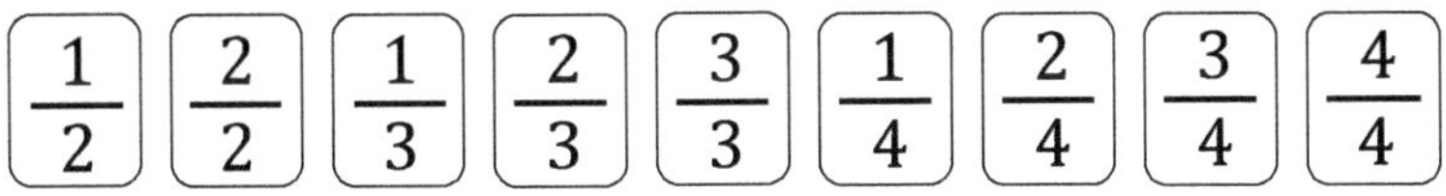

The denominators might go from 2 to 10 (54 cards) or higher. Fractions might also represent numbers greater than 1.

Fraction Number Line

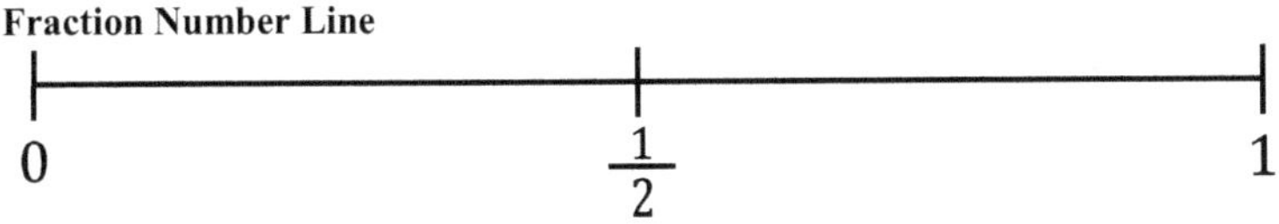

Fig. 3.1 Materials for the fraction card games

Card game 1: Solitaire. Children can get used to the materials by playing solitaire games, e.g., shuffle the deck and turn up cards one at a time then do one of the following:

- place the card where it goes on the number line,
- place the card to the left, right, or on ½ as appropriate, or
- put the cards in order (no number line needed)

Card game 2: Two-person practice game. The deck is divided as evenly as possible with the cards dealt one at a time face down. Each player places his stack of cards face down, in front of him. Players each turn up a card at the same time and the one who places it correctly on the number line first gets both cards. If they place the cards correctly at the same time, they each keep their own card. The play can proceed with or without a number line. Whoever has the most cards at the end of some time period is the winner.

Alternate play—non-competitive. Proceed as with solitaire but taking turns, show where the fraction goes on the number line (see Fig. 3.2), or arrange the cards in order (see Fig. 3.3).

Card game 3: Two-person simpler practice game. The deal proceeds as above but the goal is to place the cards, as quickly as possible, into piles: <1/2, =1/2, >1/2.

Card game 4: Fraction Game WAR. Players each turn up a card at the same time and the player with the higher card takes both cards and puts them, face down, on the bottom of his stack.

If the cards have the same value, e.g., 1/2 and 2/4, it is War. Each player turns up one card face down and one card face up. The player with the higher cards takes both piles (six cards). If the turned-up cards are again the same rank, each player places another card face down and turns another card face up. The player with the

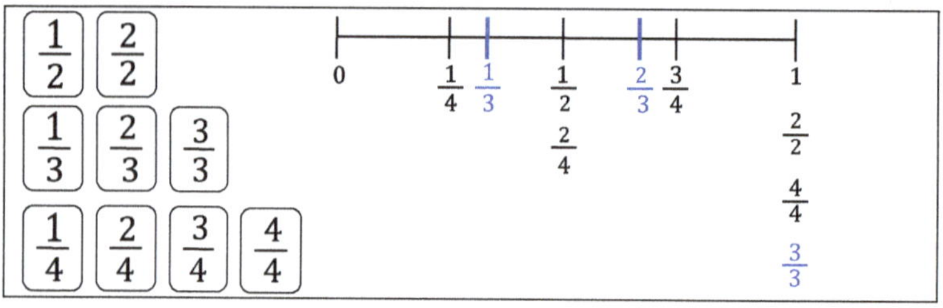

Fig. 3.2 Show where the cards go on the number line

Fig. 3.3 Arrange the cards in order

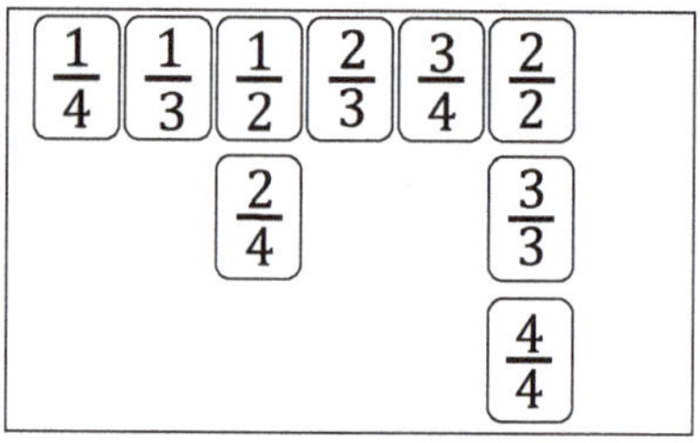

higher card takes all 10 cards, and so on. The game ends when one player has won all the cards.

Adding fractions using graphing. Another way to develop fraction sense is to use spatial representations through graphing. In these exercises, students use graph paper in two ways. They use a grid, e.g. 10 × 10 or 12 × 12, to color in regions to represent each fraction in a sum. They can also make a standard x vs. y graph to show a running sum of fractions, the x-axis showing the number of fractions added and the y-axis showing the sum so far.

Use of graphing to add fractions: convergent series. Though infinite series sounds pretty advanced, the idea of adding up fractions that follow a pattern is really pretty simple and keeping track of the sums visually provides another playful way for children to think about fractions. Note, this material is based on ideas from Cohen (1989).

Convergent series: First example. $\frac{1}{2}+\frac{1}{4}+\frac{1}{8}+\frac{1}{16}+\frac{1}{32}\cdots$ The three dots mean "and so on." Each 8 × 8 square represents one whole (see Fig. 3.4).

- Shade in the squares to show the numbers above them.
- Show the sum as a fraction.
- Show the sum as a decimal.
- Can you tell what the sum would be if we kept adding similar terms?

$$\frac{1}{2}+\frac{1}{4}+\frac{1}{8}+\frac{1}{16}+\frac{1}{32}+\frac{1}{64}+\cdots$$

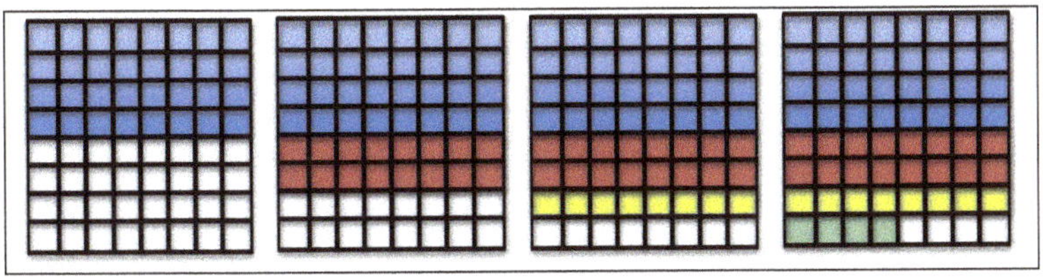

Fig. 3.4 Graphic representation of successive sums of 1/2 + 1/4 + 1/8 + 1/16

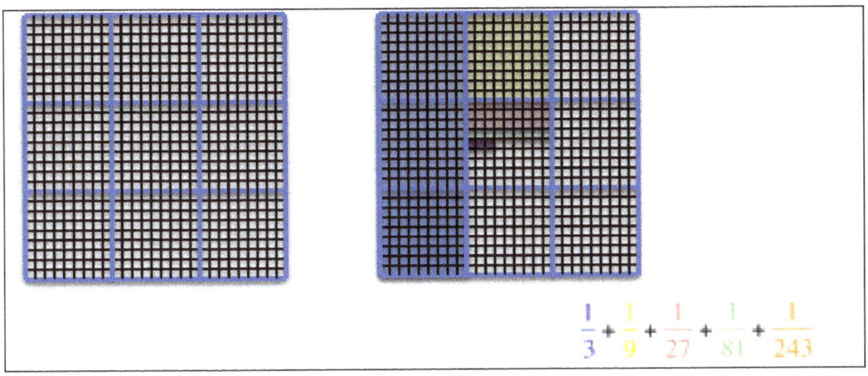

Fig. 3.5 Graphic representation of successive sums of 1/3 + 1/9 + 1/27 + 1/81 + 1/243

- What is the next term?
- What is the pattern of the denominators?

$$\frac{1}{2} \qquad \frac{1}{2}+\frac{1}{4} \qquad \frac{1}{2}+\frac{1}{4}+\frac{1}{8} \qquad \frac{1}{2}+\frac{1}{4}+\frac{1}{8}+\frac{1}{16}$$

Convergent series: Second example. $\frac{1}{3}+\frac{1}{9}+\frac{1}{27}+\frac{1}{81}+\frac{1}{243}\cdots$ This time, each denominator is three times the last one. Shade the box below (see Fig. 3.5) to show the sum of $\frac{1}{3}+\frac{1}{9}+\frac{1}{27}+\frac{1}{81}+\frac{1}{243}$. Notice that the blue lines divide the box into thirds horizontally and vertically. There are nine small boxes across and down in each blue box. Use a different color for each fraction you add on.

- What fraction represents one small black box? (answer 1/729)
- Express each of the partial sums as a single fraction, e.g., $\frac{1}{3}+\frac{1}{9}=\frac{4}{9}$ and $\frac{1}{3}+\frac{1}{9}+\frac{1}{27}=\frac{13}{27}$.
- How big do you think the sum will get if you keep on adding more similar terms?
- Will it ever get bigger than 1?
- Will it ever get bigger than 4/9?

Convergent series: Third example. For something really different, think about this simple looking sum $\frac{1}{2}+\frac{1}{3}+\frac{1}{4}+\frac{1}{5}+\frac{1}{6}\cdots$

- Does it get bigger than 1? (Yes, just add the first 3 terms.)
- Does it get bigger than 2? (Yes, if you keep adding on 1/n it grows to infinity.)
- This is not a convergent series. It does not add up to a finite sum.

Translating Verbal Descriptions into Visual Representations

In the everyday world, visualization often involves translation of information you hear or read into a mental image, and in mathematics it also involves using that mental image to reason about a solution to a problem. Recently, there has been a focus on research addressing the benefits of translating words into images for mathematics word problems. In the initial research, Hegarty and Kozhevnikov (1999) clarified that the effectiveness of image generation depends on the type of imagery used; while schematic spatial representations were associated with success in mathematical word problems, use of pictorial representations was negatively correlated with success. Schematic representations encode the spatial relations described in a word problem, while pictorial representations encode the visual appearance of the objects described in the problem. Use of schematic representations was also significantly correlated with spatial ability, while pictorial representations were not. van Garderen (2006) obtained similar findings when comparing gifted, average, and learning disabled students, with gifted students tending to use more schematic representations. When examining pathways between spatial skills and word problems in sixth graders, Boonen and associates (Boonen et al., 2013) found that 21% of the association between spatial skills and word problem solving was explained through the indirect effects of strategies involving visual-schematic representations. Thus, spatial skills can be translated into useful spatially based mathematics strategies to solve word problems through the use of visual representations.

Jitendra and colleagues (Jitendra, Nelson, Pulles, Kiss, & Houseworth, 2016) conducted a review of studies examining the benefits of using visual representation models for teaching mathematics problems with at-risk mathematics students and found a substantial benefit of using this approach. In their review of the literature, Kingsdorf and Krawec (2016) also concluded that instructional methods involving representing word problems visually in third graders have been proven effective by organizing the problem information. However, they suggest further that effectively using graphic representations requires visually representing connections between the problem parts in order to effectively link various phases of the problem-solving process.

Examples for translating verbal descriptions into visual representations with early geometry learners. As indicated earlier, the National Council of Teachers of Mathematics (NCTM, 2000) conceives of geometry problem solving as involving the ability to: spatially visualize and represent geometrical relations, hold images in spatial working memory, and mentally transform geometric shapes. This aspect of the geometry curriculum has been extensively developed within the mathematics curriculum series, *Investigations in Data, Number, and Space* (TERC, 2008). For example, throughout their curriculum, they make use of Quick Images where children are briefly shown images of quantities and shapes and asked to recognize them or reconstruct them in order to practice building and retaining such mental images. In *Taking Shape*, Joan Moss and co-authors (Moss et al., 2016) present a K-2 geometry curriculum that draws heavily on developing spatial visualization skills as a major component of mathematics education in early elementary school.

One element of these visualization skills is the ability to translate verbal into visual representations. If translating verbal information into visual representations is beneficial to later mathematics problem solving, how can educators start to develop this translation skill in young children at the outset of learning geometry? In this section, we provide examples of geometry activities and games that may be helpful in developing young students' ability to translate verbal descriptions into visual images that represent those descriptions. This proposed emphasis on translations of verbal representations into visual representations in early geometry activities may be useful later on when the spatial strategy instructional focus may be on translation of word problems into spatial representations through diagrams for older mathematics learners (e.g., Boonen et al., 2013; Jitendra et al., 2016; Kingsdorf & Krawec, 2016).

Geometry games for turning verbal or written descriptions into visual representations. Below, we present some two-person games that involve one partner giving verbal instructions to the other partner on how to draw or construct a particular representation. The objective is for the partner hearing the description to create a 2-d or 3-d representation from that description. Another set of examples uses written descriptions that have to be followed on a grid or map. These lead into learning coordinate geometry.

The Barrier Game. In the *Taking Shape* geometry curriculum, Joan Moss and co-authors (Moss et al., 2016) describe a series of "barrier games" for early elementary school that fit into this category, going from a verbal description to a spatial layout. In each game, there is a designer and a builder. There is a barrier between the two so they cannot see each other's creations as they describe and build (see Fig. 3.6). In these barrier games, children are encouraged to "Visualize" (create a visual image of what the partner is describing), "Verbalize" (ask your partner questions about where to place the tiles), and "Verify" (compare the two designs to see where they are similar and different). This visualize/verbalize/verify strategy can be useful for many types of geometry activities.

- In the simplest game, each child is given identical 7 square tiles.
- The designer creates a design with the 7 tiles.
- The designer then gives instructions to the builder so he or she can recreate the design on the other side of the barrier.

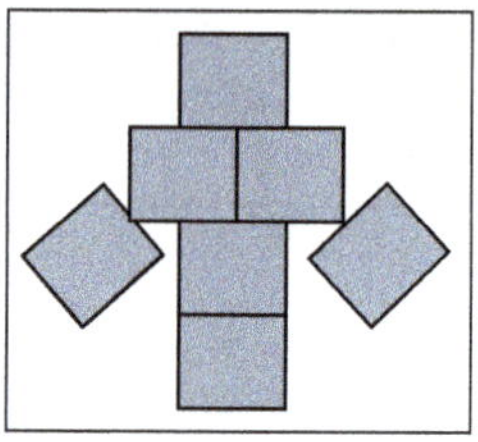

Fig. 3.6 A sample shape with 7 tiles

Fig. 3.7 Pictures from the Monster Game

- When the builder has finished, the children compare their designs.
- The children then swap roles and play again.
- The designer will have to use words and phrases like: besides, above, touching, rotate, turn, slide, left, right, etc.
- The game can be easily made more difficult by increasing the number of tiles or adding tiles with different shapes and colors.
- It could be turned into a 3-dimensional exercise by using Lego blocks.

The Monster Game. In a more open-ended activity that only requires pencil and paper, this is another two-person game to promote translation of verbal descriptions into visual representations. It even provides entertainment on long car trips. As with the barrier games, there is a designer and a builder and the children should not see each other's work until the game is finished (see Fig. 3.7).

- Each child has paper and a pencil with an eraser. Colored pencils or crayons can be used too.
- The designer draws a monster. A monster might be humanoid but they usually end up with strange sizes and numbers of arms and eyes. They can have claws in place of some hands or feet. There really are no constraints.

- The designer then gives instructions to the builder so he or she can recreate the monster on their own paper.
- When the builder has finished, the children compare their monsters.
- The children then swap roles and play again.

This is clearly very much like the barrier games, and children can also use the verbal description/drawing game to design buildings or parks or clothing as an alternative.

A mapping game based on verbal instructions. In the mathematics curriculum series, *Think Mathematics* (Educational Development Center, Inc., 2008) in grade 1, students are asked to explore direction on a map as part of their geometry section on maps, grids, and geometric figures. In one component they are asked to draw maps from verbal instructions. The example used here is from Harriet Fell's son's French Kindergarten class. The students were given a simple map of their village and then asked to draw routes, creatures, and other features according to the teacher's description. For example, after identifying their school and the village swimming pool on their maps, they drew a curve marked with arrows from school to pool. They put a duck in the pond, a goat on the farm, and waves in the pool (see Fig. 3.8). The purpose of this activity is to learn to generate visual images of maps in terms of spatial location, direction, and features on the map based on verbal directions from others. This should enable them to be more likely to use spatial strategies to solve these types of problems, such as generating mental images of maps rather than using landmarks and words.

Orienteering, drawing, and mapping routes on a grid based on verbal instructions. As part of the same unit, the *Think Mathematics* curriculum also focused on finding and following paths on a grid. By placing maps on a grid, the concept of graphing is also introduced as well as concepts involving spatial location and direction. This type of geometry activity is similar to orienteering in some ways. Orienteering is an outdoor sport where players use maps and a compass to find their way. Players may see on the map that the next goal is ¼ mile to the northeast of their current position and then use their compass to get there. In Orienteering Drawing, the player(s) is given a set of drawing directions to follow that are similar, e.g., head 3″ north. For young children, the step-by-step instructions can be given verbally and for older children it can be presented in writing (see Fig. 3.9).

Or the teacher could give a list of 2-d coordinates to make a picture. Here, the lower left-hand corner is the origin (0, 0). The child is instructed to go from coordinate to coordinate as in follow-the-dots.

$$(4,2)(4,4)(5,4)(5,5)(4,5)(4,6)(5,6)(5,7)(3,7)(3,2)(4,2)$$

For each of these methods of description, a teacher could provide the description or a child could create the description from the image.

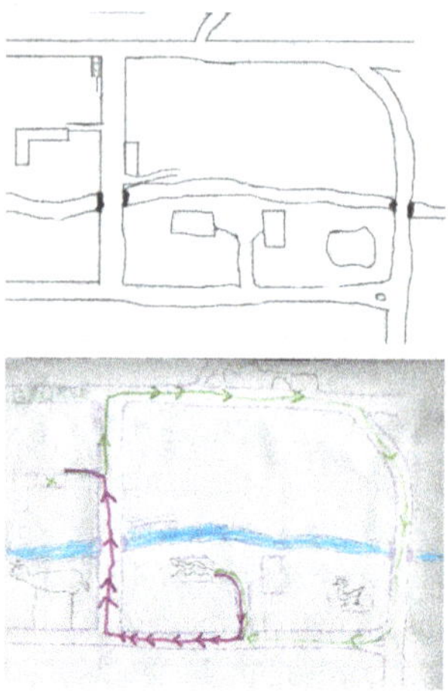

1) Color the river (Yvette) blue.
2) Put a green "X" on the L-shaped building. That's our school.
3) With your green marker, let's make a route to the swimming pool.
   a. Go along the path from the school.
   b. Turn left when you get to the road.
   c. Turn right at the stop light.
   d. Follow the road around until you can turn right.
   e. Turn right as soon as you can.
   f. Head to the left. That's the swimming pool.
4) Put waves in the swimming pool.
5) ...

Fig. 3.8 Following directions to annotate a map

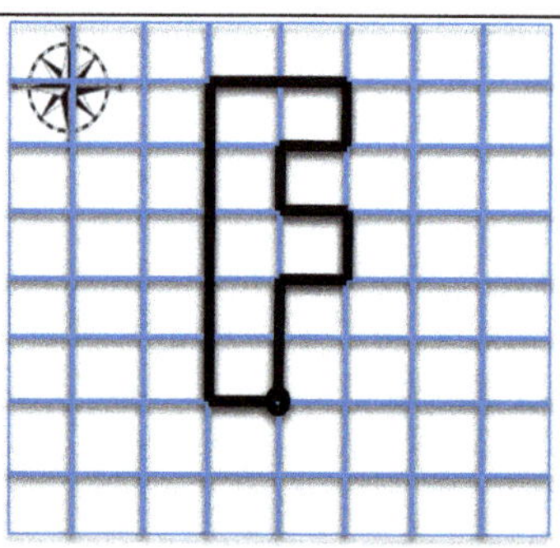

(0,0)

Assume the boxes are 1" on a side.
1. Start at the blue dot.
2. Head 2" North.
3. Head 1" East.
4. Head 1" North.
5. Head 1" West.
6. Head 1" North.
7. Head 1" East.
8. Head 1" North.
9. Head 2" West.
10. Head 5" South.
11. Head 1" East

Fig. 3.9 Orienteering Drawing

Translating 2-d Representations into 3-d Mental Images and Vice Versa

In the sciences, engineering, and in medicine, in particular, a critical skill involves being able generating mental images by moving back and forth between 2-d and 3-d representations. As reported in a literature review by Harris, Hirsh-Pasek, and

Newcombe (2013), students with high dynamic spatial transformation abilities (such as paper folding and mental rotation) were found to have greater ability to read graphs (Kozhevnikov, Motes, & Hegarty, 2007; Kozhevnikov & Thorton, 2006) and interpret diagrams (Höffler, 2010). Spatial research has been done on dynamic spatial transformations, such as mental rotation, in which the images have to be transformed and manipulated as well as generated (Mix & Cheng, 2012). A number of mathematics curricula for young children, such as *Investigations in Data, Number, and Space* (TERC, 2008) and *Taking Shape* (Moss et al., 2016) have incorporated 2-d to 3-d transformations that also require mental folding, such as identifying which 2-dimensional nets make cubes and rectangular prisms. This type of transformation is also used in origami. Boys, in particular, receive many hours of informal instruction doing these types of activities with Legos and model kits. These types of construction activities often require the ability to examine complex 2-d drawings, translate them into 3-d mental images, and ultimately produce 3-d structures.

Examples of instructional activities to develop translations of 2-d representations into 3-d images and vice versa. In the next section, we provide examples of geometry activities involving the simple translation of 2-d representations into 3-d structures and vice versa—without any dynamic transformations. Again, these activities are designed to develop spatial visualization skills and visuospatial memory, and are part of the NCTM (2000) and Common Core State Standards for Mathematics (National Governors Association Center for Best Practices, Council of Chief State School Officers State Standards for Mathematics, 2011) relating to analyzing, comparing, creating, and composing 2-d and 3-d shapes in relation to modeling shapes in the world. In order to apply dynamic transformations strategies to solve many types of geometry problems, an important first step is to be able to visualize the relation between 2-d and 3-d representations of the same figures within static images.

However, when visualizing complex shapes, simple 2-d to 3-d transformations can be quite difficult for children with poorer spatial skills, even before adding on the requirement of manipulating the stimuli dynamically. For example, it can be argued that one of the initial difficulties in solving 3-d mental rotation tasks, such as the Vandenberg task (Vandenberg & Kuse, 1978), is that the stimuli consist of complex, unfamiliar, abstract shapes that project 2-d representations of 3-d stimuli. Just transforming that 2-d picture and projecting it as a 3-d image is difficult enough, in addition to the increased cognitive processing load required by holding it working memory, and then manipulating the image in order to mentally rotate it. In an analysis of the Vandenberg items, it was found that the most difficult items on the Vandenberg were ones in which the drawings of the 3-d figures had occluded parts (Voyer & Hou, 2006).

Thus, the simple translation of 2-d representations into 3-d mental images can be quite difficult to visualize, especially when stimuli are complex. Students need practice using their visualization skills to complete this simple translation process, as they are important components of many geometry tasks that also require mental manipulation and rotations and are an important prerequisite to the mental transformations and manipulations that are needed to apply spatial strategies when solving many geometry problems.

2-d representations of the swimming pool. The first example is for younger children, and involves exercises from an École Maternelle class in France for 5-year olds where Harriet Fell's son went for kindergarten, the year the family lived in France. Since the village had a pool, swimming lessons were part of the public schools' curriculum (see map of the village shown previously). Every week, the pool was divided into different sections for the three levels of classes and different types of floatation toys were placed in the pool in each section, e.g., foam barbells, kickboards, or doughnuts. The sections and the objects within the sections varied each week. When the children returned to the school after swim class, they were asked to draw how the pool looked that week from memory (see Fig. 3.10). When asked what he did that day, George would show a different drawing of the pool, and say, "It's a drawing of the swimming pool." What was being taught here? It involves not only the ability to translate a 3-d scene into a 2-d drawing, but accurately remembering the details of the different parts, as well as showing how the different parts relate to the whole as a gestalt. We have not found examples of similar activities in US kindergarten classrooms, and this approach may be idiosyncratic to France for this age group. However, the mathematics educators in the book may be able to provide other examples.

Using modeling clay to explore surface, solids and cross-sections. Because one component of geometry, based on the NCTM Standards (2000), involves visualizing relationships between 2-d drawings and 3-d objects, clay and plasticine are great materials for exploring surfaces and solids, and have been used frequently in

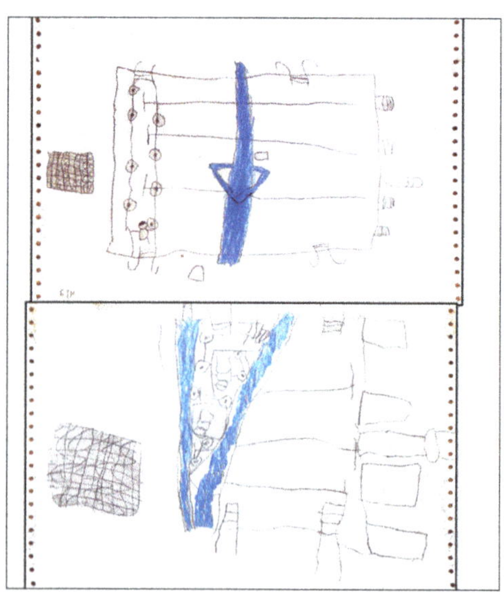

Fig. 3.10 Drawings of the swimming pool

geometry classrooms. Clay can be used to study ways of describing 3-dimensional objects in 2-dimensions. Though there is wonderful software available for visualizing three-dimensional objects, we think it is important for children to have the experience of visualizing and manipulating three-dimensional objects as in the examples that follow to fully understand the two-dimensional projections they see.

Build convex solids, given 2-d front, side, and top. Students were asked to build convex solids given the top, side, and front views (see Fig. 3.11).

Using plasticine to build objects and slicing through to see side views. Within the Common Core Standards for Mathematics at seventh grade (National Governors Association Center for Best Practices, Council of Chief State School Officers State Standards for Mathematics, 2011), students work with 3-d objects, relating them to 2-d figures by examining cross-sections. Students are expected to identify the shapes of 2-d cross-sections of 3-d object and identify 3-d objects generated by rotations of the 2-d figures. We think these types of cross-section activities can be done successfully with younger students. In the present example, given a contour map, the children are asked to draw the side views they would see if the scene were sliced along the lines, A, B, C, D, and E. They do this twice, first just from the contour map, e.g., by "walking" along the lines with their fingers, and then by using plasticine to build a model and slicing the model along the lines to see the side view (see Fig. 3.12).

Summary for Part 2. In this last half of the chapter, we provided examples of a variety of methods for teaching children to *generate images* as a critical component of developing spatial sense within the content areas of fractions, word problems, and geometry. We considered the importance of developing "spatial sense" within these mathematics content areas as a prerequisite to applying spatially based strategies in these areas.

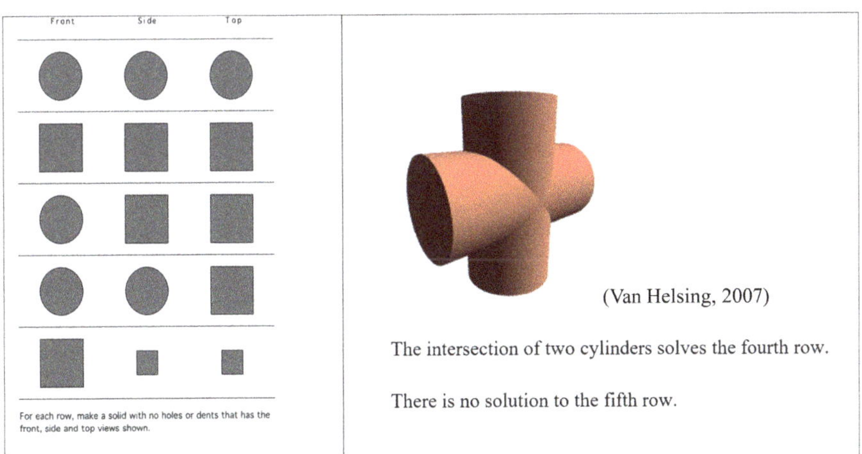

Fig. 3.11 Front, side, and top views of solids

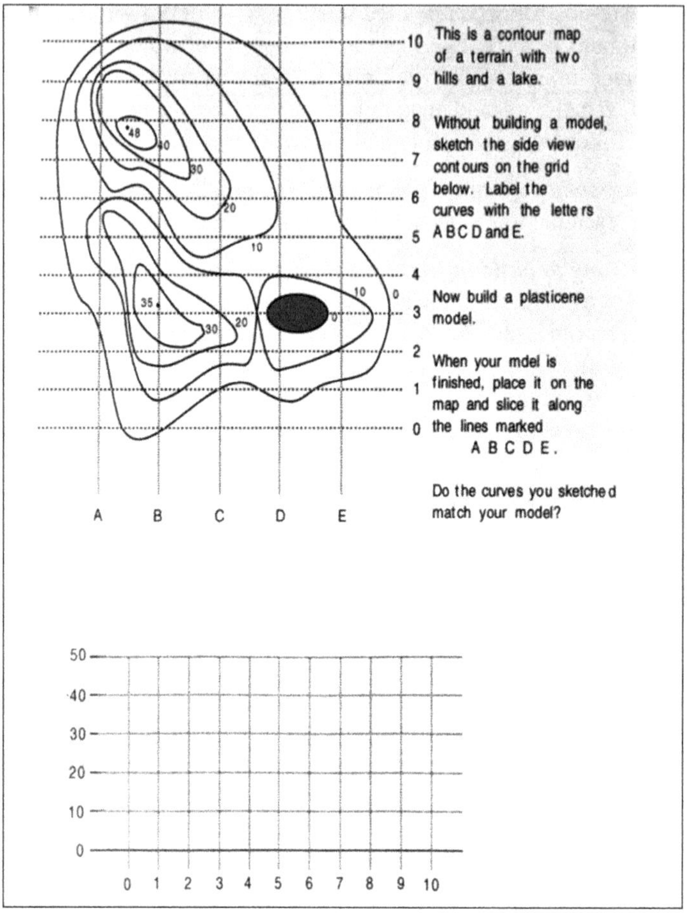

Fig. 3.12 Contour map and grid for plotting the cross-sections of the model

It is very clear from extensive research and reviews of the literature that strong spatial skills and visuospatial memory are predictive of mathematics achievement across a wide range of mathematics content areas and ages, and individuals with high spatial skills are likely to excel in mathematics (Li & Geary, 2017; Mix et al., 2016; Mix & Cheng, 2012; Wai et al., 2009). What we do not yet know is how malleable spatial skills are in terms of being able to teach students without initial high spatial ability how to: (1) acquire spatial sense, and (2) apply this knowledge by drawing on spatial strategies for mathematics problem solving. Future research on the spatial-mathematics association needs to focus on empirical research and spatial interventions to identify specific mechanisms for this association (Bailey, 2017).

Conclusion

In conclusion, we would like to propose that teaching generalized spatial skills disconnected from specific mathematics content areas may not be the best direction to go in future spatial intervention studies. Students who do not start out with strong spatial skills may need to learn to develop different types of spatial sense, specific to each type of content area, and to learn how to utilize spatial strategies based on developing spatial sense within these specific mathematics domains. Thus, the best strategy for future spatial-mathematics research may be: (1) understanding what types of spatial sense are required for different mathematics content areas, (2) conducting focused interventions for developing each type of spatial sense, (3) along with encouraging use of spatial strategies that draw upon them. If started early and done extensively across a range of mathematics content areas, students without initial spatial reasoning skills may eventually be able develop a wider mathematics spatial sense in order to approach a diverse range of mathematics problems with the added benefit of being able to draw upon critical spatial strategies from within in their problem-solving tool kit.

References

Ashcraft, M. H., & Fierman, B. A. (1982). Mental addition in third, fourth, and sixth graders. *Journal of Experimental Child Psychology, 33*, 216–234. https://doi.org/10.1016/0022-0965(82)90017-0

Baddeley, A. D., & Hitch, G. J. (1974). Working memory. In G. A. Bower (Ed.), *Recent advances in learning and motivation* (Vol. 8, pp. 47–90). New York: Academic Press.

Bailey, D. H. (2017). Causal inference and the spatial-math link in early childhood. *Monographs of the Society for Research in Child Development, 82*(1), 127–136. https://doi.org/10.1111/mono.12288

Battista, M. T. (2003). Understanding students' thinking about area and volume measurement. In D. H. Clements & G. Bright (Eds.), *Learning and teaching measurement: 2003 yearbook* (pp. 122–142). Reston, VA: National Council of Teachers of Mathematics.

Bjorklund, D. F., & Rosenblum, K. E. (2001). Children's use of multiple and variable addition strategies in a game context. *Developmental Science, 4*(2), 184–194. https://doi.org/10.1207/S15327647JCD0302_5

Boonen, A. J. H., van der Schoot, M., van Wesel, F., DeVries, M. H., & Jolles, J. (2013). What underlies successful word problem solving? A path analysis in sixth grade students. *Contemporary Educational Psychology, 38*, 271–279. https://doi.org/10.1016/j.cedpsych.2013.05.001

Carr, M., & Alexeev, N. (2011). Fluency, accuracy, and gender predict developmental trajectories of arithmetic strategies. *Journal of Educational Psychology, 103*, 617–631. https://doi.org/10.1037/a0023864

Carr, M., & Davis, H. (2002). Gender differences in mathematics strategy use: The influence of temperament. *Learning and Individual Differences, 13*, 83–95. https://doi.org/10.1006/ceps.2000.1059

Carr, M., Hettinger Steiner, H. H., Kyser, B., & Biddlecomb, B. A. (2008). Comparison of predictors of early emerging gender differences in mathematics competency. *Learning and Individual Differences, 18*, 61–75. https://doi.org/10.1016/j.lindif.2007.04.005

Carr, M., & Jessup, D. L. (1997). Gender differences in first-grade mathematics strategy use: Social and metacognitive influences. *Journal of Educational Psychology, 89*, 318–328. https://doi.org/10.1037/0022-0663.89.2.318

Casey, B. M. (2013). Spatial abilities and individual differences. In D. A. Waller & L. Nadel (Eds.), *Handbook of spatial cognition* (pp. 117–134). Washington, DC: American Psychological Association.

Casey, M. B., Andrews, N., Schindler, H., Kersh, J., Samper, A., & Copley, J. (2008). The development of spatial skills through interventions involving block-building activities. *Cognition and Instruction, 26*, 269–309. https://doi.org/10.1080/07370000802177177

Casey, M. B., Erkut, S., Ceder, I., & Mercer Young, J. (2008). Use of a storytelling context to improve girls' and boys' geometry skills in kindergarten. *Journal of Applied Developmental Psychology, 28*, 29–48. https://doi.org/10.1016/j.appdev.2007.10.005

Casey, B. M., Dearing, E., Dulaney, A., Heyman, M., & Springer, R. (2014). Young girls' spatial and arithmetic performance: The mediating role of maternal support during spatial problem solving. *Early Childhood Research Quarterly, 29*, 636–648. https://doi.org/10.1016/j.ecresq.2014.07.005

Casey, B. M., Lombardi, C., Pollock, A., Fineman, B., Pezaris, E., & Dearing, E. (2017). Girls' spatial skills and arithmetic strategies in first grade as predictors of fifth grade analytical math reasoning. *Journal of Cognition and Development, 18*(5), 530–555. https://doi.org/10.1080/15248372.2017.1363044

Casey, B. M., Pezaris, E., Fineman, B., Pollock, A., Demers, L., & Dearing, E. (2015). A longitudinal analysis of early spatial skills compared to arithmetic and verbal skills as predictors of fifth-grade girls' math reasoning. *Learning and Individual Differences, 40*, 90–100. https://doi.org/10.1016/j.lindif.2015.03.028

Cohen, D. (1989). *Calculus by and for young people*. Champaign/Urbana, IL: Don Cohen-The Mathman.

Cronin, V. (1967). Mirror-image reversal discrimination in kindergarten and first grade children. *Journal of Experimental Child Psychology, 5*, 577–585.

Cvencek, D., Metzoff, A. N., & Greenwald, A. C. (2011). Math gender stereotypes in elementary school children. *Child Development, 82*(3), 766–779. https://doi.org/10.1111/j.1467-8624.2010.01529.x

Educational Development Center, Inc. (2008). *Think math: Teacher guide, first grade*. Orlando, FL: Harcourt.

Ehrlich, S., Levine, S., & Goldin-Meadow, S. (2006). The importance of gestures in children's spatial reasoning. *Developmental Psychology, 42*(6), 1259–1268. https://doi.org/10.1037/0012-1649.42.6.1259

Fennema, E., Carpenter, T. P., Jacobs, V. R., Franke, M. L., & Levi, L. W. (1998). A longitudinal study of gender differences in young children's mathematical thinking. *Educational Researcher, 27*, 6–11.

Foley, A. E., Vasilyeva, M., & Laski, E. (2017). Children's use of decomposition strategies mediates the visuospatial memory and arithmetic accuracy relation. *British Journal of Developmental Psychology, 35*(2), 303–309. https://doi.org/10.1111/bjdp.12166

Frick, A., Ferrara, K., & Newcombe, N. S. (2013). Using a touch screen paradigm to assess the development of mental rotation between 3 ½ and 5 ½ years of age. *Cognitive Processes, 14*, 117–127. https://doi.org/10.1007/s10339-012-0534-0

Fuchs, L. S., Schumacher, R. R., Long, J., Jordan, N., Gersten, R., Cirino, P. T., … Changas, P. (2013). Improving at-risk learners' understanding of fractions. *Journal of Educational Psychology, 105*(3), 683–700. https://doi.org/10.1037/a0032446

Fuchs, L. S., Schumacher, R. F., Sterba, S. K., Long, J., Namkung, J., Malone, A., … Changas, P. (2014). Does working memory moderate the effects of fraction intervention? An aptitude–treatment interaction. *Journal of Educational Psychology, 106*, 499–514. https://doi.org/10.1037/a0034341

Gathercole, S. E., Pickering, S. J., Ambridge, B., & Wearing, H. (2004). The structure of working memory from 4 to 15 years of age. *Developmental Psychology, 40*(2), 177–190. https://doi.org/10.1037/0012/1649.40.2.177

Geary, D. C. (2011). Cognitive predictors of achievement growth in mathematics: A 5-year longitudinal study. *Developmental Psychology, 47*(6), 1539–1552. https://doi.org/10.1037/a0025510

Gunderson, E. A., Ramirez, G., Beilock, S. L., & Levine, S. C. (2012). The relation between spatial skill and early number knowledge: The role of the linear number line. *Developmental Psychology, 48*(5), 1229–1241. https://doi.org/10.1037/a0027433

Hamdan, N., & Gunderson, E. A. (2017). The number line is a critical spatial-numerical representation: Evidence from a fraction intervention. *Developmental Psychology, 53*(3), 587–596. https://doi.org/10.1037/dev0000252

Harris, J., Hirsh-Pasek, K., & Newcombe, N. S. (2013). Understanding spatial transformations: Similarities and differences between mental rotation and mental folding. *Cognitive Processing, 14*(1), 105–115.

Hawes, Z., Moss, J., Caswell, B., Naqvi, S., & MacKinnon, S. (2017). Enhancing children's spatial and numerical skills through a dynamic spatial approach to early geometry instruction: Effects of a 32-week intervention. *Cognition and Instruction, 35*(3), 236–264. https://doi.org/10.1080/07370008.2017.1323902

Hegarty, M., & Kozhevnikov, M. (1999). Types of visual-spatial representations and mathematical problem solving. *Journal of Educational Psychology, 91*(4), 684–689. https://doi.org/10.1037/0022-0663.91.4.684

Höffler, T. N. (2010). Spatial ability: Its influence on learning with visualizations—A meta-analytic review. *Educational Psychology Review, 22*(3), 245–269. https://doi.org/10.1007/s10648-010-9126-7

Hurst, M., & Cordes, S. (2016). Rational-number comparison across notation: Fractions, decimals, and whole numbers. *Journal of Experimental Psychology: Human Perception and Performance, 42*(2), 281–293. https://doi.org/10.1037/xhp0000140

Jitendra, A. K., Nelson, G., Pulles, S. M., Kiss, A. J., & Houseworth, J. (2016). Is mathematical representation of problems an evidence-based strategy for students with mathematical difficulties? *Exceptional Children, 83*(1), 8–25. https://doi.org/10.1177/0014402915625062

Joram, E. (2003). Benchmarks as tools for developing measurement sense. In D. H. Clements & G. Bright (Eds.), *Learning and teaching measurement: 2003 yearbook* (pp. 57–67). Reston, VA: National Council of Teachers of Mathematics.

Kingsdorf, S., & Krawec, J. (2016). A broad look at the literature on math word problem-solving interventions for third graders. *Cogent Education, 3*(1), 1–12. https://doi.org/10.1080/2331186X.2015.1135770

Kozhevnikov, M., Motes, M. A., & Hegarty, M. (2007). Spatial visualization in physics problem solving. *Cognitive Science, 31*, 549–579. https://doi.org/10.1080/15326900701399897

Kozhevnikov, M., & Thorton, R. (2006). Real-time data display, spatial visualization ability, and learning force and motion concepts. *Journal of Science Education and Technology, 15*(1), 111–132. https://doi.org/10.1007/s10956-006-0361-0

Krajewski, K., & Schneider, W. (2009). Exploring the impact of phonological awareness, visual-spatial working memory, and preschool quantity-number competencies on mathematics achievement in elementary school: Findings from a 3-year longitudinal study. *Journal of Experimental Child Psychology, 103*(4), 516–531. https://doi.org/10.1016/j.jecp.2009.03.009

Krawec, J. L. (2012). Problem representation and mathematical problem solving of students of varying math ability. *Journal of Learning Disabilities, 4*(2), 1–13. https://doi.org/10.1177/0022219412436976

Laski, E. V., Casey, B. M., Qingyi, Y., Dulaney, A., Heyman, M., & Dearing, E. (2013). Spatial skills as a predictor of first grade girls' use of higher-level arithmetic strategies. *Learning and Individual Differences, 23*, 123–130. https://doi.org/10.1016/j.lindif.2012.08.001

LeFevre, J., Fast, L., Skwarchuk, S., Smith-Chant, B. L., Bisanz, J., Kamawar, D., & Penner-Wilger, M. (2010). Pathways to mathematics: Longitudinal predictors of performance. *Child Development, 81*, 1753–1757. https://doi.org/10.1111/j.1467-8624.2010.01508.x

LeFevre, J., Jimenez Lira, C., Sowinski, C., Cankaya, O., Kamawar, D., & Skwarchuk, S. L. (2013). Charting the role of the number line in mathematical development. *Frontiers in Psychology, 4*(641), 1–9. https://doi.org/10.3389/fpsyg.2013.00641

Lemaire, P., & Siegler, R. S. (1995). Four aspects of strategic change: Contributions to children's learning of multiplication. *Journal of Experimental Psychology: General, 124*(1), 83–97. https://doi.org/10.1037/0096-3445.124.1.83

Levine, S. C., Foley, A., Lourenco, S., Ehrlich, S., & Ratliff, K. (2016). Sex differences in spatial cognition: Advancing the conversation. *WIREs Cognitive Science, 7*, 127–155. https://doi.org/10.1002/wcs.1380

Levine, S. C., Huttenlocher, J., Taylor, A., & Langrock, A. (1999). Early sex differences in spatial skill. *Developmental Psychology, 35*, 940–949. https://doi.org/10.1037/0012-1649.35.4.940

Li, Y., & Geary, D. C. (2013). Developmental gains in visuospatial memory predict gains in mathematics achievement. *PLoS One, 8*(7), e70160. https://doi.org/10.1371/journal.pone.0070160

Li, Y., & Geary, D. C. (2017). Children's visuospatial memory predicts mathematics achievement through early adolescence. *PLoS One, 12*(2), e0172046. https://doi.org/10.1371/journal.pone.0172046

Mix, K. S. (2010). Spatial tools for mathematical thought. In K. S. Mix, L. B. Smith, & M. Gasser (Eds.), *The spatial foundations of language and cognition*. New York: Oxford University Press.

Mix, K. S., & Cheng, U. L. (2012). The relation between space and math: Developmental and educational implications. In J. B. Benson (Ed.), *Advances in child development and behavior* (Vol. 42, pp. 199–243). Waltham, MA: Academic Press.

Mix, K. S., Levine, S. C., Cheng, Y., Young, C., Hambrick, D. Z., … Konstantopolous, S. (2016). Separate but correlated: The latent structure of space and mathematics across development. *Journal of Experimental Psychology: General, 145*(9), 1206–1227.

Miyake, A., Friedman, N. P., Rettinger, D. A., Shah, P., & Hegarty, M. (2001). How are visuospatial working memory, executive functioning, and spatial abilities related? A latent-variable analysis. *Journal of Experimental Psychology: General, 130*(4), 621–640. https://doi.org/10.1037/0096-3445.130.4.621

Moss, J., Bruce, C. D., Caswell, B., Flynn, T., & Hawes, Z. (2016). *Taking shape: Activities to develop geometric and spatial thinking grades k-2*. Toronto, ON, Canada: Pearson, Canada.

Nath, S., & Szücs, D. (2014). Construction play and cognitive skills associated with the development of mathematical abilities in 7-year-old children. *Learning and Instruction, 32*, 73–80. https://doi.org/10.1016/j.learninstruc.2014.01.006

National Council of Teachers of Mathematics. (2000). *Principles and standards for school mathematics*. Reston, VA: National Council of Teachers of Mathematics.

National Governors Association Center for Best Practices, Council of Chief State School Officers. (2011). *Common Core State Standards: Preparing America's Students for College and Career*. Retrieved from http://www.corestandards.org/.

Schneider, M., & Siegler, R. S. (2010). Representations of the magnitudes of fractions. *Journal of Experimental Psychology: Human Perception and Performance, 36*, 1227–1238. https://doi.org/10.1037/a0018170

Shen, C., Vasilyeva, M., & Laski, E. V. (2016). Here, but not there: Cross-national variability of gender effects in arithmetic. *Journal of Experimental Child Psychology, 146*, 50–65. https://doi.org/10.1016/j.jecp.2016.01.016

Shaw, J., & Pucket-Cliatt, M. (1989). Developing measurement sense. In P. Trafton & A. Schule (Eds.), *New directions for elementary school mathematics: 1989 yearbook* (pp. 149–155). Reston, VA: National Council of Teachers of Mathematics.

Siegler, R. S. (1987). The perils of averaging data over strategies: An example from children's addition. *Journal of Experimental Psychology: General, 116*(3), 250–264. https://doi.org/10.1037/0096-3445.116.3.250

Siegler, R. S. (2005). Children's learning. *American Psychologist, 60*(8), 769–778. https://doi.org/10.1037/0003-066X.60.8.769

Siegler, R. S. (2007). Cognitive variability. *Developmental Science, 10*(1), 104–109. https://doi.org/10.1111/j.1467-7687.2007.00571.x

Siegler, R. S., Duncan, G. J., Davis-Kean, P. E., Duckworth, K., Claessens, A., Engel, M., … Chen, M. (2012). Early predictors of high school mathematics achievement. *Psychological Science, 23*(7), 691–697. https://doi.org/10.1177/0956797612440101

Siegler, R. S., & Shrager, J. (1984). A model of strategy choice. In C. Sophian (Ed.), *Origins of cognitive skills* (pp. 229–293). Hillsdale, NJ: Erlbaum.

Siegler, R. S., Thompson, C. A., & Schneider, M. (2011). An integrated theory of whole number and fractions development. *Cognitive Psychology, 62*, 273–296. https://doi.org/10.1016/j.cogpsych.2011.03.001

TERC. (2008). *Investigations in number, data, and space* (2nd ed.). Boston, MA: Pearson Education.

Tzuriel, D., & Egozi, G. (2010). Gender differences in spatial ability of young children: The effects of training and processing strategies. *Child Development, 81*(5), 1417–1430. https://doi.org/10.1111/j.1467-8624.2010.01482.x

Torbeyns, J., Verschaffel, L., & Ghesquière, P. (2005). Simple addition strategies in a first-grade class with multiple strategies. *Cognition and Instruction, 23*(1), 1–21. https://doi.org/10.1207/s1532690xci2301_1

University of Chicago School Mathematics Project. (2007). *Everyday math* (3rd ed.). Boston, MA: McGraw-Hill Education.

Vandenberg, S. G., & Kuse, A. R. (1978). Mental rotations, a group test of three-dimensional spatial visualization. *Perceptual and Motor Skills, 47*(2), 599–604.

van Garderen, M. (2006). Spatial visualization, visual imagery, and mathematical problem solving of students with varying abilities. *Journal of Learning Disabilities, 39*, 496–506. https://doi.org/10.1177/00222194060390060201

Van Helsing (2007, August 7). *Bicylinder Steinmetz solid.gif.* Retrieved from https://commons.wikimedia.org/wiki/File:Bicylinder_Steinmetz_solid.gif

Vasilyeva, M., Laski, E., & Shen, C. (2015). Computational fluency and strategy choice predict individual and cross-national differences in complex arithmetic. *Developmental Psychology, 51*(1), 1489–1500. https://doi.org/10.1037/dev0000045

Verdine, B. N., Golinkoff, R. M., Hirsh-Pasek, K., Newcombe, N. S., Filipowicz, A. T., & Chang, A. (2014). Deconstructing building blocks: Preschoolers' spatial assembly performance relates to early mathematical skills. *Child Development, 85*(3), 1062–1076. https://doi.org/10.1111/cdev.12165

Voyer, D., & Hou, J. (2006). Type of items and the magnitude of gender differences on the Mental Rotations Test. *Canadian Journal of Experimental Psychology, 60*(2), 91–100. https://doi.org/10.1037/cjep2006010

Wai, J., Lubinski, D., & Benbow, C. P. (2009). Spatial ability for STEM domains: Aligning over 50 years of cumulative psychological knowledge solidifies its importance. *Journal of Educational Psychology, 101*(4), 817–835. https://doi.org/10.1037/a0016127

Chapter 4
More Space, Better Mathematics: Is Space a Powerful Tool or a Cornerstone for Understanding Arithmetic?

Krzysztof Cipora, Philipp Alexander Schroeder, Mojtaba Soltanlou, and Hans-Christoph Nuerk

Abstract Tight cognitive links between space and number processing exist. Usually, Spatial-Numerical Associations (SNAs) are interpreted causally: spatial capabilities are a cornerstone of math skill. We question this seemingly ubiquitous assumption. After presenting SNA taxonomy, we show that only some SNAs correlate with math skill. These correlations are not conclusive: (1) Their directions vary (stronger SNA relates sometimes to better, sometimes to poorer skill), (2) the correlations might be explained by mediator variables (e.g., SNA tasks involve cognitive control or reasoning), (3) the hypothetical course of causality is not resolved: For instance, contrary to conventional theories, arithmetic skills can underlie performance in some SNA tasks. However, benefits of SNA trainings on math skills seem to reinforce the claim of primary SNA role. Nevertheless, tasks used in such trainings may tap cognitive operations required in arithmetic, but not SNA representations themselves. Therefore, using space is a powerful tool rather than a cornerstone for math.

K. Cipora (✉) · M. Soltanlou
Department of Psychology, University of Tuebingen, Tuebingen, Germany

LEAD Graduate School and Research Network, University of Tuebingen, Tuebingen, Germany
e-mail: krzysztof.cipora@uni-tuebingen.de

P. A. Schroeder
Department of Psychology, University of Tuebingen, Tuebingen, Germany

Department of Psychiatry and Psychotherapy, University of Tuebingen, Tuebingen, Germany

H.-C. Nuerk (✉)
Department of Psychology, University of Tuebingen, Tuebingen, Germany

Leibniz-Institut für Wissensmedien, Tuebingen, Germany

LEAD Graduate School and Research Network, University of Tuebingen, Tuebingen, Germany
e-mail: hc.nuerk@uni-tuebingen.de

K. S. S. Mix, M. T. Battista (eds.), *Visualizing Mathematics*, Research in Mathematics Education, https://doi.org/10.1007/978-3-319-98767-5_4

Keywords Spatial-Numerical Associations (SNA) · Extension Spatial-Numerical Associations · Directional Spatial-Numerical Associations · SNARC effect · Arithmetic skills · Cognitive skills · SNA Taxonomy · Multi-digit number processing · Compatibility effect · Grounded cognition · Embodied cognition · Situated cognition · Embodied math trainings · Number Line Estimation (NLE) · Cardinality · Ordinality · Place identification · Place-value activation · Place-value computation · Place-value integration

Numbers and Space: A Long-Lasting Relation

The idea that numerical magnitudes could be represented on a directed mental line appeared very early in the history of mathematics. Dating back to the Greek philosophers of Aristotle's era, it was known that numerical magnitudes may be represented by a geometric line. In the Middle Ages, Campanus of Novara argued that a ratio found in one type of continuum can also be found in another. Subsequently, the medieval mathematician Nicole Oresme became a pioneer of quantifying space in terms of a multi-axis coordinate system (see Grant, 1972). That idea was later popularized and developed by René Descartes, after whom the widely used perpendicular coordinate system was named.

The mathematical concept of the number line (and coordinate systems in particular) and the possibility to transfer abstract numerical quantities onto space substantially influenced the development of mathematics. Nevertheless, the mapping of numbers onto space (and quantifying space by means of numbers) is not a totally abstract or arbitrary invention. There is little doubt that spatial and numerical representations can be tightly and bidirectionally associated; there are numerous examples that this can happen both voluntarily and relatively automatically (e.g., Shaki & Fischer, 2014). It was also demonstrated that non-numerical spatial abilities and mathematics abilities are correlated (Mix et al., 2016), which may also indicate common underlying cognitive mechanisms.

In the present book chapter, we use the term Spatial-Numerical Associations (SNAs; see Cipora, Patro, & Nuerk, 2015, for elaboration) to refer to a broad range of different behavioral phenomena (see Box 4.1). We shall see that there is no agreement about the origins of SNAs, i.e., whether they are innate or shaped by culture (including exposure to conventional representations of numbers on rulers, graphs, and so on). Here, we argue that one possible reason for such disagreements is that SNAs consist of different phenomena, differing in their origins, their general characteristics and their propensity to be changed by situated influences. We also discuss how particular SNAs are related to school mathematics achievement and outline that some SNAs are not predictive of later arithmetic performance, whereas others are fundamentally necessary. SNA trainings have been shown to transfer to other arithmetic skills; we will discuss why this may be the case, and in particular, whether a spatial-numerical representation per se is improved by training, or whether such training use space and its potential to be associated with number as a powerful tool to train other numerical skills and representations. For each SNA type, we discuss whether and how it informs mathematics education with regard to trainings and other possibilities for intervention.

Space and Numbers Live Next to Each Other

One of the first scientific inquiries on how humans represent numbers revealed that thinking of numbers includes some spatial components, at least in some people. In a paper published in 1880, Sir Francis Galton (Charles Darwin's cousin) described several reports of individuals who claimed to possess very vivid spatial visualizations of numbers (Galton, 1880). These usually took complex curvilinear forms, and according to the reports of people who experienced them, specific number representations remained precise and stable over time. Such explicit spatial number forms (i.e., directly available in self-reports) are pronounced in a considerable proportion of the general population (estimates vary from 2.2 to 29.0% of the population), and are referred to as synaesthetic visuo-spatial forms (e.g., Simner, Mayo, & Spiller, 2009). Nevertheless, number-space synaesthetes (and their variable visualizations) are to some degree exceptional, and there has been a debate as to how far their spatial-numerical representations can inform us about the general population (e.g., Cohen Kadosh & Henik, 2007).

Moyer and Landauer (1967) were the first to describe the numerical distance effect (see Box 4.1). This effect refers to the behavioral finding of longer response times on a comparison task, when the numerical difference decreases between the target stimulus (e.g., the numbers 1 or 4) and the referent number (e.g., 5, to which 1 or 4 are compared). The authors explained their finding with the claim that numerical magnitudes are converted to analogue magnitudes, which are then compared (and thus it takes longer to compare numbers that are closer together). While the distance effect has been interpreted in a spatial framework, the assumption of a spatial organization of number magnitude is not necessary to account for the effect. Following up on Moyer and Landauer's (1967) work, Restle (1970) developed the concept of the Mental Number Line (MNL) as an analog system, which organizes the representation of all numbers by distinctive markers placed on a visual line. Making numerical judgment requires the participants to "zoom in" on the MNL close enough so that numbers to be compared are located in different regions. Thus, the smaller the difference between numbers to be compared, the more "zooming in" operations need to be carried out. Restle's concept of the number line has seen differentiations and extensions. For instance, it was suggested that multiple number lines are activated for multi-symbol numbers and not only one analog number line (Nuerk, Moeller, & Willmes, 2015).

SNA: Not a Single Melting Pot

Space is not only related to number magnitude representation, but also to other mathematical representations (numerical intervals, ordinality, mathematic functions; see M. H. Fischer & Shaki, 2014). However, it seems that the SNA term describes a relatively general property of cognition, which needs to be further specified. Although the vast majority of studies have been sound and conclusive, their results unfortunately cannot be combined easily to provide the big picture of how numbers are associated with space. Furthermore, the very general yet reasonable question of whether and to which extent SNAs are important for arithmetic (or even more broadly mathematics) learning is highly dependent on the type of SNA under study. Differences in SNAs will be elaborated on in detail in subsequent parts of the chapter.

Box 4.1 A. Key Terms and Concepts of Experimental Psychology Used in This Chapter

Attention (see Raz & Buhle, 2006). Attention is a mechanism of information selection. One of the most influential models of attention outlined by Posner (see Petersen & Posner, 2012) postulates three independent networks: alerting, orienting, and executive. The alerting system controls the arousal level and sustained vigilance. The orienting system prioritizes sensory input from a particular sensory modality and/or location in space. Executive attention is responsible for inhibition of irrelevant responses, conflict monitoring, and switching between tasks. In this chapter, two components of attention are of particular importance. Orienting (related to spatial shifts of attentional focus) and executive, especially components related to inhibition and conflict monitoring, will be discussed in the context of the role the domain-general factors play in SNAs.

Compatibility effect (see Kornblum et al., 1990). This term describes the fact that some tasks are easier or more difficult depending on stimuli and response sets used. For instance, it is easier to respond if a meaning of the stimulus (e.g., left/right pointing arrow) corresponds to the required response (e.g., left/right located key) than when it does not (stimulus-response compatibility). Task difficulty also differs depending on the meaning of objects being presented (stimulus-stimulus compatibility; e.g., Verbruggen, Liefooghe, Notebaert, & Vandierendonck, 2005). Namely, the task is easier if irrelevant stimuli presented together with the target stimulus are associated with the same response, than when they are associated with another response. Differences in task difficulty are reflected in reaction times, reaction time variances, error rates (all larger for incompatible trials), as well as in neural correlates of cognitive processing (differences in electrical or metabolic brain responses). This is because in case of incompatible trials, competing cognitive processes involved in initiating different responses need to be resolved by

(continued)

Box 4.1 (continued)

cognitive control. Compatibility effects are widely used in the domain of numerical cognition in investigating Spatial Numerical Associations (SNAs) or the multi-digit number processing.

Domain-specific and domain-general processes (see Hohol, Cipora, Willmes, & Nuerk, 2017). Domain-general processes can be viewed as mental general-purpose tools, which can be used to process a very broad range of information (e.g., numerical, spatial, verbal, and so on). On the contrary, domain-specific processes are those specialized in processing particular type of information (e.g., quantity). One of the most important debates in numerical cognition is the interplay and mutual relationships between these two types of processes and the roles they play.

B. Key Terms and Phenomena Investigated Within Numerical Cognition

Estimation (see Gallistel & Gelman, 2000). The ability to provide an approximate number of elements in a collection; possessed by humans and several other species. The cognitive system responsible for this process is called the Approximate Number System (ANS). The accuracy of estimation decreases with an increasing number of elements; however, the minimal ratio enabling successful discrimination between sets is constant within an individual (the Weber fraction; see Piazza & Izard, 2009). In other words, the larger is the absolute size of the set, the greater the absolute difference between sets must be (so that this ratio remains constant) in order to enable successful discrimination.

Mental Number Line (MNL; Restle, 1970). A theoretical construct or a metaphor on how numbers are mentally represented. According to this view, numbers are represented as the points on a directional number line. It is claimed that the MNL is logarithmically compressed (i.e., distances between small numbers are larger than distances between large numbers).

Numerical distance effect (Moyer & Landauer, 1967). This term denotes the observation that the time required to perform number comparisons increases with a decreasing difference between the numbers to be compared. It can be observed for both non-symbolic (dot patterns) and symbolic (e.g., Arabic numbers) notation. The numerical distance effect can be split into absolute (e.g., in 85_53, is 85 − 53 = 32), decade (e.g., in 85_53, is 8 − 5 = 3), and unit distance (e.g., in 85_53, is 5 − 3 = 2) aspects. Importantly, similar effects in the context of comparing physical objects (e.g., line length) were described earlier in classical Gestalt psychology literature (see Cohen Kadosh, Lammertyn, & Izard, 2008).

Ratio effect (see Lyons, Nuerk, & Ansari, 2015). This term denotes the relationship between numerical ratio (i.e., the ratio of numbers to be compared) and comparison task performance: the performance decreases as the ratio approaches "1." The ratio effect is very similar to the distance effect; however, it considers

(continued)

Box 4.1 (continued)

the absolute magnitude of numbers to be compared. Thus, to some extent the ratio effect takes into account both distance and size effects.

Size effect (see Brysbaert, 1995). It denotes the observation that the time needed to make numerical judgments increases with increasing magnitude. As in the case of the numerical distance effect, the size effect was first observed in tasks requiring comparison of physical properties of objects.

Spatial-Numerical Association of Response Codes (SNARC; Dehaene, Bossini, & Giraux, 1993). SNARC denotes an observation that numerical magnitudes are associated with space: in left-to-right readers, small magnitude numbers are associated with the left hand side, whereas large magnitude numbers are associated with the right hand side.

Spatial-Numerical Associations (SNA; see Cipora et al., 2015). A broad range of phenomena demonstrating that numbers (especially their magnitudes) are bidirectionally associated with space. It can be observed by means of behavioral or neural signatures.

Subitizing (see Piazza, Mechelli, Butterworth, & Price, 2002). From the Latin word *subitius* (sudden). An elementary capacity to quickly and effortlessly determine the number of elements in small sets (i.e., no larger than 4 elements) possessed by humans and other species.

Unit-decade compatibility effect (Nuerk, Weger, & Willmes, 2001). Denotes the fact that while comparing two two-digit numbers, responses are faster and more accurate if the number of units in the numerically larger number is larger than the number of units in the numerically smaller number (compatible trial, e.g., 23_69), than in a situation when the number of units in the larger number is smaller than in the smaller number (incompatible trial, e.g., 29_63).

Place-value processing (Nuerk et al., 2015). The place-value concept refers to the value of a digit according to its position within a sequence of digits in the Arabic number system. The value of a digit increases by a power of 10 (base-10 system) with each step going from the right digit to the left digit in a multi-digit Arabic number (e.g., $257 = [2] \times 10^2 + [5] \times 10^1 + [7] \times 10^0$). Nuerk et al. (2015) proposed a three-level categorization of componential processing for place-value understanding: place identification, place-value activation, and place-value computation.

Numbers and Space: Fundamental Principles and Questions

An overview of tasks used to measure SNAs is provided in Box 4.2. Tight relationships between space and number can be observed in varied tasks, age groups, and even species.

Grounded, Embodied, and Situated Influences on SNA

One implicit assumption about SNAs is that they change during development, but are rather stable across different situations (i.e., similar to personality characteristics). However, an increasing amount of evidence converges to show that SNAs may also be subject to situated influences (see M. H. Fischer, 2012 for theoretical justification and introduction to this term; Wasner, Moeller, Fischer, & Nuerk, 2014 for applications in other areas).

The common assumption regarding embodied (in a general sense) influences is that sensory and motor experiences present during the acquisition of knowledge (such as semantic number magnitude) are re-activated during retrieval and when operations are performed. Within these general embodied influences, according to Fischer (2012), SNAs can be influenced by grounded, embodied, and situated influences, which we will briefly explain in the following section:

- *Grounded* principles are reflected in universal rules of number semantics such as the fact that larger numerosities imply physically "more" of something, including parts of smaller numerosities (M. H. Fischer, 2012), or the vertical association of larger numerosities with higher physical space (Wiemers, Bekkering, & Lindemann, 2017).
- *Embodied influences* (in a narrower sense) refer to bodily influences and cultural sensorimotor experiences that influence cognition even though they might not be immediately relevant to a situation. For instance, the SNARC effect (see Box 4.1) is moderated by the reading direction of a language (e.g., Shaki, Fischer, & Petrusic, 2009).
- *Situated influences* are nested in the empirical context or experimental situation: For instance, situated influences on SNAs are shown when the SNARC effect in bilinguals changes from one experimental situation in which words are presented in a left-to-right written language, to another experimental situation in which words are presented in a right-to-left written context (M. H. Fischer, Shaki, & Cruise, 2009). In a similar vein, finger counting habits (generally considered a directional embodied influence) were remarkably different for participants that were simply told to either use their fingers for counting, to hold their hands in front of them to count aloud, or to indicate their counting habits themselves in a questionnaire (i.e., their hands were occupied; Wasner et al., 2014). The difference between embodied and situated influences is that embodied influences are culturally learnt and may not be induced by the experimental situation, but nev-

ertheless modulate cognition between different cultural groups. Situated influences are specific to the particular experimental or empirical situation in which an effect or an underlying representation is assessed.

It is important to note that situated influences can be further distinguished (Cipora, Patro, & Nuerk, 2018). A more exhaustive overview and a taxonomy on situated influences on SNA is beyond the scope of this chapter.

Several empirical studies have tested grounded, embodied, and situated modulations of the directional SNAs from brief interventions to long-term trainings, either to study their fundamental features and demonstrate their underlying mechanisms or to study potential interventions. In fact, several factors allow for re-training the shape of SNAs, yet the impact of these interventions or training on arithmetic is unclear. Notably, effects of embodied cultural differences due to reading direction have been established in different experimental studies (Moeller, Shaki, Göbel, & Nuerk, 2015; Shaki et al., 2009), which indicate that culturally experienced sensorimotor interactions with the environment, such as eye movements in reading a language, can shape SNAs.

Correlations of SNA and Arithmetic Skill and Potential Underlying Mechanisms

The question of whether SNAs are related to arithmetic skill is in our view much too broad to be answered adequately, since it depends on which SNA is considered. Some studies report that stronger or more adequate SNAs are related to better mathematics skills (e.g., Siegler & Ramani, 2009). Other studies indicate no such effects (Cipora & Nuerk, 2013), whereas a third category of studies show that stronger SNAs are related to poorer mathematics skills (Hoffmann, Mussolin, Martin, & Schiltz, 2014). With such varied results, it seems essential to consider different SNA types separately. Furthermore, mediating variables, such as domain general cognitive factors (Hohol et al., 2017) and knowledge of formal principles of mathematics should be taken into account.

First, several SNA types can equally be considered compatibility effects (see Box 4.1). Like compatibility effects, the SNA indexes the extent to which *irrelevant* information influences processing of the currently relevant numerical or spatial information. In order to successfully perform the task, one in fact needs to refrain from processing the interfering information in half of the trials (assuming that the task consists 50% compatible and 50% incompatible trials). Therefore, domain-general processes (see Box 4.1) need to be involved to inhibit the irrelevant aspect of the stimuli/response, such as physical size, distance between numbers, or implicit mapping of numerical magnitudes onto space.

Some of these operations seem to be governed especially by executive functions, which themselves have been shown to correlate with mathematics skill level (Cragg & Gilmore, 2014; Nemati et al., 2017) or with directional SNA (Hoffmann, Pigat,

& Schiltz, 2014). Thus, it seems that interference-based SNA should either not correlate or correlate negatively with mathematics achievement. In fact, this view is supported by several studies (see Cipora et al., 2015 for review).

Knowledge of formal rules of mathematics may not only influence arithmetic skill directly, but also mediate the relation between SNA and arithmetic skill. SNAs related to explicit counting direction in children can serve as an example: Through numerical development, it is very important that the child realizes and understands that the direction of counting elements is in fact irrelevant to the counting operation (Gelman & Gallistel, 1978). Namely, this means that it does not matter where the sequence is started for counting. Thus, successful acquisition and use of numerical knowledge in fact require a cognitive flexibility that would undermine the SNAs related to a particular direction. Sometimes the formal knowledge counters SNA in ways that are even more specific. When solving calculation problems, the physical size of digits does not directly influence their meaning. On the other hand, in problems such as "$2^5 + 5^2 = ?$", differences in physical sizes carry arbitrary semantic information. Namely, the smaller size of the superscript does not mean that these numbers are either numerically smaller or less important, but instead provides information about the required operation. In that case, proper calculation requires (1) knowledge of the arbitrary rule on power notation and (2) inhibition of SNA, which would misleadingly associate smaller extension in the superscript with smaller number magnitude.

On the contrary, the potential correlation between performance in explicit number line estimation tasks (see Box 4.2) and arithmetic seems theoretically justified. The accurate mapping of numbers onto spatial locations requires an understanding of numerical magnitude and relations between numbers. In this case, the SNA is not reflected by the interference effect and does not go against formal mathematics knowledge. On the other hand, such correlation is not very surprising because the very strategies that underlie good number line estimation rely on arithmetic skills (Barth, Starr, & Sullivan, 2009; Link, Huber, Nuerk, & Moeller, 2014; Link, Nuerk, & Moeller, 2014).

Even if correlations are observed, one still needs to be cautious in their interpretation. This is particularly the case in children's studies, because developmental changes can naturally produce changes in both arithmetic and NLE abilities separately. Especially in studying predictors of later arithmetic achievement, it is thus essential to include multiple measures and control variables in order to avoid flagging a correlation as meaningful, when it is actually driven by shared covariation due to age or another aspect of development. For example, conceptually unrelated abilities such as the speed of running and mathematics ability both increase from grade 1 to grade 2 of schooling. Thus, if both measures were assessed at grade 1 and grade 2, a positive correlation would reflect shared variance due to age, but would not indicate that running speed could predict later achievements in mathematics. Of course, the mediation of a correlation by another variable would be more critical to assess in potentially meaningful predictors of arithmetic achievement, such as counting abilities, working memory, or something else. Thus, a correlation can be

Box 4.2: Measuring Spatial-Numerical Associations

Several tasks have been developed in order to tap how humans associate numbers with space. They will be briefly discussed below. This overview shows that numbers seem to be inherently associated with space, irrespective of the type of task used to measure it.

Extension SNA: Approximate:

Interference tasks measuring the size congruity effect (Henik & Tzelgov, 1982). The participants are presented with pairs of numbers differing both in physical size and numerical magnitude. The participant's task is to indicate either the physically or numerically larger one. Difference in reaction times between congruent (physically larger number is also numerically larger) and incongruent (physically larger number is numerically smaller) trials are measured.

Following cardinality across modalities (de Hevia & Spelke, 2010). Used in infant studies. Children are familiarized with either ascending or descending sequences of one type of stimuli (e.g., a line becoming longer and longer). Subsequently, the test probe (either an ascending or descending sequence of another type of stimuli, e.g., an increasing number of dots) is presented. Looking times are compared between trials in which the order in the test probe was familiar or novel.

Line bisection task (Fischer, 2001). Participant's task is to indicate the midpoint of the line. The line is either constructed out of numbers (e.g., 99999999999999; Fischer, 2001) or flanked with numbers presented either in symbolic or nonsymbolic format (e.g., 2-----------7; de Hevia & Spelke, 2009). In the first case midpoint is biased to the left/right when line is built of small/large magnitude numbers. In the second case, the midpoint is located closer to the larger magnitude number.

Spatial arrangement of operation order (Landy & Goldstone, 2010). Participants are presented with arithmetic operations in which following the operation order is essential. The physical spacing between numbers can be either congruent (e.g., 2*2 + 2) or incongruent (e.g., 2 * 2+2) with the operation order. Reaction times are compared between conditions.

Extension SNA: Exact

Number line estimation (Siegler & Opfer, 2003). There are two major variations of this task, namely bounded and unbounded versions (Cohen & Blanc-Goldhammer, 2011). In the bounded number line estimation the participant is presented with a line marked with numbers on both of its ends (e.g., 0 and 100). In the unbounded task, only the left side of the number is marked with a number. Additionally there is another short line indicating one unit. In both cases the task is to mark the position of a given number on the line. Recently another variation of the task was developed in which touchscreen technology

(continued)

Box 4.2 (continued)
is used not only to track the response given by the participant but also the exact movement trajectory.

Directional SNA: implicit coding—cardinality
Parity judgment task (e.g., Dehaene et al., 1993). The participant is to make binary decisions on whether the presented number is odd or even. Two lateralized response keys are used. In the mid-experiment, the response-to-key assignment is switched so that both right- and left-hand responses are collected for each number.

Magnitude classification task (see Wood, Willmes, Nuerk, & Fischer, 2008). The participant is to make binary decisions on whether the presented number is smaller or larger than a criterion value (which is constant across the experiment). Two lateralized response keys are used. Similarly to parity judgment, response-to-key assignment is flipped in the mid-experiment.

Magnitude comparison task (see Wood et al., 2008). The criterion value according to which the numerical value of the target number is compared changes from trial to trial.

Other binary decisions on numerical magnitude (e.g., Fias, Lauwereyns, & Lammertyn, 2001). The participants are deciding on other characteristics of the number (e.g., whether it contains a particular phoneme or whether it is written upright or with italics).

Directional SNA: implicit coding—place-value
Physical comparison task (Ganor-Stern, Tzelgov, & Ellenbogen, 2007): A variation of a typical two-digit number comparison task. While one number is shown in large font, the other is shown in small font, irrelevant to the magnitude of the numbers. In half of the trials the magnitude and physical size of the number pairs are compatible but not in the other half. Participants are to decide which number is physically larger, while ignoring numerical magnitudes.

Directional SNA: implicit coding—ordinality
Parity judgment task with working memory load. This is a variation of the typical parity judgment task in which the participant is first asked to memorize a sequence of numbers and then performs the typical parity task. The participant is to react only to numbers, which were present in the sequence. After a block of such trials, it is checked whether the participant can correctly recall the sequence.

Directional SNA: implicit coding—functions
Spatial biases in mental arithmetic (Knops, Viarouge, & Dehaene, 2009). Participants are to perform operations (additions and subtractions) on either symbolic or nonsymbolic material. It is tested whether performing such operations influences concurrent spatial activity (directional hand movements, oculomotor activity, lateral attentional shifts).

(continued)

Box 4.2 (continued)

Arithmetic sign and space (Pinhas, Shaki, & Fischer, 2014). The participants decide whether the presented sign is "+" or "−" using two response keys.

Directional SNA: explicit coding—place value

Two-digit number comparison (Nuerk et al., 2001): Two two-digit numbers are presented simultaneously on the screen with the same font size, while they differ in magnitude. Participants are asked to select the numerically larger number.

Directional SNA: explicit coding—ordinality

Counting tasks (e.g., Shaki, Fischer, & Göbel, 2012). Such tasks are used mostly in cross cultural children's studies. Participants are presented with a row or a matrix of tokens and are asked to count them. The direction of counting is checked.

The tasks may involve several response formats. Binary decisions can be made bimanually, unimanually using two different fingers or with a single finger in a pointing task. Responses may also be given using feet or directional eye movements. In other setups responses may be given orally, by pointing, grasping, or the whole movement trajectory can be measured (Fischer & Shaki, 2014 for review).

explained (partialled out) by another variable, which correlates with both variables of the original correlation.

To sum up, there is no general and consistent pattern of correlations between SNA and arithmetic skill, as some SNAs are consistently related to arithmetic skill (number line estimation) while others (SNARC) are not consistently related to arithmetic skill. In the following sections we present evidence for relationships (or lack thereof) between particular SNA types and arithmetic skill.

Causal Relations: What Can Be Derived from Training Studies? Implications and Caveats

Apart from practical implications, the development of efficient trainings can shed light on underlying theories and causal relationships as well. Namely, the fact that a given training is efficient provides evidence for a causal nature of relations between constructs of interest. However, some caveats apply.

Even if a training is successful, it often consists of several modules or representations. For instance, in spatial-numerical trainings, one might enhance spatial representations and processes, numerical representations and processes, or their relations.

Even within numerical cognition, it might not be trivial to pinpoint the aspects that directly or indirectly relate to arithmetic abilities. What is more, additional variables such as gender or socioeconomic status may render trainings more effective in different learners. Thus, training programs may be tailored to a particular group: for instance, embodied trainings might be successful for typically developing children, but may not function for children with learning disabilities, because the instructions for the motor action in embodied training place too high a demand on working memory or other cognitive resources. Or vice versa, an embodied training might not further help typically developing children, but may facilitate learning in disabled children for whom normal instruction is not sufficient. In sum, causal implications from successful or unsuccessful trainings are not as straightforward as one might presume.

In this chapter, we focus on spatial-numerical trainings. However, we wish to make explicit that these are not the only successful numerical trainings. Training on other tasks, such as the MNL task or non-symbolic magnitude comparison, could also change mathematical performance either directly or indirectly by improving domain-general abilities.

By introducing the fundamental principles and questions in SNA research, we have now laid the groundwork for a new and extended taxonomy for SNAs.

A New Taxonomy for SNAs: How Different SNAs Have to Be Differentiated

Patro, Nuerk, Cress, and Haman (2014) outlined the first proposal of a systematic SNA taxonomy, which considered only phenomena that can be observed in preliterate children. Subsequently, Cipora et al. (2015) extended this taxonomy by including SNAs observed in adults. Here, we partially clarify and extend it further by adding place-value processing as an instance of directional SNA considering both implicit and explicit coding components. The graphic summary of the taxonomy is presented in Fig. 4.1.

The taxonomy will be discussed in the following paragraphs together with results demonstrating relationships (or a lack thereof) between particular SNA types and arithmetic skill. Furthermore, wherever such evidence exists, we will discuss situated influences on a given SNA type. In line with Cipora et al. (2018), we will classify manipulations of situated influences into categories depending on which stage of information processing was affected (perceptual, representational, action) and when the manipulation was applied (pre-experimental or intra-experimental). Note that this taxonomy does not consider the numerical distance effect or the size effect. Assuming the analog system of numerical magnitude does not necessarily imply existence of a spatial component. The MNL does not need to be spatially mapped/oriented in order to explain the numerical distance effect or the size effect (see Bonato, Fabbri, Umiltà, & Zorzi, 2007; Cipora et al., 2015). These two fascinating phenomena fall outside the scope of the taxonomy.

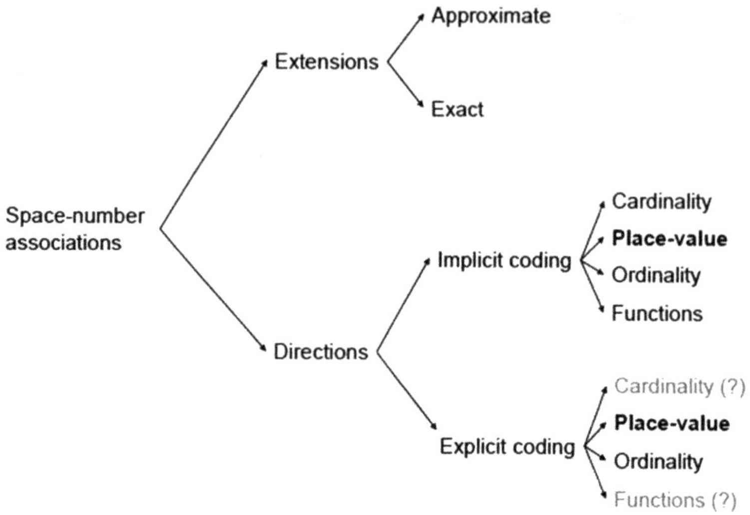

Fig. 4.1 Overview of the Spatial-Numerical Associations (SNAs) taxonomy

Overview of the taxonomy

The primary taxonomy proposal introduced by Patro et al. (2014) includes the distinction between extension SNA and directional SNA. Within extension SNA there are two subcategories: (a) approximate (formerly cardinality) and (b) exact (formerly interval). Cipora et al. (2015) further subdivided directional SNAs into two categories: (1) SNAs related to implicit coding of numerical magnitude, and (2) those related to explicit coding. Within each subcategory, Patro et al. (2014) differentiated SNA types related to cardinality and ordinality, and Cipora et al. (2015) added a third category of functions.

Category Extension SNA: Subcategory Approximate

Type and Paradigms

This SNA type might be observed very early in development (e.g., de Hevia & Spelke, 2010), in non-human animals (Tudusciuc & Nieder, 2007) as well as in adults (Henik & Tzelgov, 1982). For example, the *size congruity effect* is present when numerals of different physical sizes are shown, and responses are facilitated if semantic and physical size information are matched (Henik & Tzelgov, 1982). A more exhaustive overview of experimental paradigms is presented in Box 4.2. Importantly, the relation between space and numbers is bidirectional. Spatial aspects of the stimuli also affect numerical judgments (e.g., widely spaced numbers are judged to be more numerically distant; Lonnemann, Krinzinger, Knops, & Willmes, 2008). Interference between space and numbers seems not to be restricted to simple numerical judgment, but also affects calculation efficiency. Facing problems like "2 * 2+2" vs. "2*2 + 2" (i.e., the problem is identical but spacing differs), participants are faster and more accurate in the first problem as the spatial arrangement corresponds to proper operation order (Landy & Goldstone, 2010). We have previously called this category cardinality, but now term it approximate, because the above examples are not about exact relations between the physical and numerical space (like intervals in the number line estimation task described further on), but rather about larger magnitudes in one dimension being related to magnitudes or functions in another dimension, while the exact relation is usually not important and unspecified.

Terming this subcategory approximate was also inspired by our proposal to take studies of the *approximate number system* (ANS, see Box 4.1) into this subcategory. ANS studies are not usually considered a SNA; however, based on more recent research, we postulate that approximate extension SNAs are in fact an important factor which needs to be considered when analyzing results of these studies (Dietrich, Huber, & Nuerk, 2015; Gebuis & Reynvoet, 2012). In ANS studies, non-symbolic sets are judged and spatial parameters such as extension, density, size, and others have been found to interfere with numerosity (see, e.g., Leibovich & Henik, 2013). Thus, visual properties of the stimuli presented in the typical ANS task are either positively or negatively correlated with the actual number of elements present within each set. This means that larger spatial extension (or another parameter) is either consistent with larger numerosity or interferes with it. If the association is consistent, performance is usually better (Szűcs, Nobes, Devine, Gabriel, & Gebuis, 2013).

Situatedness

Studies on situated influences on this SNA category are rather scarce. Usually, researchers were interested in demonstrating the phenomena and sometimes in looking for correlations between SNAs and other cognitive characteristics. One notable exception is the experiment by Fornaciai, Cicchini, and Burr (2016), which documented that numerosities of dot collections are systematically underestimated when the dots are connected by task-irrelevant lines. This result is supported by different patterns of psychometric functions of numerosity that adapt depending on whether the lines are present or not in the display. Nevertheless, this field requires further exploration.

Correlations with Arithmetic Skill

The relation between approximate extension SNA and arithmetic skill has mostly been looked at with ANS studies. The general rationale is that an understanding of non-symbolic magnitude, e.g., of set sizes in visual dot patterns, would constitute the deep basis of any magnitude-related activities, including formal mathematics (Feigenson, Dehaene, & Spelke, 2004). Despite these strong theoretical foundations, usually the observed correlations were either non-existent (e.g., Sasanguie & Reynvoet, 2014), or very small, around $r = 0.2$, and decrease with age (Schneider et al., 2017 for meta-analysis).

Even these relatively low (but theoretically sound) correlations need to be treated with caution. As we mentioned above, when one has to compare two sets of dots, apart from numerosity they also differ by spatial features (Szűcs et al., 2013). In so-called compatible trials, physical features (e.g., convex hull, overall area covered by the elements, size of elements) correlate with the number of elements. In so-called incompatible trials, the visual features correlate negatively with the number of elements. In this case, if the task is to judge numerosities, strong SNA would be beneficial for compatible trials but detrimental in incompatible trials. In some studies, the correlation between performance in non-symbolic comparison and arithmetic skill was not present any more when the interference component was controlled for (Cragg & Gilmore, 2014). In fact, children seem to be often misguided by spatial components when solving numerical tasks (Stavy & Tirosh, 2000) such that they tend to follow the principle "More A–More B." Despite being useful in everyday life (i.e., physical and temporal properties of objects are usually correlated with numbers, for instance a larger pile comprises more elements than a smaller one), formal mathematics requires abstracting from physical properties, e.g., despite having the same physical size, numbers refer to different magnitudes (Bueti & Walsh, 2009).

Contemporary conceptions of the ANS include the multimodal processing of spatial and quantity information (Leibovich, Katzin, Harel, & Henik, 2017), although it is important to highlight that one should always expect interactions between domain-general factors and a domain-specific factor such as the ANS (Hohol et al., 2017).

There are no genuine correlations between other instances of this SNA type and the arithmetic skill level as well. In particular, there were no correlations between the size congruity effect and mathematics performance (Bugden & Ansari, 2011; Rodic et al., 2015). Lonnemann et al. (2008) observed that boys (8- to 9-year-olds) who exhibited stronger SNA in judging numerical and spatial distances performed mathematics better. However, this effect was not present in girls. In general, it seems that this SNA category is not genuinely (or only weakly) correlated with arithmetic skill.

Trainings

Studies that use non-symbolic ANS tasks as training yield inconsistent results. Positive effects have been reported: for instance, two experiments on training in non-symbolic addition and subtraction improved performance in symbolic operations (Park & Brannon, 2013). Another positive outcome was observed for children that engaged primitive quantities on exact arithmetic problems (Hyde, Khanum, & Spelke, 2014). However, no such cross-over effect was observed in another large-scale study with children randomly assigned to different groups including training on exact numerosities (Obersteiner, Reiss, & Ufer, 2013). Another recent study corroborated this negative result that extensive arithmetic training and substantial improvements in arithmetic performance were not reflected in matching ANS acuity changes (Lindskog, Winman, & Poom, 2016). A possible indication of these inconsistent results would be to focus on the processes involved in mathematical operations and not only on the assumed numerical representation.

Category Extension SNA: Subcategory Exact

Type and Paradigms

Previously, we called this subcategory "Intervals." However, the difference between "approximate" and "exact" categories is the requirement of an exact (vs. approximate) match between spatial interval or a specific magnitude and the numerical interval of the magnitude. Performance is usually measured as a deviation from the exact match.

The number line estimation (NLE) task (see Box 4.2) involves associating number intervals with respective exact spatial extensions. The task itself is very easy to explain to the participants, including small children (e.g., 5- to 6-year-olds; Siegler & Booth, 2004), which itself can be treated as an argument that mapping numerical magnitude onto spatial extensions is natural. This can be illustrated in a classical example where children always think that numerical magnitude maps onto spatial extension, and cannot detach from mapping extension to numerical quantity when distinct objects are presented; i.e., the Piagetian Number Conservation experiment (Gelman & Gallistel, 1978). Performance in the NLE task has been linked to the internal representation of numerical magnitude (e.g., Siegler, 2009). It was claimed that with training, internal magnitude representation changes from a logarithmic format (i.e., large magnitude numbers are compressed) into a linear one (with equal distances between numbers). Nevertheless, the log-to-linear change in mental representation of the magnitude (Siegler & Opfer, 2003) was challenged. It was shown that improvement of proportional judgment skill may explain the results better (i.e., typical bounded NLE is in fact solved by means of proportional judgment; Barth & Paladino, 2011). The other challenging view was that performance in the task instead reflects place-value integration (Moeller, Pixner, Zuber, Kaufmann, & Nuerk, 2011), which will be discussed in subsequent parts of this chapter, or the ability to integrate familiar and unfamiliar numerical ranges (Ebersbach, Luwel, Frick, Onghena, & Verschaffel, 2008). In general, these alternative explanations of non-linear response pattern in the NLE refer to some lack of thorough understanding of the numerical magnitude.

To identify the processes underlying the NLE, another version of the task was developed—the unbounded NLE (Cohen & Blanc-Goldhammer, 2011; see Box

4.2). This task variant was proposed to eliminate the proportional judgment component as well as more complex reasoning which also plays a role in case of bounded NLE (one might set up virtual anchor points in the middle of the line, then at quartile points, and adjust the exact estimation relative to these points, e.g., to mark 77 one divides the line into halves and quarters and puts the estimation a little bit to the right of the virtual division point marking 75). Interestingly, the link between number line performance and arithmetic skill seems stronger for the bounded version of the task than for the unbounded one.

However, recently, Kim and Opfer (2017) argued that the log-to-linear change in the representation can provide a unifying framework for NLE. They used both bounded and unbounded NLE tasks. The mixed log-linear model (i.e., both components were included in one model) accounted for performance in both bounded and unbounded NLE. Performance in both tasks was strongly correlated ($r = 0.73$). Interestingly, the overall performance accuracy was higher in bounded NLE compared to the supposedly easier (i.e., allowing summation strategy only) unbounded task. According to their interpretation, these results seem to support the log-to-linear shift. However, they are not in line with divergent validities, because the bounded number line correlates with arithmetic skill, while the unbounded number line does not (Link, Huber, et al., 2014; Link, Nuerk, & Moeller, 2014). Link, Huber, et al. (2014) suggested that bounded NLE allows the participants to use a wider range of available strategies. Undoubtedly, this issue is far from being resolved and we can expect intense discussion on the topic.

Finally, it is often observed in developmental studies that a linear correlation between numerical magnitude and space in the number line tasks takes different shapes in children, typical adults, and mathematics-deficient adults. More precisely, children and mathematics-deficient adults often assign larger inter-digit distances to magnitude steps in smaller ranges, which leads to a different shape of the correlation (Moeller, Pixner, Kaufmann, & Nuerk, 2009). Often, it has been assumed that this behavior reflects a shift from logarithmically compressed number representations to a linear representation, mimicking one model fit for these patterns of results. However, it may be also possible that the behavior of children and mathematics-deficient adults stems from a certain strategy for solving the task (Nuerk et al., 2001). More precisely, the logarithmic pattern observed in the correlation (see Figures 1 and 2 in Moeller et al., 2009) can also be almost perfectly accounted for by two linear regressions, i.e., a bilinear fit for single- vs. multi-digit processing. In any case, proportional reasoning is always required, and thus we can infer that a benefit to other mathematical operations (that also often include proportional judgment) is also derived from this moderator variable. When solving the task and assigning space to numbers, such bilinear patterns would result from different strategies for mapping number magnitudes up to a certain point (e.g., 10) and then assigning the remaining space of the visually presented line to the remaining number magnitudes (e.g., up to 100). Educated adults that are asked to assign spatial distances on the number line task from one thousand to one billion (incorrectly) assign the half of the line to a landmark for one million (Landy, Silbert, & Goldin,

2013), which reflects the bilinear strategy observed in children and points to a purely verbal (but not logarithmic compressed) strategy.

The representations underlying different task types of the NLE are still under heavy debate. Contradictory to many previous studies, Dietrich, Huber, Dackermann, Moeller, and Fischer (2016) argue that NLE might not be related to a spatial representation of numbers. The authors suggest that place-value understanding of the MNL task is most probably driven by the relation between MNL and arithmetic performance (Booth & Siegler, 2006). In line with this finding, a discontinuity of the MNL in very large numbers was reported, which was interpreted as reflecting a limitation of the magnitude perception system in humans (Landy, Charlesworth, & Ottmar, 2014, 2017). All in all, although NLE is probably the task most frequently cited as providing evidence for the relation between space and arithmetic, recent studies have started to cast doubt on this relation and suggest some mediator processes between them. Finally, finger-tracking technologies can be used to study number line mappings (e.g., Pinheiro-Chagas, Dotan, Piazza, & Dehaene, 2017) or place-value structures (Bloechle, Huber, & Moeller, 2015), which may allow for better targeting and investigation of the relevant cognitive processes underlying different number-space mappings.

Situatedness

Huber, Moeller, and Nuerk (2014) demonstrated that providing feedback after each response may successfully train participants to perform the NLE task according to one of several models (linear, exponential, logarithm, sigmoid, inverted sigmoid). The training for each mapping consisted of only 30 trials. All participants participated in all conditions in counterbalanced order. Thus, it seemed as if that brief training successfully influenced NLE performance without changing the participants' numerical long-term representation (i.e., involving such a change for five times within one experimental session). Astonishingly, the follow-up study detected a general deficit in adults with developmental dyscalculia (severe mathematics impairments) with less precision in the NLE irrespective of the underlying shape (Huber, Sury, Moeller, Rubinsten, & Nuerk, 2015). Individuals with developmental dyscalculia had severe problems using benchmark points in this non-linear, but bounded NLE, again corroborating the idea that performance in the NLE is not only driven by spatial representation of number, but also by arithmetic strategies.

Correlations with Arithmetic Skill

The vast majority of studies on the relation between this SNA type and arithmetic skill have reported zero order correlations (i.e., without an additional control variable partialled out). However, the overall picture gets much more complicated when one aims to investigate the nature of this relation in a more detailed way. The extent to which linear function reflects a child's performance is correlated with his/her mathematics skill (see Siegler, Thompson, & Opfer, 2009 for an overview). Lefevre et al. (2013) used a longitudinal experimental design to investigate the causality of these correlations. Their results showed that NLE performance could not predict future arithmetic performance better than arithmetic performance could predict future NLE performance. However, number system knowledge predicted future NLE performance. Thus, as the authors conclude, NLE is more related to mathematics than to spatial performance.

Link, Nuerk, and Moeller (2014) observed that only performance in the bounded NLE correlated significantly with arithmetic performance in fourth graders. This might suggest that fluency in making proportional judgments (i.e., relatively complex mathematics reasoning) correlates with arithmetic performance. In other words, the results of the bounded NLE suggest that one type of mathematics understanding correlates with another type of mathematics understanding, which is in fact not very surprising. On the other hand, SNA itself (indexed with the unbounded task) seems not to be genuinely correlated with arithmetic performance.

In sum, we tend to agree with the conclusion of Dackermann, Fischer, Nuerk, and Cress (2017) that the correlation between NLE and arithmetic skills may not mean that SNAs assessed by the NLE are (causally) important for arithmetic skills, but that vice versa, good arithmetic skills are causally important for good performance and application of helpful strategies in the NLE.

Trainings

Opfer and Siegler (2007) showed that presenting feedback rapidly and strongly improves NLE performance in second graders. Providing a single instance of feedback on accuracy in the task led to considerable improvements in performance (i.e., a log-to-linear change).

Potential beneficial effects of NLE tasks and training on numerical learning and arithmetic skill were obtained in respective studies. For instance, playing a board game designed to resemble the MNL for only 1 h improved the numerical understanding of low-income preschoolers (Siegler & Ramani, 2009). More precisely, when children were assigned to play a linear board game—as opposed to either a circular board game or a numerical control condition (including number counting, object counting, and number naming)—the researchers observed a steeper and more linear performance in the NLE task, more accurate performance in magnitude comparison tasks, and in addition problems. Additional experimentation showed the potential of extending the number board game to the 1–100 range, and that positive effects are distinct from those of control training with mere counting (Laski & Siegler, 2014). The training is also effective in small group teaching setups (e.g., classroom context) and without extensive training of the teachers themselves (Ramani, Siegler, & Hitti, 2012).

Throughout the last decade, several board games and computerized games targeting the link between number and space were developed and reported upon in different studies. Some aspects of instruction principles (e.g., adaptive increase of difficulty depending on recent performance) proved highly relevant to target the individual level of arithmetic skill, especially in computerized tasks (Wilson et al., 2006). In contrast, other aspects such as constructivism and intrinsic feedback are yet to be fully explored (Laurillard, 2016). Such interventions—including the renowned scientific games Rescue Calcularis (Kucian et al., 2011), The Number Race, and The Number Catcher[1]—are particularly thought to improve numerical skills in neurodevelopmental disorders such as dyscalculia (Kucian et al., 2011).

Regarding evaluations, the most comprehensive evaluation of a single number training game to date was performed in a randomized controlled trial on the Number Race (Sella, Tressoldi, Lucangeli, & Zorzi, 2016). In this trial, children were randomly assigned to play either the Number Race or an alternative computer-based activity, so that the comparison group was active in a different scope. Interestingly, Sella et al. (2016) observed large improvements in mental calculation, number-space mappings, and smaller improvements in semantic processes, rendering an optimistic view on the training validity. In another study, when a more recent tablet version of the NLE game was contrasted with a comparison training in kindergarten children, both training methods elicited distinct and common learning effects (Maertens, De Smedt, Sasanguie, Elen, & Reynvoet, 2016).

Accumulating evidence supports the broad theoretical notion that cognitive processes are embedded in corporeal experiences. This includes very basic interactions such as pointing and grasping objects and attributing numerical distance to spatial distance, as well as the use of finger counting. With accumulating sensory and motor interactions and learning transitions between these immediate physical experiences and the concepts of numerosity and magnitude, a rich and multimodal network evolves to allow for flexible representations and arithmetical procedures.

By implication, providing active opportunities for physical interactions could foster the development of abstract numerical or even arithmetic abilities. Some first evidence is available to support the embodied learning approach. For example, an intervention with first-graders included a NLE task with full-body involvement, i.e., walking to an estimated location displayed on the floor. Children showed more improvements with this intervention than a control group of children who solved the task without full-body involvement, not only in number line representations, but also in non-trained mathematics-related tasks (Link, Moeller, Huber, Fischer, & Nuerk, 2013). Furthermore, in kindergarten children, and compared to another numerical training without an active component, specific improvements were documented in embodied magnitude training (U. Fischer, Moeller, Bientzle, Cress, & Nuerk, 2011). In this training study, full-body movements were incorporated for mathematical learning by use of a digital dance mat in magnitude classification tasks. The control group also performed magnitude comparisons in the training sessions with the same stimuli, but there was no presentation of a spatial number line and they solved the task on a tablet PC. Results showed enhanced mathematical performance following the active sensorimotor training mediated by mental number

[1] www.thenumberrace.com; http://www.thenumbercatcher.com/

line representations. Such studies show the promising potential of vivid spatial-numerical activities for the acquisition of numerical concepts. Nevertheless, more research is required to particularly compare the novel training intervention to default procedures in large groups.

Category: Directional SNA with Implicit Coding: Subcategory Cardinalities

Type and Paradigms

In this SNA category the link between space and number is not highlighted in any instruction, and is not relevant to the task itself. Most often, space and numbers are highlighted separately in respective tasks, for instance, when magnitude classifications are indicated by left-hand vs. right-hand key presses.

The most important phenomenon that falls into this category is the *SNARC effect* (Spatial-Numerical Associations of Response Codes; Dehaene et al., 1993; see Box 4.1) which refers to the faster speed of responses with the left hand than the right hand for small vs. large single-digit numbers.

Over past 25 years, the SNARC effect has been replicated numerous times and tested with varied participant groups (healthy individuals and clinical samples, of varied age), stimuli sets (both symbolic and non-symbolic), tasks (response criterion either referring or not referring to numerical magnitude), and response formats (see M. H. Fischer & Shaki, 2014 for a current review; Wood et al., 2008 for a meta-analysis). Another SNA within this category is revealed by biases in random number generation (the participant is asked to generate random numbers) caused by either head (Loetscher, Schwarz, Schubiger, & Brugger, 2008) or whole body movements (Schroeder & Pfister, 2015; Shaki & Fischer, 2014), and biases in numerical estimations caused by changes in body posture (Eerland, Guadalupe, & Zwaan, 2011). Cultural influences such as reading habits in right-to-left reading Arabic countries can produce reverse effects (Shaki et al., 2009), although this was not observed in native speakers of Hebrew (but see Zohar-Shai, Tzelgov, Karni, & Rubinsten, 2017).

Situatedness

Contextual cues that signal a different space-to-number assignment can modulate the SNARC effect (M. H. Fischer, Mills, & Shaki, 2010), but also previous episodes of responding with the incongruent mapping can instantly reduce regular SNAs in the next trial (Pfister, Schroeder, & Kunde, 2013). It is also possible to establish regular or reversed direction of SNARC effects by targeted training in children. For instance, children were asked to play a non-numerical spatial game on a touch screen, containing a frog that was moved across a pond either toward the left or right side of the screen (Patro, Fischer, Nuerk, & Cress, 2016). The direction of a SNARC effect depended on the direction of the training. This finding demonstrates that simple activities that are not necessarily related to numerical knowledge still impact associations between numbers and space. Conversely, this and related findings from situated numerical cognition, showing flexibility of SNAs in adults (e.g., Pfister et al., 2013), advise us to carefully gauge the value of exploring SNARC effects for arithmetic skill or as an effect of lifelong culturally dependent practice.

Correlations with Arithmetic Skills

Several positions exist in the contemporary literature concerning the relation between SNARC and arithmetic skill in adults and children. It was suggested that adults skilled in mathematics can be characterized by a weaker SNARC effect; however, the several results did not reach statistical significance (Dehaene et al., 1993, Exp. 1; Bonato et al., 2007; Bull, Cleland, & Mitchell, 2013; Cipora & Nuerk, 2013). On the other hand, a relation was found when individuals with mathematics difficulties (Hoffmann, Hornung, Martin, & Schiltz, 2013) and professional mathematicians (Cipora et al., 2016) were considered. The SNARC effect seemed to decrease with increasing mathematics proficiency. It suggests that differences in the SNARC effect get more pronounced in the case of extreme mathematics skill groups but are either nonexistent or very weak at a typical mathematics skill level.

The same seems to be true in the case of children (Gibson & Maurer, 2016) and adolescents (Schneider, Grabner, Zurich, & Paetsch, 2009). On the other hand, Georges, Hoffmann, and Schiltz (2017b) reported a correlation. Interestingly, contrary to tendencies in adult studies, children who were characterized as having a stronger SNARC effect scored higher on a standardized mathematics ability test. Bachot, Gevers, Fias, and Roeyers (2005) observed that 7- to 12-year-olds with visuospatial disorder and developmental dyscalculia, contrary to a control group, do not reveal the SNARC effect. On the other hand, Crollen and Noël (2015) did not find such an effect in fourth graders with poor visuospatial abilities.

Together, these findings suggest that there is no consistent and genuine relation between directional implicitly coded SNA for cardinalities and arithmetic or mathematics, and the possible direction of the relation is not consistent across development. However, one must take into account that across different studies mathematics skill was operationalized in varied ways, ranging from performance in (speeded) calculation tests to categorical descriptions of individuals' field of study/occupation. This is important because of double dissociations between calculation proficiency and mathematics expertise (Pesenti, 2005). According to Cipora et al. (2016), professional mathematicians who do not demonstrate the SNARC effect deal mostly with highly abstract constructs and for that reason their SNAs may be either nonexistent or highly flexible.

To the best of our knowledge, there have been no studies investigating the relation between mathematics skill and other forms of this SNA subcategory. In our view, this should also be pursued in future studies.

Trainings

To the best of our knowledge there have been no training studies aimed at using this SNA type in order to improve mathematics skills.

Category: Directional SNA with Implicit Coding: Subcategory—Ordinalities

Type and Paradigms

Directional SNAs can emerge due to the quantity present in numerical cardinality as well as due to the ordinal structure in numerical and non-numerical stimuli (such as series of objects, or sequentially arranged structures such as weekdays). Thus, with number symbols in the SNARC effect described above, it is ambiguous whether cardinal or ordinal information becomes spatially arranged. Complementing the influence of reading habits in left-to-right and right-to-left directed Western (English) and Eastern (Hebrew) languages, researchers tested these interrelations by means of ambiguous symbols that depict both ordinal series and quantity. For different tasks, either type of information was considered relevant, but interestingly, only the SNARC effect based on ordinality was found to follow participants' right-to-left reading direction. When the task emphasized the symbols' cardinality, a regular left-to-right direction was observed in the Hebrew-speaking participants (Shaki & Gevers, 2011). In recent experiments, this apparent dissociation of spatial associations that are based on number and order was further investigated in Western culture. First, it was observed that the correlation between different SNARC effects was remarkably low for different cardinal (1–5) and ordinal stimuli (Monday-Friday sequence; Schroeder, Nuerk, & Plewnia, 2017a) but also for the same numerical stimuli in different magnitude and parity classification tasks (Georges, Hoffmann, & Schiltz, 2017a). Finally, spatial associations of ordinal and cardinal nature were clearly dissociated with transcranial brain stimulation (Schroeder, Nuerk, & Plewnia, 2017b). Thus, it is important to differentiate ordinal and cardinal aspects of SNAs also with respect to other behavioral and neurocognitive studies (Huber, Klein, Moeller, & Willmes, 2016) and multiple cognitive processes seem to underlie the emergence of spatial associations (Schroeder et al., 2017b). Nevertheless, the current data converge in that ordinal SNAs represent a separable and distinct directional SNA category.

Situatedness

In principle, it is plausible that situated influences can shape the direction and intensities of SNAs based on ordinality and cardinality alike. Memorizing a sequence of elements (i.e., ordinality information) evokes SNARC-like effects. When the participants were instructed to memorize randomized sequences of numbers (e.g., 9, 2, 5, 6, 3) and then performed parity judgments to classify these numbers, the SNARC effect reflected the order of numbers in a sequence, not their magnitudes (van Dijck & Fias, 2011). Namely, elements from the beginning of the list were associated with the left hand side and those from the end, with the right hand side. Thus, it seemed that situated influence (memorizing a sequence carrying ordinal information) may cover up the (long-term or short-term) SNA related to cardinality of numbers constituting the sequence. Nevertheless, subsequent studies showed that in the case of number sequences, the SNARC effect can depend on both number magnitude and a position in a sequence (Huber, Klein, et al., 2016).

Correlations with Arithmetic Skill

So far, there no studies have demonstrated correlations of this SNA type with arithmetic skill. Nevertheless, it could be explored whether serial order working memory and ordinality deficits constitute different types of mathematics deficit (e.g., Rubinsten & Sury, 2011). Focusing on this area could help disambiguate between contradictory findings on the relation between cardinal SNA and arithmetic skill, as in most cases both ordinal and cardinal properties of numbers are tightly connected.

Trainings

To the best of our knowledge there have been no training studies aimed at using this SNA type in order to improve mathematics skills.

Category: Directional SNA with Implicit Coding: Subcategory Operations

Type and Paradigms

Not only basic numerical concepts are associated with space: so are arithmetic operations such as addition and subtraction. Even single "−" and "+" signs are related to the left and right sides (operation sign spatial association—OSSA; Pinhas et al., 2014), but the spatial association also holds for entire arithmetic operations (see Box 4.1). This was described at different levels such as movements along the number line (Pinhas & Fischer, 2008), eye movements (Klein, Huber, Nuerk, & Moeller, 2014; Masson, Letesson, & Pesenti, 2018), attentional shifts (Masson & Pesenti, 2014), and arm movements (Wiemers, Bekkering, & Lindemann, 2014). This collection of phenomena is referred to as *Operational Momentum.* However, this term initially referred only to over-/under-estimation of the results of addition, multiplication/subtraction, division (McCrink, Dehaene, & Dehaene-Lambertz, 2007).

Situatedness

To date, studies on spatial biases in mental arithmetic focused on demonstrating its various forms depending on modality (symbolic vs. non-symbolic) and type of spatial bias. To the best of our knowledge there have been no studies demonstrating situated influences on this effect.

Correlations with Arithmetic Skill

We are also not aware of studies demonstrating a relation between spatial associations with arithmetic functions/operations and arithmetic skill level.

Trainings

To the best of our knowledge, there have been no specific training studies taking advantage of SNA related to implicit coding of arithmetic operations.

Category Directional SNA with Explicit Coding: Subcategory—Ordinalities

Type and Paradigms

This SNA set refers to biases evoked *explicitly* by the task instruction. In tasks measuring this SNA type, participants are asked to count objects from a predetermined set. The direction is the dependent measure. Objects may be arranged linearly (either in a horizontal or vertical line) thus providing only two possibilities (e.g., Shaki et al., 2012) or objects may form an array providing more possibilities (e.g., Göbel, 2015). A special case pertains to finger counting (Lindemann, Alipour, & Fischer, 2011).

In several instances such studies demonstrated that the counting direction was influenced by reading habits in a given culture. Moreover, culture-driven preference patterns become more pronounced during reading acquisition (Shaki et al., 2012). In general, in Western cultures both children and adults prefer counting from left to right. Individuals who use right-to-left script (e.g., Arab speakers) count from right to left. This suggests that both explicit and implicit SNAs are culturally modulated and subject to embodied moderations.

Situatedness

This SNA type is very prone to situated influences. In Cantonese (a Chinese language), speakers can use either vertical or horizontal script. Reading a short paragraph written vertically may affect Cantonese participants' counting directions (Göbel, 2015).

Moreover, remarkably different proportions of right-starters (i.e., participants starting the finger counting sequence from their right hand) vs. left-starters emerged depending on whether participants were (a) asked to spontaneously count, (b) presented with a schematic drawing of hands, or (c) asked when their dominant hand was occupied (Wasner et al., 2014). All these results suggest that there is huge space for situated influences in this SNA category (at least for finger counting).

Correlations with Arithmetic Skill

To the best of our knowledge, there is no data showing correlations between this SNA type and arithmetic skill. Studies showing a relation between finger *gnosis* and arithmetic skill in children (e.g., Penner-Wilger & Anderson, 2013) are not instructive on this issue because they only demonstrate that an adequate representation of fingers seems to be related to arithmetic performance.

Trainings

Interventions directly related to this SNA type are hard to find. Of course, some finger counting interventions were related to finger gnosis, counting, or cardinality (Gracia-Bafalluy & Noël, 2008, but see J. P. Fischer, 2010 for a methodological critique). Improvement of finger gnosis or finger cardinality or finger counting was related to improvements in arithmetic skill. However, since these interventions were not aiming at SNAs, nor was the spatially organized finger counting routine exclusively trained, they cannot be considered directly related to SNAs.

Category Directional SNA and Explicit Coding: Subcategory Cardinalities and Operations

The taxonomy proposed by Cipora et al. (2015) postulated the existence of those two SNA types, but nevertheless, they have not been reported in empirical studies so far.

SNA with Multi-digit Numbers

For operations on symbolic numbers, it may be intuitive to simply extend the concept of SNA. However, as we show in the following section, additional processes need to be considered. These processes may include parallel computation or even show distinct effects of integrating the values and their perceptual positions in place-value processing (Nuerk, Moeller, Klein, Willmes, & Fischer, 2011). Usually, multi-symbol number processing is not seen as a SNA. However, in the Arabic place-value processing, in fact it is one by many definitions. Therefore, we will discuss multi-symbol number processing in the remainder of this chapter from the SNA perspective.

Multi-symbol Processing as an Integration of Space and Magnitude

The ability to understand multi-digit numbers such as 83, 5729, and 1,000,000 involves understanding the digits and the values imposed by their spatial positions. However, numbers are not only comprised of different digits, but also of other symbols like the "–" sign for negative number, the "." for decimal numbers, or the "/" or "_" for fractions (Huber, Nuerk, Willmes, & Moeller, 2016). Because we also refer to such numbers, which are comprised of digits and other symbols in certain spatial positions, we refer to multi-symbol numbers. It is important to note that not only the spatial positions of digits (29 vs. 92), but also the spatial positions of symbols (e.g., 2.93 vs. 29.3) change the value of the multi-symbol number. While this may appear obvious at first to educated adults, the perceptual and cognitive consequences of these highly overlearned characteristics can be even more informative regarding multi-digit processing. In the following, we will explain the relation between the spatial and multi-symbol processing and domain-specific factors as a part of mathematics.

Place-Value System

Multi-digit Arabic numbers rely on both value (number magnitude), e.g., 9 is larger than 4, and place (unit, decade, hundred, etc.), e.g., 4 as a decade is larger than 9 as a unit. Therefore, in order to process multi-digit numbers, one needs to integrate the value and the place (see Nuerk et al., 2015).

The place-value system simplifies calculation and allows fast, automatic calculation using typical algorithms. Based on the place-value system, Nuerk et al. (2015) suggested a new theoretical framework for multi-digit number processing, which comprises three distinctions: place identification, place-value activation, and place-value computation (cf. Box 4.1). In the following, we will discuss these three steps along with language influence, which is essential in multi-digit number processing.

Place Identification

The first perceptual step in multi-digit number processing is to correctly identify the position of digits and symbols. As mentioned above, digits are arranged in a specific manner in multi-digit Arabic numbers, which leads to a differentiation between the two magnitudes in different places (unit, decade, hundred, etc.). Therefore, this processing level does not include semantic processing of the digits. However, language properties facilitate or interfere with place identification (Nuerk et al., 2015, for overview). For instance, Miura et al. (1994) observed that Japanese, Chinese, and Korean children perform better in distinguishing units and decades compared to their counterparts from Western countries, i.e., the US, Sweden, and France. This difference is because Asian children learn Arabic multi-digit numbers based on a transparent number-word system (e.g., 38 is three ten eight), in contrast to Western children. Furthermore, interference between the place identification and the syntactic structure of number-words across languages may worsen performance (Nuerk et al., 2005). For example, in the German language, decade and unit digits are stated inversely (38 is read or said as "eight and thirty"), which leads to transcoding errors in half of the German-speaking 7-year-olds (Zuber, Pixner, Moeller, & Nuerk, 2009). This might be even more pronounced in three-digit numbers, which are read in the order of hundreds (left digit), units (right digit), and then decades (middle digit). This observation exemplifies how verbal representation, as a linguistic feature, influences place identification.

Place-Value Activation

In the next step of multi-digit number processing, the integration of places and magnitudes of the digits is highlighted. For instance, to successfully compare two-digit numbers, individuals need to know to first compare the decades, which requires processing of both place and value (Nuerk et al., 2001). The unit-decade compatibility effect (cf. Box 4.1) in symbolic multi-digit number comparison is an index of place-value activation. In unit-decade compatible pairs (e.g., 21_89) in a symbolic number comparison task, the unit comparison leads to the same response as the decade comparison. However, in unit-decade incompatible pairs (e.g., 29_93), the unit comparison leads to an error. Therefore, the place-value system has a critical influence on incompatible trials. Depending on the unit-decade compatibility, the place-value activation might have a more or less important influence on performance (Miura et al., 1994). It is important to note that the unit magnitude does not have to be processed when two two-digit numbers are to be compared. However, the compatibility effect automatically interferes with processing a digit when its identity and magnitude are irrelevant for the comparison in question.

Linguistic features such as inversion also influence the unit-decade compatibility effect, which leads to a more pronounced effect for instance in German speakers.

Pixner and colleagues (2009) observed that the interference effect of unit magnitude is larger in German-speaking children compared to Italian-speaking children.

Furthermore, the interaction of the unit-decade compatibility effect with decade distance and unit distance requires place-value activation as well (Nuerk et al., 2001). What is important is that the distance effect (cf. Box 4.1) for two-digit numbers may be a better diagnostic marker than the distance effect for single-digit numbers. For instance, Ashkenazi, Mark-Zigdon & Henik (2009) observed that dyscalculic children differed from controls in two-digit distance effects, but not in the single-digit distance effect. A possible reason is that single-digit numbers are highly overlearned, so that their processing does not differentiate between groups at different skill levels. Another explanation is that compared to single-digit numbers, multi-digit numbers involve SNAs and automatic place-value activation, which differ between dyscalculics and controls. However, while some studies have observed that number magnitude, i.e., a larger decade distance effect, negatively correlated with arithmetic performance (e.g., De Smedt, Verschaffel, & Ghesquière, 2009), other studies documented a paradoxical positive relation between the numerical distance effect and arithmetic performance (e.g., Moeller et al., 2011).

While most of the studies regarding three levels of place-value system involve explicit coding—for instance, all of the above-mentioned studies—very few studies have investigated implicit place-value activation. In the study by Kallai and Tzelgov (2012), while participants were explicitly asked to compare only the value of two numbers including several zeros (e.g., 050 vs. 007), the place-value activation was documented. In another study, Ganor-Stern et al. (2007) observed that the place-value activation interacts with the size congruity effect, while participants were not explicitly asked to decide about both place and value. The size congruity effect is related to processing of the (irrelevant) magnitude dimension when the task is to compare the (relevant) physical dimension of the decades. They suggested that the compatibility effect is automatic as it interacts with the size congruity effect.

Place-Value Computation

Place-value computation is one step further beyond place identification and place-value activation in multi-digit number processing. It refers to a manipulation of place-values, such as in carry operations. For instance, to solve 28 + 65, the added magnitude of the units needs to be carried over to the decades (80 + 13). Hence, the magnitudes of the units and decades in relation to each other and to their places need to be taken into account. Surprisingly, language has an influence on this higher level of place-value computation as well. Göbel, Moeller, Pixner, Kaufmann, and Nuerk (2014) investigated addition problems with and without carry operations in German-speaking and Italian-speaking children. They reported a larger carry effect in the language with inversion (German) than without (Italian). As might be expected, because the added magnitude of the units needs to be kept in working memory, working memory mediated the carry effect. Complementary to these findings,

Colome, Laka, and Sebastian-Galles (2010) reported the role of full base-20 in the Basque number word system. Basque participants responded faster to the addition problems containing 20 and a teen number (e.g., 37 is said "20 + 17") rather than any other addition problems with similar results but other summands than 20. Altogether, we conclude that understanding different levels of the place-value system is essential for multi-digit number processing and therefore, for numerical and arithmetic development.

Training Multi-digit Numbers And Predicting Future Performance

Multi-digit number processing is a special case regarding trainings, because in contrast to other SNAs, its understanding is an explicit goal in education. As one of the basic numerical competencies, the place-value system develops during childhood and is clearly related to mathematics achievement. In a longitudinal study, Moeller et al. (2011) observed that place-value understanding in first grade was a reliable predictor of arithmetic performance in third grade and suggested that it should be understood as a developmental process. Ho and Cheng (1997) observed that place-value training improved arithmetic performance in first graders. Altogether, place-value processing, which is the spatial arrangement of the digits in multi-digit numbers that leads to different evaluation of a digit in different places, helps us to learn and apply mathematical knowledge in multi-digit numbers in a less effortful way.

Conclusions

The leading question of this chapter was: is space a cornerstone or a powerful tool for learning and for understanding arithmetic? As often in science, there is no simple unitary answer on this question. First, there is not "THE" spatial-numerical association, but there are multiple SNAs, which differ in their spatial dimensions (e.g., extension, directionality) and their numerical dimensions (cardinality, ordinality, operations, place-value integration, etc.). For some of them, stable correlations with arithmetic skill have been reported (e.g., extension, interval-scaled like in the NLE, place-value processing), while for others correlations are low, scarce and inconsistent (e.g., SNARC effect). Although exceptions apply, a general rule of thumb seems to be that SNAs related to the directionality in space (e.g., small numbers left, large numbers right) are usually not important for arithmetic skill.

However, when it comes to specific SNAs related to exact extension or (place-value) position, the picture is different. For both extension and positions, we see relatively stable correlations with arithmetic skill and indications that training these processes and representations successfully transfers to other arithmetic skills. It is unclear whether the spatial-numerical representation of numbers per se is trained (e.g., linear or logarithmic) or whether spatial-numerical tasks are a powerful tool to train associated number representations like magnitude representations or the internal power system underlying our base-10 place-value notation in Arabic numbers.

Here, we want to add a disclaimer in order not to be misunderstood. In our view, it is absolutely essential to have a functioning magnitude representation and a functioning representation of the base-10-system. There is no doubt that these systems can be mapped to space and that participants skillful in mathematics can use space as a powerful tool to facilitate their performance. Furthermore, there is convincing evidence that spatial visualizations are a powerful tool to train more complex internal mathematical representations. All these points are essentially undisputed. The reasons and the causality for these findings, however, are disputed. We still question whether a particular spatial-numerical representation is trained, which underlies all our mathematical understanding, or whether a spatial mapping can enhance our numerical representations and skills.

Our point can be illustrated by comparing spatial-numerical mappings to mnemonic techniques and methods in memory research. There is no question that they work, or that people who perform well in memory tasks can usually use them; there are training studies showing that they can improve memory performance. But does this mean that mnemonic representations underlie our memory system? Of course, every analogy has its limitations, but here we illustrate the general point that correlations and successful trainings do not automatically indicate an underlying representation. This is an open question for us.

Nevertheless, there is no doubt that space is at least an extremely powerful tool for learning and understanding arithmetic and probably the most important association with numbers and arithmetic besides language. We believe that it is impossible to fully understand arithmetic without understanding its relation to space in learning and operating numerical skills. So, although we believe that the case of particular underlying spatial representations is not as straightforward as it is sometimes assumed in the literature, we are convinced that SNAs deserve much more thorough and in-depth research in the future. However, in our view, the direction should be to understand the nature and impact of each different SNA rather than to demonstrate yet another SNA without particular reference to its function and impact.

Acknowledgments KC and MS are supported by a DFG grant [NU 265/3-1] to HCN. KC, MS, and HCN are further supported by the LEAD Graduate School & Research Network [GSC1028], which is funded within the framework of the Excellence Initiative of the German federal and state governments. We thank Julianne Skinner for proofreading the manuscript.

References

Ashkenazi, S., Mark-Zigdon, N., & Henik, A. (2009). Numerical distance effect in developmental dyscalculia. *Cognitive Development, 24*(4), 387–400. https://doi.org/10.1016/j.cogdev.2009.09.006

Bachot, J., Gevers, W., Fias, W., & Roeyers, H. (2005). Number sense in children with visuospatial disabilities: Orientation of the mental number line. *Psychology Science, 47*(1), 172–183.

Barth, H. C., & Paladino, A. M. (2011). The development of numerical estimation: Evidence against a representational shift. *Developmental Science, 14*(1), 125–135. https://doi.org/10.1111/j.1467-7687.2010.00962.x

Barth, H. C., Starr, A., & Sullivan, J. (2009). Children's mappings of large number words to numerosities. *Cognitive Development, 24*(3), 248–264. https://doi.org/10.1016/j.cogdev.2009.04.001

Bloechle, J., Huber, S., & Moeller, K. (2015). In touch with numbers: Embodied and situated effects in number magnitude comparison. *Journal of Cognitive Psychology, 27*(4), 478–489. https://doi.org/10.1080/20445911.2014.1001760

Bonato, M., Fabbri, S., Umiltà, C., & Zorzi, M. (2007). The mental representation of numerical fractions: Real or integer? *Journal of Experimental Psychology: Human Perception and Performance, 33*(6), 1410–1419. https://doi.org/10.1037/0096-1523.33.6.1410

Booth, J. L., & Siegler, R. S. (2006). Developmental and individual differences in pure numerical estimation. *Developmental Psychology, 41*(6), 189–201. https://doi.org/10.1037/0012-1649.41.6.189

Brysbaert, M. (1995). Arabic number reading: On the nature of the numerical scale and the origin of phonological recoding. *Journal of Experimental Psychology: General, 124*(4), 434–452. https://doi.org/10.1037/0096-3445.124.4.434

Bueti, D., & Walsh, V. (2009). The parietal cortex and the representation of time, space, number and other magnitudes. Philosophical Transactions of the Royal Society of London. *Series B, Biological Sciences, 364*(1525), 1831–1840. https://doi.org/10.1098/rstb.2009.0028

Bugden, S., & Ansari, D. (2011). Individual differences in children's mathematical competence are related to the intentional but not automatic processing of Arabic numerals. *Cognition, 118*(1), 32–44. https://doi.org/10.1016/j.cognition.2010.09.005

Bull, R., Cleland, A. A., & Mitchell, T. (2013). Sex differences in the spatial representation of number. *Journal of Experimental Psychology: General, 142*(1), 181–192. https://doi.org/10.1037/a0028387

Cipora, K., Hohol, M., Nuerk, H.-C., Willmes, K., Brożek, B., Kucharzyk, B., & Nęcka, E. (2016). Professional mathematicians differ from controls in their spatial-numerical associations. *Psychological Research, 80*(4), 710–726. https://doi.org/10.1007/s00426-015-0677-6

Cipora, K., & Nuerk, H.-C. (2013). Is the SNARC effect related to the level of mathematics? No systematic relationship observed despite more power, more repetitions, and more direct assessment of arithmetic skill. *Quarterly Journal of Experimental Psychology, 66*(10), 1974–1991. https://doi.org/10.1080/17470218.2013.772215

Cipora, K., Patro, K., & Nuerk, H.-C. (2018). Situated influences on spatial-numerical associations. In T. Hubbard (Ed.), *Spatial biases in perception and cognition*. (pp. 41–59). Cambridge, UK: Cambridge University Press.

Cipora, K., Patro, K., & Nuerk, H.-C. (2015). Are Spatial-Numerical Associations a Cornerstone for Arithmetic Learning? The Lack of Genuine Correlations suggests: No. *Mind, Brain, & Education, 9*(4), 190–207. https://doi.org/10.1111/mbe.12093

Cohen, D. J., & Blanc-Goldhammer, D. (2011). Numerical bias in bounded and unbounded number line tasks. *Psychonomic Bulletin & Review, 18*(2), 331–338. https://doi.org/10.3758/s13423-011-0059-z

Cohen Kadosh, R., & Henik, A. (2007). Can synaesthesia research inform cognitive science? *Trends in Cognitive Sciences, 11*(4), 177–184. https://doi.org/10.1016/j.tics.2007.01.003

Cohen Kadosh, R., Lammertyn, J., & Izard, V. (2008). Are numbers special? An overview of chronometric, neuroimaging, developmental and comparative studies of magnitude representation. *Progress in Neurobiology, 84*(2), 132–147. https://doi.org/10.1016/j.pneurobio.2007.11.001

Colome, A., Laka, I., & Sebastian-Galles, N. (2010). Language effects in addition: How you say it counts. *Quarterly Journal of Experimental Psychology, 63*(5), 965–983. https://doi.org/10.1080/17470210903134377

Cragg, L., & Gilmore, C. (2014). Skills underlying mathematics: The role of executive function in the development of mathematics proficiency. *Trends in Neuroscience and Education, 3*(2), 63–68. https://doi.org/10.1016/j.tine.2013.12.001

Crollen, V., & Noël, M. P. (2015). Spatial and numerical processing in children with high and low visuospatial abilities. *Journal of Experimental Child Psychology, 132*, 84–98. https://doi.org/10.1016/j.jecp.2014.12.006

Dackermann, T., Fischer, U., Nuerk, H.-C., & Cress, U. (2017). Applying embodied cognition: From useful interventions and their theoretical underpinnings to practical applications. *ZDM, 49*(4), 545–557. https://doi.org/10.1007/s11858-017-0850-z

Dehaene, S., Bossini, S., & Giraux, P. (1993). The mental representation of parity and number magnitude. *Journal of Experimental Psychology: General, 122*(3), 371–396. https://doi.org/10.1037/0096-3445.122.3.371

de Hevia, M. D., & Spelke, E. S. (2009). Spontaneous mapping of number and space in adults and young children. *Cognition, 110*(2), 198–207. https://doi.org/10.1016/j.cognition.2008.11.003

de Hevia, M. D., & Spelke, E. S. (2010). Number-space mapping in human infants. *Psychological Science, 21*(5), 653–660. https://doi.org/10.1177/0956797610366091

De Smedt, B., Verschaffel, L., & Ghesquière, P. (2009). The predictive value of numerical magnitude comparison for individual differences in mathematics achievement. *Journal of Experimental Child Psychology, 103*(4), 469–479. https://doi.org/10.1016/j.jecp.2009.01.010

Dietrich, J. F., Huber, S., Dackermann, T., Moeller, K., & Fischer, U. (2016). Place-value understanding in number line estimation predicts future arithmetic performance. *British Journal of Developmental Psychology, 34*(4), 502–517. https://doi.org/10.1111/bjdp.12146

Dietrich, J. F., Huber, S., & Nuerk, H.-C. (2015). Methodological aspects to be considered when measuring the approximate number system (ANS)—A research review. *Frontiers in Psychology, 6*, 295. https://doi.org/10.3389/fpsyg.2015.00295

Ebersbach, M., Luwel, K., Frick, A., Onghena, P., & Verschaffel, L. (2008). The relationship between the shape of the mental number line and familiarity with numbers in 5- to 9-year old children: Evidence for a segmented linear model. *Journal of Experimental Child Psychology, 99*(1), 1–17. https://doi.org/10.1016/j.jecp.2007.08.006

Eerland, A., Guadalupe, T. M., & Zwaan, R. A. (2011). Leaning to the left makes the Eiffel Tower seem smaller: Posture-modulated estimation. *Psychological Science, 22*(12), 1511–1514. https://doi.org/10.1177/0956797611420731

Feigenson, L., Dehaene, S., & Spelke, E. (2004). Core systems of number. *Trends in Cognitive Sciences, 8*(7), 307–314. https://doi.org/10.1016/j.tics.2004.05.002

Fias, W., Lauwereyns, J., & Lammertyn, J. (2001). Irrelevant digits affect feature-based attention depending on the overlap of neural circuits. *Cognitive Brain Research, 12*(3), 415–423. https://doi.org/10.1016/S0926-6410(01)00078-7

Fischer, J. P. (2010). Numerical performance increased by finger training: A fallacy due to regression toward the mean? *Cortex, 46*(2), 272–273. https://doi.org/10.1016/j.cortex.2008.06.010

Fischer, M. H. (2001). Number processing induces spatial performance biases. *Neurology, 57*(5), 822–826. https://doi.org/10.1212/WNL.57.5.822

Fischer, M. H. (2012). A hierarchical view of grounded, embodied, and situated numerical cognition. *Cognitive Processing, 13*(Suppl 1), S161–S164. https://doi.org/10.1007/s10339-012-0477-5

Fischer, M. H., Mills, R. A., & Shaki, S. (2010). How to cook a SNARC: Number placement in text rapidly changes spatial-numerical associations. *Brain and Cognition, 72*(3), 333–336. https://doi.org/10.1016/j.bandc.2009.10.010

Fischer, M. H., & Shaki, S. (2014). Spatial associations in numerical cognition—From single digits to arithmetic. *Quarterly Journal of Experimental Psychology, 67*(8), 1461–1483. https://doi.org/10.1080/17470218.2014.927515

Fischer, M. H., Shaki, S., & Cruise, A. (2009). It takes just one word to quash a SNARC. *Experimental Psychology, 56*(5), 361–366. https://doi.org/10.1027/1618-3169.56.5.361

Fischer, U., Moeller, K., Bientzle, M., Cress, U., & Nuerk, H.-C. (2011). Sensori-motor spatial training of number magnitude representation. *Psychonomic Bulletin & Review, 18*(1), 177–183. https://doi.org/10.3758/s13423-010-0031-3

Fornaciai, M., Cicchini, G. M., & Burr, D. C. (2016). Adaptation to number operates on perceived rather than physical numerosity. *Cognition, 151*, 63–67. https://doi.org/10.1016/j.cognition.2016.03.006

Gallistel, C. R., & Gelman, R. (2000). Non-verbal numerical cognition: From reals to integers. *Trends in Cognitive Sciences, 4*(2), 59–65. https://doi.org/10.1016/S1364-6613(99)01424-2

Galton, F. (1880). Visualised numerals. *Nature, 21*(533), 252–256. https://doi.org/10.1038/021494e0

Ganor-Stern, D., Tzelgov, J., & Ellenbogen, R. (2007). Automaticity and two-digit numbers. *Journal of Experimental Psychology: Human Perception and Performance, 33*(2), 483–496. https://doi.org/10.1037/0096-1523.33.2.483

Gebuis, T., & Reynvoet, B. (2012). The interplay between nonsymbolic number and its continuous visual properties. *Journal of Experimental Psychology: General, 141*(4), 642–648. https://doi.org/10.1037/a0026218

Gelman, R., & Gallistel, C. R. (1978). *The child's understanding of number*. Cambridge, MA: Harvard University Press.

Georges, C., Hoffmann, D., & Schiltz, C. (2017a). How and why do number-space associations co-vary in implicit and explicit magnitude processing tasks? *Journal of Numerical Cognition, 3*(2), 182–211. https://doi.org/10.5964/jnc.v3i2.46

Georges, C., Hoffmann, D., & Schiltz, C. (2017b). Mathematical abilities in elementary school: Do they relate to number–space associations? *Journal of Experimental Child Psychology, 161*, 126–147. https://doi.org/10.1016/j.jecp.2017.04.011

Gibson, L. C., & Maurer, D. (2016). Development of SNARC and distance effects and their relation to mathematical and visuospatial abilities. *Journal of Experimental Child Psychology, 150*, 301–313. https://doi.org/10.1016/j.jecp.2016.05.009

Göbel, S. M. (2015). Up or down? Reading direction influences vertical counting direction in the horizontal plane—A cross-cultural comparison. *Frontiers in Psychology, 6*, 228. https://doi.org/10.3389/fpsyg.2015.00228

Göbel, S. M., Moeller, K., Pixner, S., Kaufmann, L., & Nuerk, H.-C. (2014). Language affects symbolic arithmetic in children: The case of number word inversion. *Journal of Experimental Child Psychology, 119*(1), 17–25. https://doi.org/10.1016/j.jecp.2013.10.001

Gracia-Bafalluy, M., & Noël, M. P. (2008). Does finger training increase young children's numerical performance? *Cortex, 44*(4), 368–375. https://doi.org/10.1016/j.cortex.2007.08.020

Grant, E. (1972). Nicole Oresme and the medieval geometry of qualities and motions. A treatise on the uniformity and difformity of intensities known as "tractatus de configurationibus qualitatum et motuum": Marshall Clagett (ed. and tr.), edited with an introduction (...). *Studies in History and Philosophy of Science Part A, 3*(2), 167–182. https://doi.org/10.1016/0039-3681(72)90022-2

Henik, A., & Tzelgov, J. (1982). Is three greater than five: The relation between physical and semantic size in comparison tasks. *Memory & Cognition, 10*(4), 389–395. https://doi.org/10.3758/BF03202431

Ho, C. S.-H., & Cheng, F. S.-F. (1997). Training in place-value concepts improves children's addition skills. *Contemporary Educational Psychology, 22*(4), 495–506. https://doi.org/10.1006/ceps.1997.0947

Hoffmann, D., Hornung, C., Martin, R., & Schiltz, C. (2013). Developing number-space associations: SNARC effects using a color discrimination task in 5-year-olds. *Journal of Experimental Child Psychology, 116*(4), 775–791. https://doi.org/10.1016/j.jecp.2013.07.013

Hoffmann, D., Mussolin, C., Martin, R., & Schiltz, C. (2014). The impact of mathematical proficiency on the number-space association. *PLoS One, 9*(1), e85048. https://doi.org/10.1371/journal.pone.0085048

Hoffmann, D., Pigat, D., & Schiltz, C. (2014). The impact of inhibition capacities and age on number-space associations. *Cognitive Processing, 15*(3), 329–342. https://doi.org/10.1007/s10339-014-0601-9

Hohol, M., Cipora, K., Willmes, K., & Nuerk, H.-C. (2017). Bringing back the balance: Domain-general processes are also important in numerical cognition. *Frontiers in Psychology, 8*, 499. https://doi.org/10.3389/fpsyg.2017.00499

Huber, S., Klein, E., Moeller, K., & Willmes, K. (2016). Spatial-numerical and ordinal positional associations coexist in parallel. *Frontiers in Psychology, 7*, 438. https://doi.org/10.3389/fpsyg.2016.00438

Huber, S., Moeller, K., & Nuerk, H.-C. (2014). Dissociating number line estimations from underlying numerical representations. *Quarterly Journal of Experimental Psychology, 67*(5), 991–1003. https://doi.org/10.1080/17470218.2013.838974

Huber, S., Nuerk, H.-C., Willmes, K., & Moeller, K. (2016). A general model framework for multisymbol number comparison. *Psychological Review, 123*(6), 667–695. https://doi.org/10.1037/rev0000040

Huber, S., Sury, D., Moeller, K., Rubinsten, O., & Nuerk, H.-C. (2015). A general number-to-space mapping deficit in developmental dyscalculia. *Research in Developmental Disabilities, 43–44*, 32–42. https://doi.org/10.1016/j.ridd.2015.06.003

Hyde, D. C., Khanum, S., & Spelke, E. S. (2014). Brief non-symbolic, approximate number practice enhances subsequent exact symbolic arithmetic in children. *Cognition, 131*(1), 92–107. https://doi.org/10.1038/jid.2014.371

Kallai, A. Y., & Tzelgov, J. (2012). The place-value of a digit in multi-digit numbers is processed automatically. *Journal of Experimental Psychology: Learning, Memory, and Cognition, 38*(5), 1221–1233. https://doi.org/10.1037/a0027635

Kim, D., & Opfer, J. E. (2017). A unified framework for bounded and unbounded numerical estimation. *Developmental Psychology, 53*(6), 1088–1097. https://doi.org/10.1037/dev0000305

Klein, E., Huber, S., Nuerk, H.-C., & Moeller, K. (2014). Operational momentum affects eye fixation behaviour. *Quarterly Journal of Experimental Psychology, 67*(8), 1614–1625. https://doi.org/10.1080/17470218.2014.902976

Knops, A., Viarouge, A., & Dehaene, S. (2009). Dynamic representations underlying symbolic and nonsymbolic calculation: Evidence from the operational momentum effect. *Attention, Perception & Psychophysics, 71*(4), 803–821. https://doi.org/10.3758/APP

Kornblum, S., Hasbroucq, T., & Osman, A. (1990). Dimensional overlap: cognitive basis for stimulus-response compatibility—a model and taxonomy. *Psychological Review, 97*(2), 253–270. https://doi.org/10.1037//0033295x.97.2.253

Kucian, K., Grond, U., Rotzer, S., Henzi, B., Schönmann, C., Plangger, F., … von Aster, M. (2011). Mental number line training in children with developmental dyscalculia. *NeuroImage, 57*(3), 782–795. https://doi.org/10.1016/j.neuroimage.2011.01.070

Landy, D., Charlesworth, A., & Ottmar, E. (2014). Cutting in line: Discontinuities in the use of large numbers in adults. In *Proceedings of the 36th Annual Conference of the Cognitive Science Society* (pp. 815–820).

Landy, D., Charlesworth, A., & Ottmar, E. (2017). Categories of large numbers in line estimation. *Cognitive Science, 41*(2), 326–353. https://doi.org/10.1111/cogs.12342

Landy, D., & Goldstone, R. L. (2010). Proximity and precedence in arithmetic. *Quarterly Journal of Experimental Psychology, 63*(10), 1953–1968. https://doi.org/10.1080/17470211003787619

Landy, D., Silbert, N., & Goldin, A. (2013). Estimating large numbers. *Cognitive Science, 37*(5), 775–799. https://doi.org/10.1111/cogs.12028

Laski, E. V., & Siegler, R. S. (2014). Learning from number board games: You learn what you encode. *Developmental Psychology, 50*(3), 853–864. https://doi.org/10.1037/a0034321

Laurillard, D. (2016). Learning number sense through digital games with intrinsic feedback. *Australasian Journal of Educational Technology, 32*(6), 32–44. https://doi.org/10.14742/ajet.3116

LeFevre, J. A., Lira, C. J., Sowinski, C., Cankaya, O., Kamawar, D., & Skwarchuk, S. L. (2013). Charting the role of the number line in mathematical development. *Frontiers in Psychology, 4*, 641. https://doi.org/10.3389/fpsyg.2013.00641

Leibovich, T., & Henik, A. (2013). Magnitude processing in non-symbolic stimuli. *Frontiers in Psychology, 4*, 375. https://doi.org/10.3389/fpsyg.2013.00375

Leibovich, T., Katzin, N., Harel, M., & Henik, A. (2017). From "sense of number" to "sense of magnitude"—The role of continuous magnitudes in numerical cognition. *Behavioral and Brain Sciences, 40*, e164. https://doi.org/10.1017/s0140525x16000960

Lindemann, O., Alipour, A., & Fischer, M. H. (2011). Finger counting habits in middle eastern and western individuals: An online survey. *Journal of Cross-Cultural Psychology, 42*(4), 566–578. https://doi.org/10.1177/0022022111406254

Lindskog, M., Winman, A., & Poom, L. (2016). Arithmetic training does not improve approximate number system acuity. *Frontiers in Psychology, 7*, 1364. https://doi.org/10.3389/fpsyg.2016.01634

Link, T., Huber, S., Nuerk, H.-C., & Moeller, K. (2014). Unbounding the mental number line-new evidence on children's spatial representation of numbers. *Frontiers in Psychology, 4*, 1021. https://doi.org/10.3389/fpsyg.2013.01021

Link, T., Moeller, K., Huber, S., Fischer, U., & Nuerk, H.-C. (2013). Walk the number line—An embodied training of numerical concepts. *Trends in Neuroscience and Education, 2*(2), 74–84. https://doi.org/10.1016/j.tine.2013.06.005

Link, T., Nuerk, H.-C., & Moeller, K. (2014). On the relation between the mental number line and arithmetic competencies. *Quarterly Journal of Experimental Psychology, 67*(8), 1597–1613. https://doi.org/10.1080/17470218.2014.892517

Loetscher, T., Schwarz, U., Schubiger, M., & Brugger, P. (2008). Head turns bias the brain's internal random generator. *Current Biology, 18*(2), R60–R62. https://doi.org/10.1016/j.cub.2007.11.015

Lonnemann, J., Krinzinger, H., Knops, A., & Willmes, K. (2008). Spatial representations of numbers in children and their connection with calculation abilities. *Cortex, 44*(4), 420–428. https://doi.org/10.1016/j.cortex.2007.08.015

Lyons, I. M., Nuerk, H.-C., & Ansari, D. (2015). Rethinking the implications of numerical ratio effects for understanding the development of representational precision and numerical processing across formats. *Journal of Experimental Psychology: General, 144*(5), 1021–1035. https://doi.org/10.1037/xge0000094

Maertens, B., De Smedt, B., Sasanguie, D., Elen, J., & Reynvoet, B. (2016). Enhancing arithmetic in pre-schoolers with comparison or number line estimation training: Does it matter? *Learning and Instruction, 46*, 1–11. https://doi.org/10.1016/j.learninstruc.2016.08.004

Masson, N., & Pesenti, M. (2014). Attentional bias induced by solving simple and complex addition and subtraction problems. *Quarterly Journal of Experimental Psychology, 67*(8), 1514–1526. https://doi.org/10.1080/17470218.2014.903985

Masson, N., Letesson, C., & Pesenti, M. (2018). Time course of attentional shifts in mental arithmetic: Evidence from gaze metrics. *The Quarterly Journal of Experimental Psychology, 71*(4), 1009–1019. https://doi.org/10.1080/17470218.2017.1318931

McCrink, K., Dehaene, S., & Dehaene-Lambertz, G. (2007). Moving along the number line: Operational momentum in nonsymbolic arithmetic. *Perception & Psychophysics, 69*(8), 1324–1333. https://doi.org/10.3758/BF03192949

Miura, I. T., Okamoto, Y., Kim, C. C., Chang, C.-M., Steere, M., & Fayol, M. (1994). Comparisons of children's cognitive representation of number: China, France, Japan, Korea, Sweden, and the United States. *International Journal of Behavioral Development, 17*(3), 401–411. https://doi.org/10.1177/016502549401700301

Mix, K., Levine, S., Cheng, Y.-L., Young, C., Hambrick, D., Ping, R., & Konstantopoulos, S. (2016). Separate but correlated: The latent structure of space and mathematics across development. *Journal of Experimental Psychology: General, 145*(9), 1206–1227. https://doi.org/10.1037/xge0000182

Moeller, K., Pixner, S., Kaufmann, L., & Nuerk, H.-C. (2009). Children's early mental number line: Logarithmic or decomposed linear? *Journal of Experimental Child Psychology, 103*(4), 503–515. https://doi.org/10.1016/j.jecp.2009.02.006

Moeller, K., Pixner, S., Zuber, J., Kaufmann, L., & Nuerk, H.-C. (2011). Early place-value understanding as a precursor for later arithmetic performance—A longitudinal study on numerical development. *Research in Developmental Disabilities, 32*(5), 1837–1851. https://doi.org/10.1016/j.ridd.2011.03.012

Moeller, K., Shaki, S., Göbel, S. M., & Nuerk, H.-C. (2015). Language influences number processing—A quadrilingual study. *Cognition, 136*, 150–155. https://doi.org/10.1016/j.cognition.2014.11.003

Moyer, R. S., & Landauer, T. K. (1967). Time required for judgements of numerical inequality. *Nature, 215*(5109), 1519–1520. https://doi.org/10.1038/2151519a0

Nemati, P., Schmid, J., Soltanlou, M., Krimly, J.-T., Nuerk, H.-C., & Gawrilow, C. (2017). Planning and self-control, but not working memory, directly predict multiplication performance in adults. *Journal of Numerical Cognition, 3*(2), 441–467. https://doi.org/10.5964/jnc.v3i2.61

Nuerk, H.-C., Moeller, K., Klein, E., Willmes, K., & Fischer, M. H. (2011). Extending the mental number line: A review of multi-digit number processing. *Zeitschrift Für Psychologie/Journal of Psychology, 219*(1), 3–22. https://doi.org/10.1027/2151-2604/a000041

Nuerk, H.-C., Weger, U., & Willmes, K. (2005). Language effects in magnitude comparison: Small, but not irrelevant. *Brain and Language, 92*(3), 262–277. https://doi.org/10.1016/j.bandl.2004.06.107

Nuerk, H.-C., Moeller, K., & Willmes, K. (2015). Multi-digit number processing: Overview, conceptual clarifications, and language influences. In R. Cohen Kadosh & A. Dowker (Eds.), *The Oxford handbook of numerical cognition*. Oxford: Oxford University Press (pp. 106–139). https://doi.org/10.1027/2151-2604/a000040

Nuerk, H.-C., Weger, U., & Willmes, K. (2001). Decade breaks in the mental number line? Putting the tens and units back in different bins. *Cognition, 82*(1), B25–B33. https://doi.org/10.1016/S0010-0277(01)00142-1

Obersteiner, A., Reiss, K., & Ufer, S. (2013). How training on exact or approximate mental representations of number can enhance first-grade students' basic number processing and arithmetic skills. *Learning and Instruction, 23*(1), 125–135. https://doi.org/10.1016/j.learninstruc.2012.08.004

Opfer, J. E., & Siegler, R. S. (2007). Representational change and children's numerical estimation. *Cognitive Psychology, 55*(3), 169–195. https://doi.org/10.1016/j.cogpsych.2006.09.002

Park, J., & Brannon, E. M. (2013). Training the approximate number system improves math proficiency. *Psychological Science, 24*(10), 2013–2019. https://doi.org/10.1177/0956797613482944

Patro, K., Fischer, U., Nuerk, H.-C., & Cress, U. (2016). How to rapidly construct a spatial-numerical representation in preliterate children (at least temporarily). *Developmental Science, 19*(1), 126–144. https://doi.org/10.1111/desc.12296

Patro, K., Nuerk, H.-C., Cress, U., & Haman, M. (2014). How number-space relationships are assessed before formal schooling: A taxonomy proposal. *Frontiers in Psychology, 5*, 419. https://doi.org/10.3389/fpsyg.2014.00419

Penner-Wilger, M., & Anderson, M. L. (2013). The relation between finger gnosis and mathematical ability: Why redeployment of neural circuits best explains the finding. *Frontiers in Psychology, 4*, 877. https://doi.org/10.3389/fpsyg.2013.00877

Pesenti, M. (2005). Calculation abilities in expert calculators. In J. I. D. Campbell (Ed.), *Handbook of mathematical cognition* (pp. 413–430). New York, NY: Psychology Press.

Petersen, S. E., & Posner, M. I. (2012). The attention system of the human brain: 20 years after. *Annual review of neuroscience, 35*, 73–89. https://doi.org/10.1146/annurev-neuro-062111-150525

Pfister, R., Schroeder, P. A., & Kunde, W. (2013). SNARC struggles: Instant control over spatial-numerical associations. *Journal of Experimental Psychology: Learning, Memory, and Cognition, 39*(6), 1953–1958. https://doi.org/10.1037/a0032991

Piazza, M., & Izard, V. (2009). How humans count: Numerosity and the parietal cortex. *The Neuroscientist, 15*(3), 261–273. https://doi.org/10.1177/1073858409333073

Piazza, M., Mechelli, A., Butterworth, B., & Price, C. J. (2002). Are subitizing and counting implemented as separate or functionally overlapping processes? *NeuroImage, 15*(2), 435–446. https://doi.org/10.1006/nimg.2001.0980

Pinhas, M., & Fischer, M. H. (2008). Mental movements without magnitude? A study of spatial biases in symbolic arithmetic. *Cognition, 109*(3), 408–415. https://doi.org/10.1016/j.cognition.2008.09.003

Pinhas, M., Shaki, S., & Fischer, M. H. (2014). Heed the signs: Operation signs have spatial associations. *Quarterly Journal of Experimental Psychology, 67*(8), 1527–1540. https://doi.org/10.1080/17470218.2014.892516

Pinheiro-Chagas, P., Dotan, D., Piazza, M., & Dehaene, S. (2017). Finger tracking reveals the covert stages of mental arithmetic pedro. *Open Mind, 1*(1), 30–41. https://doi.org/10.1162/opmi

Pixner, S., Moeller, K., Zuber, J., & Nuerk, H.-C. (2009). Decomposed but parallel processing of two-digit numbers in 1st graders. *The Open Psychology Journal, 2*, 40–48. https://doi.org/10.2174/1874350100902010040

Ramani, G. B., Siegler, R. S., & Hitti, A. (2012). Taking it to the classroom: Number board games as a small group learning activity. *Journal of Educational Psychology, 104*(3), 661–672. https://doi.org/10.1037/a0028995

Raz, A., & Buhle, J. (2006). Typologies of attentional networks. *Nature Reviews Neuroscience, 7*(5), 367–379. https://doi.org/10.1038/nrn1903

Restle, F. (1970). Speed of adding and comparing numbers. *Journal of Experimental Psychology, 83*(2, Pt. 1), 274–278. https://doi.org/10.1037/h0028573

Rodic, M., Zhou, X., Tikhomirova, T., Wei, W., Malykh, S., Ismatulina, V., ... Kovas, Y. (2015). Cross-cultural investigation into cognitive underpinnings of individual differences in early arithmetic. *Developmental Science, 18*(1), 165–174. https://doi.org/10.1111/desc.12204

Rubinsten, O., & Sury, D. (2011). Processing ordinality and quantity: The case of developmental dyscalculia. *PLoS One, 6*(9), e24079. https://doi.org/10.1371/journal.pone.0024079

Sasanguie, D., & Reynvoet, B. (2014). Adults' arithmetic builds on fast and automatic processing of Arabic digits: Evidence from an audiovisual matching paradigm. *PLoS One, 9*(2), e87739. https://doi.org/10.1371/journal.pone.0087739

Schneider, M., Beeres, K., Coban, L., Merz, S., Susan Schmidt, S., Stricker, J., & De Smedt, B. (2017). Associations of non-symbolic and symbolic numerical magnitude processing with mathematical competence: A meta-analysis. *Developmental Science, 20*(30), e12372. https://doi.org/10.1111/desc.12372

Schneider, M., Grabner, R. H., Zurich, E., & Paetsch, J. (2009). Mental number line, number line estimation, and mathematical achievement: Their interrelations in grades 5 and 6. *Journal of Educational Psychology, 101*(2), 359–372. https://doi.org/10.1037/a0013840

Schroeder, P. A., Nuerk, H.-C., & Plewnia, C. (2017a). Space in numerical and ordinal information: A common construct? *Journal of Numerical Cognition, 3*(2), 164–181. https://doi.org/10.5964/jnc.v3i2.40

Schroeder, P. A., Nuerk, H.-C., & Plewnia, C. (2017b). Switching between Multiple Codes of SNARC-Like Associations: Two Conceptual Replication Attempts with Anodal tDCS in Sham-Controlled Cross-Over Design. *Frontiers in Neuroscience, 11*, 654. https://doi.org/10.3389/fnins.2017.00654

Schroeder, P. A., & Pfister, R. (2015). Arbitrary numbers counter fair decisions: Trails of markedness in card distribution. *Frontiers in Psychology, 6*, 240. https://doi.org/10.3389/fpsyg.2015.00240

Sella, F., Tressoldi, P., Lucangeli, D., & Zorzi, M. (2016). Training numerical skills with the adaptive videogame "The Number Race": A randomized controlled trial on preschoolers. *Trends in Neuroscience and Education, 5*(1), 20–29. https://doi.org/10.1016/j.tine.2016.02.002

Shaki, S., & Fischer, M. H. (2014). Random walks on the mental number line. *Experimental Brain Research, 232*(1), 43–49. https://doi.org/10.1007/s00221-013-3718-7

Shaki, S., Fischer, M. H., & Göbel, S. M. (2012). Direction counts: A comparative study of spatially directional counting biases in cultures with different reading directions. *Journal of Experimental Child Psychology, 112*(2), 275–281. https://doi.org/10.1016/j.jecp.2011.12.005

Shaki, S., Fischer, M. H., & Petrusic, W. M. (2009). Reading habits for both words and numbers contribute to the SNARC effect. *Psychonomic Bulletin & Review, 16*(2), 328–331. https://doi.org/10.3758/PBR.16.2.328

Shaki, S., & Gevers, W. (2011). Cultural characteristics dissociate magnitude and ordinal information processing. *Journal of Cross-Cultural Psychology, 42*(4), 639–650. https://doi.org/10.1177/0022022111406100

Siegler, R. S. (2009). Improving the numerical understanding of children from low-income families. *Child Development Perspectives, 3*(2), 118–124. https://doi.org/10.1111/j.1750-8606.2009.00090.x

Siegler, R. S., & Booth, J. L. (2004). Development of numerical estimation in young children. *Child Development, 75*(2), 428–444. https://doi.org/10.1111/j.1467-8624.2004.00684.x

Siegler, R. S., & Opfer, J. E. (2003). The development of numerical estimation: Evidence for multiple representations of numerical quantity. *Psychological Science, 14*(3), 237–243. https://doi.org/10.1111/1467-9280.02438

Siegler, R. S., & Ramani, G. B. (2009). Playing linear number board games—But not circular ones—Improves low-income preschoolers' numerical understanding. *Journal of Educational Psychology, 101*(3), 545–560. https://doi.org/10.1037/a0014239

Siegler, R. S., Thompson, C. A., & Opfer, J. E. (2009). The logarithmic-to-linear shift: One learning sequence, many tasks, many time scales. *Mind, Brain, and Education, 3*(3), 143–150. https://doi.org/10.1111/j.1751-228X.2009.01064.x

Simner, J., Mayo, N., & Spiller, M. J. (2009). A foundation for savantism? Visuo-spatial synaesthetes present with cognitive benefits. *Cortex, 45*(10), 1246–1260. https://doi.org/10.1016/j.cortex.2009.07.007

Stavy, R., & Tirosh, D. (2000). *How students (mis-) understand science and mathematics: Intuitive rules*. New York: Teachers College Press.

Szűcs, D., Nobes, A., Devine, A., Gabriel, F. C., & Gebuis, T. (2013). Visual stimulus parameters seriously compromise the measurement of approximate number system acuity and comparative effects between adults and children. *Frontiers in Psychology, 4*, 444. https://doi.org/10.3389/fpsyg.2013.00444

Tudusciuc, O., & Nieder, A. (2007). Neuronal population coding of continuous and discrete quantity in the primate posterior parietal cortex. *PNAS, 104*(36), 14513–14518. https://doi.org/10.1073/pnas.0705495104

van Dijck, J.-P., & Fias, W. (2011). A working memory account for spatial-numerical associations. *Cognition, 119*(1), 114–119. https://doi.org/10.1016/j.cognition.2010.12.013

Verbruggen, F., Liefooghe, B., Notebaert, W., & Vandierendonck, A. (2005). Effects of stimulus-stimulus compatibility and stimulus-response compatibility on response inhibition. *Acta Psychologica, 120*(3), 307–326. https://doi.org/10.1016/j.actpsy.2005.05.003

Wasner, M., Moeller, K., Fischer, M. H., & Nuerk, H.-C. (2014). Aspects of situated cognition in embodied numerosity: The case of finger counting. *Cognitive Processing, 15*(3), 317–328. https://doi.org/10.1007/s10339-014-0599-z

Wiemers, M., Bekkering, H., & Lindemann, O. (2014). Spatial interferences in mental arithmetic: Evidence from the motion-arithmetic compatibility effect. *Quarterly Journal of Experimental Psychology, 67*(8), 1557–1570. https://doi.org/10.1080/17470218.2014.889180

Wiemers, M., Bekkering, H., & Lindemann, O. (2017). Is more always up? Evidence for a preference of hand-based associations over vertical number mappings. *Journal of Cognitive Psychology, 29*(5), 642–652. https://doi.org/10.1080/20445911.2017.1302451

Wilson, A. J., Dehaene, S., Pinel, P., Revkin, S. K., Cohen, L., & Cohen, D. (2006). Principles underlying the design of "The Number Race", an adaptive computer game for remediation of dyscalculia. *Behavioral and Brain Functions, 2*, 19. https://doi.org/10.1186/1744-9081-2-19

Wood, G., Willmes, K., Nuerk, H.-C., & Fischer, R. (2008). On the cognitive link between space and number: A meta-analysis of the SNARC effect. *Psychology Science Quarterly, 50*(4), 489–525.

Zohar-Shai, B., Tzelgov, J., Karni, A., & Rubinsten, O. (2017). It does exist! A left-to-right spatial–numerical association of response codes (SNARC) effect among native Hebrew speakers. *Journal of Experimental Psychology: Human Perception and Performance, 43*(4), 719–728. https://doi.org/10.1037/xhp0000336

Zuber, J., Pixner, S., Moeller, K., & Nuerk, H.-C. (2009). On the language specificity of basic number processing: Transcoding in a language with inversion and its relation to working memory capacity. *Journal of Experimental Child Psychology, 102*(1), 60–77. https://doi.org/10.1016/j.jecp.2008.04.003

Chapter 5
What Processes Underlie the Relation Between Spatial Skill and Mathematics?

Christopher Young, Susan C. Levine, and Kelly S. S. Mix

Abstract In this chapter, we review approaches to modeling a connection between spatial and mathematical thinking across development. We critically evaluate the strengths and weaknesses of factor analyses, meta-analyses, and experimental literatures. We examine those studies that set out to describe the nature and number of spatial and mathematical abilities and specific connections among these abilities, especially those that include children as participants. We also find evidence of strong spatial-mathematical connections and transfer from spatial interventions to mathematical understanding. Finally, we map out the kinds of studies that could enhance our understanding of the mechanism by which spatial and mathematical processing are connected and the principles by which mathematical outcomes could be enhanced through spatial training in educational settings.

Keywords Process modeling · Cognitive processes · Factor analysis · Spatial skills · Spatial cognition · Cognitive development · Mathematical concepts · Latent structure · Spatial visualization · Cognitive science · Education · Spatial ability · Mathematical ability · Individual differences · Intelligence · Number concepts · Common Core State Standards for Mathematics · Exploratory factor analysis · Confirmatory factor analysis · Multidimensionality · Meta-analysis

---

C. Young (✉)
Consortium on School Research, University of Chicago, Chicago, IL, USA

S. C. Levine
Departments of Psychology, and Comparative Human Development and Committee on Education, University of Chicago, Chicago, IL, USA

K. S. S. Mix
Department of Human Development and Quantitative Methodology, University of Maryland, College Park, MD, USA
e-mail: youngcj@uchicago.edu

K. S. S. Mix, M. T. Battista (eds.), *Visualizing Mathematics*, Research in Mathematics Education, https://doi.org/10.1007/978-3-319-98767-5_5

Introduction

Many parents, teachers, and members of the public at large believe that learning mathematics is primarily focused on remembering arithmetic facts. This is despite a general push from professional mathematics organizations and advisory committees, like the National Council of Teachers of Mathematics, arguing that mathematics instruction should incorporate more spatial thinking, with less focus on the solving routine number problems and less teaching largely rote skills (e.g., CCSSI, 2010; NCTM, 2000; NRC, 2005). Teachers' spatial skills are correlated with their motivation and interest in teaching mathematics and students' spatial skills are correlated with their persistence in learning mathematics (Edens & Potter, 2013). Moreover, teachers are amenable to teaching spatial skills when they are informed about their importance (Krakowski, Ratliff, Gomez, & Levine, 2010). A recent meta-analysis shows that spatial skills can be improved via a variety of interventions (e.g., Uttal et al., 2013). Improving spatial skills pays off in the longer term; an individual's spatial skill predicts the likelihood he or she will enter a Science, Technology, Engineering and Mathematics (STEM) field beyond both verbal and mathematical abilities (Casey, Nuttall, & Pezaris, 2001; Casey, Nuttall, Pezaris, & Benbow, 1995; Wai, Lubinski, & Benbow, 2009). Adding spatial skills to our conventionally number-focused mathematics instruction may be a way to increase students' mathematics understanding, and prompting teachers to support students' spatial thinking may be an attainable and effective way to improve mathematics achievement.

In this chapter, we describe how spatial thinking relates to mathematical thinking. To do so we will need to address several questions along the way. First, what are spatial and mathematical abilities, and which skills comprise them? Second, what are the strengths and limitations of the factor analytic approach used to describe the interconnected structures of spatial skills and of mathematical skills? Third, how can cognitive science help us to understand the connections between spatial skill and mathematics? Finally, what are the educational implications of these connections? In answering these questions, we will illustrate a pathway for future research and provide guiding principles for the design of future studies and implementation of effective educational practices that leverage the connection between spatial skills and mathematics instruction.

What Are Spatial and Mathematical Abilities and What Skills Comprise Them?

We begin by briefly discussing the methods psychologists have used to analyze the relations among cognitive abilities. Early psychologists first attempted to define "intelligence" by analyzing the "structure of the intellect," specifically whether certain cognitive components were irreducible and unique (Spearman, 1927; Thurstone,

1938; Guilford, 1967). This work relied on factor analysis, a procedure that identifies "latent" factors that can account for co-variation in many tasks. This method provides a simple, quantitative solution to the question we, and earlier theorists, ask: what are the fewest and most important skills needed to describe an ability?

In the next section, we describe past efforts to describe and analyze spatial and mathematical abilities. There have been multiple attempts to isolate independent skills in each ability, as well as meta-analyses that have looked across many studies in order to validate the strongest theories about the nature of those skills. Understanding the constellation of unique skills and how they relate to one another is a critical first step to understanding how spatial thinking may be advantageous when thinking of mathematics problems. Specifically, these relations may guide hypotheses about the best candidate spatial skills to strengthen in order to improve specific aspects of mathematical thinking.

Skills Making Up the Spatial Domain

Spatial skill is most broadly defined as "how individuals deal with materials presented in space-whether in one, two or three dimensions, or how individuals orient themselves" (Carroll, 1993). More concretely, "spatial ability" has been defined as the "ability to generate, retain, retrieve, and transform well-structured visual images" (Lohman, 1994). The NSF Spatial Intelligence and Learning Center proposed a 2 × 2 framework that categorized spatial skills by whether the transformation was dynamic or static and whether it occurs within an object or between multiple objects (e.g., a dynamic-within object spatial problem is imagining an image turning clockwise and a static-between object spatial problem is reading a map, Newcombe & Shipley, 2015). Spatial skills vary widely in terms of the stimuli that they operate on and the type of transformation that is performed, leading to the widely held belief that there are multiple, distinct spatial processes (Linn & Petersen, 1985; Voyer, Voyer & Bryden, 1995; Hegarty & Waller, 2005). In many cases, similar spatial tasks vary in the spatial skills they require (e.g., the dissociation in imagining a different perspective vs. imagining an object rotating, Hegarty & Waller, 2004) and as abilities change throughout development they take on different characteristics (e.g., sex differences in two-dimensional but not three-dimensional mental rotation, Neubauer, Bergner, & Schatz, 2010). In short, spatial ability is a broad domain and relating spatial skills to mathematical ones is a complicated task (for a more thorough review of a variety of spatial skills and mathematical skills, see Mix & Cheng, 2011).

Multiple large-scale factor analyses were conducted over the twentieth century with the aim of differentiating specific intelligences from "*g*" or general intelligence. L.L. Thurstone investigated the issue primarily by factor analyzing a battery of cognitive tasks. He described the resulting factors as "primary mental abilities," two of which were "spatial visualization" and "number facility" (Thurstone, 1933; Thurstone, 1938). Following his efforts, theorists began to assume that spatial intelligence exists, and attempted to better describe its characteristics, and how spatial skills might further explain variance across a variety of spatial measures.

Of four large-scale factor analyses in the latter part of the twentieth century that included a wide range of spatial measures, all four found evidence for a spatial visualization factor, generally dealing with imagistic transformations (Carroll, 1993; Lohman, 1988; McGee, 1979; Michael, Guilford, Fruchter, & Zimmerman, 1957). This factor consistently included tasks such as *paper folding*, a task measuring the ability to predict the result of a series of folds to a piece of paper and a hole punch, and *form board*, which measured the ability to pick the shapes needed to assemble a larger shape (Carroll, 1993; Lohman, 1988; McGee, 1979; Michael et al., 1957). Another task, *cube comparisons*, which measured the ability to determine whether a set of drawings of dice-like stimuli are two different views of the same object, loaded on the spatial visualization factor in two of the four studies, but in earlier studies loaded on separate spatial relations and orientation factor, which generally captured tasks that required perceiving the relative positions and angle of nearby objects. The other factors that were extracted across these studies were interpreted as various forms of specialized perceptual and motoric factors (kinesthetic imagery, Michael et al., 1957; closure speed, flexibility of closure, perceptual speed, and visual memory, Carroll, 1993). Thus, these initial studies all found a similar spatial visualization factor, but differed in terms of other factors they extracted, likely because they included different tasks in the analysis.

More recent factor analytic research has investigated how spatial skills relate to other cognitive abilities. For instance, Miyake, Friedman, Rettinger, Shah, and Hegarty (2001) examined the connection of spatial skills to working memory, the ability to remember and manipulate information, and executive functioning, the ability to monitor one's own behavior and to select among choices to achieve specific goals. The three spatial skills tested were spatial visualization, which involves complex mental manipulations of objects, spatial relations, which involve simpler, speeded two-dimensional transformations, and visuospatial perceptual speed, which involves quickly perceiving and making judgments about stimuli, such as whether a particular shape is present in a complex image. Each of these skills differ in the time-scale they act on and the type of transformations they require, rather than being defined by content or whether they rely on visual or kinesthetic information. These skills not only loaded on separate factors, but each also related differentially to working memory and executive function (i.e., both spatial visualization and spatial relations were significantly related to executive functioning, while visuospatial perceptual speed alone was significantly related to working memory, Miyake et al., 2001).

Evidence that each of the spatial skills have divergent connections to other skills provides external validation that they are in fact separate. It is also useful to contextualize the different kinds of spatial skill with respect to their potential roles in mathematics problems; some skills seem to be more basic and could have important role in imagining the transformations signified by arithmetic operations, while others might be more related to choosing an effective, spatially grounded strategy to solve problems involving numbers. Thus, developing an accurate understanding of our primary spatial abilities is critical if we hope to improve our understanding and use of the pathway from spatial skill to mathematics achievement. In the next section, we look at characteristics of mathematical skills, as a way of identifying potential skill to skill connections between spatial and mathematical thinking.

Skills Making Up Mathematical Domain

Mathematical ability, like spatial ability, is highly complex and multifaceted; humans must learn and use a variety of concepts, from how to differentiate and represent approximate magnitudes (Feigenson, Dehaene & Spelke, 2004) to a basic understanding of what a "natural number" is and how it relates to fundamental principles of arithmetic, such as commutativity (Rips, Bloomfield & Asmuth, 2008). Mathematical problems also vary widely in the property of numerical magnitudes involved in the problem (e.g., parity, rationality, size) and the operation applied to these magnitudes. In addition, even for specific magnitudes and operations, there are marked differences in the efficiency of the strategies used to solve these problems (Siegler, 1999). Further, different numerical skills take on greater importance in schooling across development, and in some cases the connection of basic skills to applied, mathematical reasoning depends on the exact way a mathematical problem is framed (Libertus, Feigenson, & Halberda, 2013; Landy, Brookes, & Smout, 2011).

The effort to extract primary mental abilities that was applied to spatial abilities was also applied to mathematical skills, resulting in a factor that was dubbed "facility with numbers" (Thurstone, 1938). Follow-up factor analyses carried out on purely mathematical measures over various ages during development extracted factors that seemed to be less-than-pure mathematical factors (e.g., deductive reasoning and adaptability to a new task, in a study of tenth grade students understanding of algebra, Kline, 1960; abstraction, analysis, application, in a study of elementary school students mathematical reasoning, Rusch, 1957). These early results are also notable in that many theorists found evidence of a spatial factor in mathematics (e.g., Kline, 1960; Werdelin, 1966) or else argued that there was a spatial sensorimotor intelligence factor important to mathematical reasoning (Aiken Jr, 1970; Coleman, 1960; Skemp, 1961).

Few studies have examined mathematical measures broadly enough to reveal separate skills. Yet this examination is vital because mathematics is frequently divided by differences in content rather than skills. For instance, the recently adopted Common Core State Standards for Mathematics (CCSS-M, 2010) in first grade creates a domain called "Counting and Cardinality," which includes performance standards that are nominally connected but that actually require a variety of different skills and conceptual understandings. Memorizing and reciting the count list is quite different from an active process of "counting on" from a number besides one. Further, understanding that the count list is used to determine the exact the number of items in a set requires more than knowledge of the count list, and is achieved well after young children can count fluently from 1 to 10 or higher (e.g., Sarnecka & Carey, 2008). These three skills are also distinct from the ability to identify and interpret numerals, yet these are placed in the same domain (CCSS-M, 2010).

On the whole, the factor analytic approach has not identified the kinds of distinct mathematical skills that have emerged from cognitive science research, which makes it difficult to identify skill-to-skill connections between spatial skill and particular aspects of mathematics. As summarized above, factor analytic studies of spatial tasks researchers have found evidence of multiple spatial skills, albeit with some inconsistency from study to study. Attempts to identify the structure of mathematical skills have been less successful, with some studies revealing factors that are related to solving mathematics problems, like deductive reasoning, and others revealing factors related to generic cognitive functions, like adaptability to a new task (Kline, 1960). In addition, in both the spatial and mathematical domains, researchers have sometimes found evidence of only a single, domain-wide factor (Mix et al., 2016; Mix et al., 2017). The absence of strong evidence of distinct skills and ambiguity of the results highlights the limitations of the factor analytic approach. Nonetheless, this approach does have some strengths—in particular providing a way to delimit hypotheses about how skills within and across domains relate to each other. In the next section, we further explore these strengths and limitations.

Strengths and Limitations of the Factor Analytic Approach

As we have seen, researchers have relied on factor analysis to define precisely what we mean by spatial skills, mathematical skills, and their overlap. This information is potentially important for educators as it could guide the development of effective approaches to improving mathematics learning, but can we trust the results these analyses yield?

The central strength of the factor analytic approach is that it explains the covariance in scores from a large set of correlated tasks, omitting random error and variance associated with each task, unlike correlational and regression approaches (Bollen, 1989). Factor analysis also has a built-in method of rejecting unnecessary skills (e.g., the Kaiser rule, only skills that explain significant covariation are kept, Kaiser, 2016). It also provides a way to decide among competing theories about the nature of factors (comparing models with different skills statistically is implicit in any factor analysis; Tomarken & Waller, 2005).

A second strength of factor analysis is that it allows a researcher to choose whether the skills can be related to one another not. Specifically, one can choose whether skills are allowed to be correlated rather than totally distinct to each other (geomin rotation, e.g., in a case where skills are thought to rely on common cognitive resources) or that tasks should only load strongly on one factor and not others (varimax rotation, e.g., in a case where skills are thought to be wholly distinct, Browne, 2010). While the number and overall strength of each skill will not change, rotation may cause a change in which tasks will be related to each skill. This strength requires that the researcher make a principled choice about the domain being considered because multiple rotation methods will fit the data equally well, and each will lead to different interpretations.

A third strength of the approach is the simplicity of checking preliminary exploratory factor analysis (EFA) with a confirmatory factor analysis (CFA). While many other methods could be used to replicate a finding, the use of an initial EFA, to identify a pattern of loadings, followed by CFA, where that specific pattern is tested, has a proven track record (e.g., Gerbing & Hamilton, 1996). This well-worked path of substantiating claims about the structure of skills is particularly useful given that investigating the large number of tasks that measure spatial and numerical reasoning is a costly effort, and a variety of different, and at times conflicting, factor structures emerged from earlier studies that did not use this approach (e.g., three broad factors in Michael et al., 1957 versus seven specific factors in Carroll, 1993).

Despite these benefits, the factor analytic approach also has limitations—practical, statistical, and interpretive. Practically, researchers must select enough tasks to cover the domain of interest, but also be selective about those tasks to ensure that they are reliable. Oversampling tasks that are closely related is problematic because it may cause a spurious skill to be extracted because of the similarities in the tasks, or even lead to multiple skills being extracted when absent oversampling there would be only one (as noted by Hegarty & Waller, 2005). Conversely, including tasks that are multidimensional (e.g., that rely on both spatial and non-spatial reasoning), will make loadings more difficult to interpret. Similarly, selecting tasks that are strongly affected by the way participants respond, such as those that include multiple choice questions, may result in extracting "methods" factors that capture variance not as a result of similar underlying processes but rather as a result of using the same test format (e.g., Maul, 2013).

Factor analyses are also limited by statistical power and sample considerations. Researchers must collect data concurrently on a large number of participants, as smaller sample factor analyses on few participants do not always recover stable factor structure in empirical studies (MacCallum, Widaman, Zhang, & Hong, 1999) or true factors in simulation studies (Preacher & MacCallum, 2002). Approximately 20 observations per task are needed to achieve adequate power (Hair Jr, Anderson, Tatham, & Black, 1995). Missing data in any one task requires more complicated statistical procedures, like imputation (Little & Rubin, 1989) or relying on the remaining data to reveal the factor structure through maximum likelihood estimation (Tucker & Lewis, 1973). Even a stable factor structure may arise from random sampling error, which would lead to the same pattern of loadings not being found in a second sample (Cliff & Pennell, 1967). In short, factor analysis reveals true skills only when a large sample of representative and complete data is collected.

Finally, the interpretation of a factor analysis is a complicated issue. The nature of each factor in a solution is decided by the researcher based on the tasks that load on that factor as well as based on the tasks that do not load on that factor (Rummel, 1970). These types of decisions are highly subjective and the researcher's biases may lead them to interpret random patterns as meaningful (Armstrong & Soelberg,

1968). The researcher must decide how and why the tasks that load on a skill could be logically connected and in most cases the data are correlational and cannot reveal a causal relation. In the next section, we discuss how factor analysis can be used in concert with a cognitive science informed and process-oriented view of spatial and numerical ability to better understand their connection.

How Can Cognitive Science Help Us to Understand the Connections Between Spatial Skill and Mathematics?

An alternative to an approach that relies solely on the measurement of skills through factor analysis is a multi-faceted cognitive science approach that focuses on mental processes. This approach asks more specific questions using a variety of tools. "Which theories explain differences within spatial and mathematical skills?" can be addressed through meta-analysis of many studies. "What skills are used in both spatial and mathematical problems?" can be answered by a cross-domain factor analysis. "What components of a spatial skill allows a child to solve a mathematics problem?" can be addressed with process models of skills.

We apply this approach to the present question, first by examining theoretical distinctions among characteristics of skills through meta-analytic studies. We will then discuss a recent analysis of the connection between spatial and mathematical tasks using factor analysis, and the kinds of process models that follow from this analysis. Finally, we examine how the components of spatial skills might influence mathematical reasoning by training spatial skills and examining the benefits to mathematical problem solving.

Which Theories Explain Differences Among Spatial and Mathematical Skills?

Meta-analysis and Divisions Among Skills

Although factor analytic studies have provided insights into the relations of various skills, there are considerations that factor analysis can miss because it is a primarily data-driven approach. Cognitive science counters this deficiency by incorporating more theory-driven research that takes into account the expertise of content experts as well as consistent findings in the field to articulate theories about the nature of skills and then makes predictions and tests critical assumptions. These assumptions are best tested with as much evidence as possible, which requires aggregating multiple studies, even some that were not conducted to explicitly test the theory in question, in a meta-analysis. With regard to examining the relation of spatial and numerical skills, one important way in which this theory testing has taken place is in the form of resolving factor analysis with frequently reported differences associated with gender (e.g., Casey et al., 1995) and with prevailing theories about the structure of each domain.

Gender differences in spatial skills. Linn and Petersen (1985) tested differences in three spatial skills that were chosen to test the theory that there were gender differences in particular kinds of spatial thinking: spatial perception, in which subjects must determine spatial relations in spite of distractors, mental rotation, the ability to quickly rotate two- or three-dimensional objects, and spatial visualization, which included multistep, complex spatial transformations, excluding mental rotation. Linn and Petersen's categorization schema was validated by their analysis of age and gender-based differences between categories: while mental rotation showed sex differences across the lifespan, spatial perception never did, and spatial visualization showed differences only in samples older than 18 years (Linn & Petersen, 1985). Similarly, Voyer, Voyer, and Bryden (1995) organized spatial skills in the same way and found that both mental rotation and spatial perception showed much more frequent, and larger, effect sizes based on gender than did spatial perception. These results suggest that there may be gender differences in spatial skills, from genetic and/or environmental causes (Levine, Foley, Lourenco, Ehrlich, & Ratliff, 2016), which may be relevant to the design of effective interventions that address gender gaps in mathematics achievement (Casey et al., 1995).

Differentiating spatial skills: Static or dynamic? Within or between objects?. Uttal and colleagues (2016) conducted a theory-driven meta-analysis, beginning with a process-oriented account reflected by a typology with orthogonal dimensions that had support from behavioral (Newcombe & Shipley, 2015) and neurological evidence (e.g., Chatterjee, 2008). Specifically, Uttal et al. (2016) argued that spatial relations between objects are processed differently than processing the spatial properties of the objects themselves, known as the intrinsic-extrinsic division. Further, they argued that spatial information that is accessible from stationary frames is processed differently than spatial information that involves movement and change, the static-dynamic division. This theoretical frame relies on direct evidence of dissociations, such as occurs based on whether tasks are large or small scale (e.g., Hegarty, Montello, Richardson, Ishikawa, & Lovelace, 2006) or imagining movement of an object versus changing perspective in relation to a scene containing multiple objects (Hegarty & Waller, 2004; Huttenlocher & Presson, 1973). The results of the meta-analysis showed that spatial training leads to unique transfer both within each cell of the typology (e.g., from mental to rotation training to other dynamic, intrinsic measures), but also transfer between cells (e.g., from mental rotation training to perspective taking), suggesting that the underlying processes in the various cells are not entirely distinct.

While the typology's clear dimensions might be intuitively useful as they are clearer than highly interpreted factors, they may miss the mark in terms of empirical support. For instance, a line of research regarding the cognitive styles of children and adults has provided evidence that the extrinsic-intrinsic dimension is useful for understanding how individuals process spatial information, but also adds to it an algorithmic, verbal form of processing spatial information, and does not distinguish between dynamic and static imagery (e.g., Kozhevnikov, Kozhevnikov, Yu, & Blazhenkova, 2013). A direct test of the typology seems to confirm this, with evidence for the extrinsic/intrinsic continuum but not the static/dynamic one (Mix, Hambrick, Satyam, Burgoyne, & Levine, 2018). Similarly, Atit, Shipley, and Tikoff (2013) tested the dimensions in the typology by measuring adults on a variety of mental transformation tasks, including mental rotation, paper folding, and a "breaking" test, and found that an additional orthogonal axis, between rigid and non-rigid transformations was needed to explain their results. Thus, there appear to be a variety of competitors to the original 2 × 2 typology of spatial skill categories, and it seems that neither the factor-analytical nor the theory-driven approach has arrived at a final solution about the number and types of spatial skills. The addition of this theoretical framing to the original data-driven factor analysis provides new ways to describe and explain differences spatial reasoning skill, which we can probe using both experiments and modeling.

Theoretical approach to mathematical skills. While the factor analytic approach applied to mathematics resulted in multiple skills, different studies contradicted one another regarding the nature of those skills. Several theoretically motivated lines of research have used other methods to determine how understanding of mathematical concepts and basic representation of numerical magnitude interact with each other, effectively providing more process-oriented dimensions and a more detailed hierarchy of basic and composite numerical skills. This work focuses our attention on skills that are likely to have the greatest effect on mathematical achievement over time. We review this literature by highlighting a few illustrative studies that will help us to focus our efforts.

Mathematics as concepts vs. procedures. Researchers have for years debated whether teaching students to be fluent in mathematical procedures is helpful for improving their understanding of mathematics or if it merely causes children to demonstrate rote-learning skills (Schoenfeld, 2014). While some research suggests that promoting a conceptual grounding of mathematics is of greatest important (e.g., Star, 2005) others argue for a more measured approach, wherein concepts and procedures are more mutually supportive (e.g., Baroody, 2003; Rittle-Johnson, Siegler, & Alibali, 2001). Recent research suggests that certain mathematical concepts that are often overlooked early on might have an outsized role in later mathematics achievement, e.g., patterning skill in early elementary school predicts mathematics achievement later, above and beyond more procedural acts, like counting (Rittle-Johnson, Fyfe, Hofer, & Farran, 2016). The relative importance of teaching concepts versus procedures to mathematical achievement is an important issue to raise because certain spatial skills might be more important when learning numerical concepts than for learning numerical procedures, and vice versa (e.g., certain gestures highlight conceptual groupings for young children's addition, Goldin-Meadow, Cook, & Mitchell, 2009, while undergrads benefit more from abstract than concrete examples when learning modulo Kaminski, Sloutsky, & Heckler, 2009).

Thinking symbolically vs. non-symbolically. Recent debate has also focused on how mathematics achievement relates to our basic sense of numerical magnitude, frequently indexed by non-symbolic magnitude comparisons and ordering (e.g., Barth, Beckmann, & Spelke, 2008; Mazzocco, Feigenson, & Halberda, 2011; Mundy & Gilmore, 2009; Rousselle & Noël, 2007). While some argue that this number sense is strongly related to our ability to represent number in more complex number skills (Mazzocco et al., 2011), recent large-scale studies have also shown that our symbolic sense of number is the strongest predictor of mathematics achievement across the elementary school years (comparing in first grade, ordering in sixth grade, Lyons, Price, Vaessen, Blomert, & Ansari, 2014). Meta-analysis confirmed a significantly larger relation of our symbolic number sense to our ability to understand more complex mathematics subject areas non-symbolic number sense (Schneider et al., 2016) and thus may be a better target for training. However, the question of whether spatial skills training can be used to train symbolic and/or non-symbolic number sense remains open, particularly if certain mathematical skills, like understanding fractional magnitude depend on earlier developing non-symbolic ratio or proportional reasoning (Matthews, Lewis, & Hubbard, 2016; Möhring, Newcombe, & Frick, 2015).

These strands of research together suggest that there are many important divisions in spatial and mathematical skills that have not been detected by the basic factor analyses of each domain, but nevertheless could play a role in the relation between spatial and mathematical thinking. In the next section, we consider what can be learned from studies that examine the connection of spatial and mathematical skills. Specifically, we look at recent studies that use factor analysis on both mathematical and spatial domains, to determine whether there is evidence for hybrid spatial-mathematical skills and/or evidence for specific spatial skills that are closely related to general mathematical skills. We then outline how process models of spatial skills can help build out theories about how spatial skills may support mathematical achievement.

What Skills Are Used in Both Spatial and Mathematical Problems?

A number of studies have examined the relation between specific spatial skills and mathematical achievement (for a review, see Mix & Cheng, 2012: Visuospatial working memory, Raghubar, Barnes, & Hecht, 2010; Mental Rotation, Kyttälä, et al., 2003; Block Design, Markey, 2009; Patterning, Rittle-Johnson et al., 2016). Some other spatial skills, often those less researched, have not shown the same connection to mathematics, despite in some cases clear areas where it seems like there might be overlap. For instance, we might expect interpreting maps or solving problems of scale (e.g., DeLoache, Uttal, & Pierroutsakos, 1998; Huttenlocher, Newcombe, & Vasilyeva, 1999), which both involve symbolic thinking, to be useful for understanding numerical symbols. We might also expect that those individuals skilled in disembedding shapes from visual scenes might be better able at analyzing charts and graphs (Clark III, 1988). However, such connections have not been frequently reported.

An obstacle to identifying mechanisms that connect specific mathematical and spatial skills is the high degree of interrelation among skills. Even in rare cases where multiple measures of spatial skills are included in studies with mathematical outcomes, it can be difficult to interpret the result because all spatial skills are correlated with mathematical outcomes. This type of evidence fails to provide support for the theory that certain specific spatial skills are important for mathematics achievement nor how they enable better performance and learning of mathematical skills.

Cross-Domain Factor Analysis of Spatial Reasoning and Mathematical Reasoning

In this section, we focus on two questions central to the goal of this chapter: first, are the spatial and mathematically domains connected generally or by specific skills, and second, is there evidence for shared processes used in both domains? We highlight a pair of studies conducted by the authors that addressed these questions by using factor analysis to determine whether skills in the spatial and mathematical domain load on a single or multiple factors across children's kindergarten to sixth grade education. To our knowledge, there have not been studies that have systematically examined how spatial and mathematical skills, and their interconnections, change over developmental time.

Surveying the field of spatial and mathematical connections. Mix and colleagues conducted a two-stage, exploratory and confirmatory factor analysis of data collected over the 2012–2013 and 2013–2014 school years (Mix et al., 2016; Mix et al., 2017). The goal of the studies was to examine what latent factors explain covariation in age-appropriate mathematics measures and spatial measures. In each study, tasks that had the greatest likelihood of showing spatial-mathematical connections based on the existing literature were included, e.g., between spatial visualization and complex mathematical relations, between form perception and symbolic reasoning, and between spatial scaling and a number line representations (Thompson, Nuerk, Moeller, & Kadosh, 2013; Landy & Goldstone, 2010; Slusser, Santiago & Barth, 2013, respectively). As shown in Fig. 5.1, by design these spatial skills fall into different places along the dimensions described by the spatial typology (e.g., Uttal, et al., 2016), which should allow us to pick up on differential connections between, for instance, extrinsic versus intrinsic spatial skills and mathematical skills.

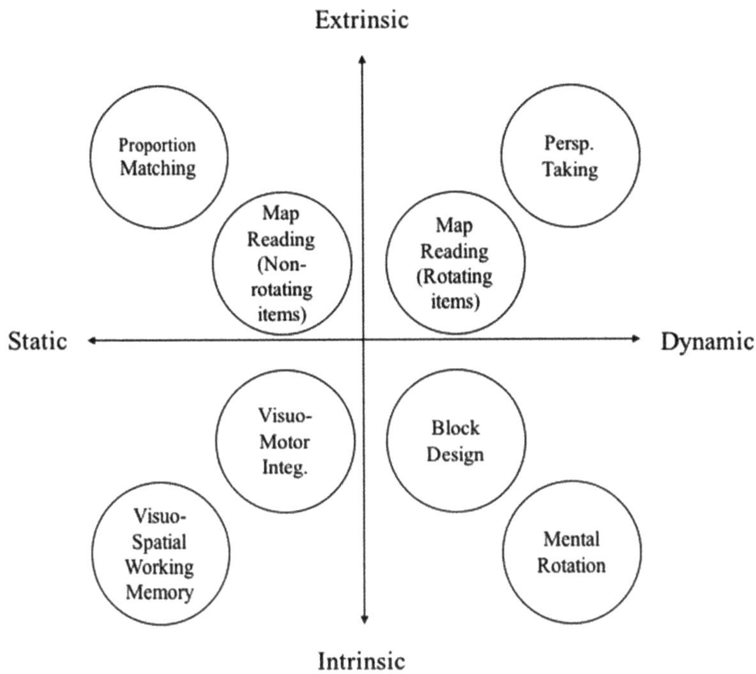

Fig. 5.1 Measures included in EFA/CFA within Spatial Typology (Uttal et al., 2013)

Selection of measures of spatial and mathematical skill. Each task, with some grade appropriate modifications, was administered to kindergarten, third and sixth grade children, which allowed for the possibility of detecting developmental shifts in the relation of spatial and mathematical skills. In addition, the data were collected from students across a wide variety of school settings to ensure the results were generalizable. The specific tasks and their sources are shown in Table 5.1.

Separate but correlated spatial and mathematical factors. In both studies spatial and mathematical processing, as measured by their latent factors, were found to be separate but highly correlated from kindergarten through sixth grade, controlling for general cognitive ability as measured by a vocabulary test. The processes that are accessed when performing a broad range of spatial tasks are highly related to those accessed when performing a broad range of mathematical tasks across development. It was perhaps surprising that separate, domain-specific factors were obtained, given that the covariance among tasks might have been based on one of many other shared task characteristics. For instance, factors may have instead tracked to the way in which children responded to tasks (e.g., productive vs. receptive), or to form of stimuli (symbolic vs. non-symbolic), or to the cognitive resources required (high vs. low executive function), or, as some have previously theorized, we might have found no differentiation between spatial and mathematical skills at all (e.g., a single factor that all measures loaded on). These results show that spatial

Table 5.1 Skills measured in Mix et al. (2016)

Skill	Description, kindergarten and third grade–sixth grade variants	Reference
Mental rotation	Select 2 scrambled letters that match a target with mirror distractors/ Select 2 block figures, that match a target with mirror distractors	Neuburger et al. (2011)/Peters et al. (1995)
Block design	Recreate a complex pattern with, multisided, multicolored blocks	Wechsler et al. (2004)
Visual spatial working memory	Recall positions of an increasingly large array of objects	Kaufman and Kaufman (1983)
Visuomotor integration	Copy images of geometric forms	Beery and Beery (2004)
Perspective taking	Select photo matching view from other's perspective/ Draw arrow showing the direction from object 1 to object 2 when facing object 3	Frick, Möhring, and Newcombe (2014)/Hegarty and Waller (2004); Kozhevnikov and Hegarty (2001)
Map reading	Identify a location on a model using a scale map, sometimes from a rotated map/ Identify a location on a map from photographs	Liben and Downs (1989)
Place value/rational numbers	Compare, order, & interpret multidigit numerals, match numerals to expanded equivalents/ Interpret and translate between different numerical formats (e.g. decimals, tractions)	Novel/Hresko, Schlieve, Herron, Swain, and Sherbenou (2003)
Word problems/ problem solving	Answer word problems testing age appropriate math concepts/ Answer word problems testing age appropriate math concepts	Ginsburg and Baroody (2003)
Calculation	Solve arithmetic problems (K: Addition &. Subtraction, 3rd: Operations through Division)/ Solve arithmetic problems (Operations through Division, more digits)	Novel/Hresko et al. (2003)
Missing terms/algebra	Solve arithmetic problems with missing addends, minuends or subtrahends/ Solve problems involving algebraic concepts and procedures	Novel/Hresko et al. (2003)

(continued)

Table 5.1 (continued)

Skill	Description, kindergarten and third grade–sixth grade variants	Reference
Number line estimation	Estimate position of numbers on a line(K: (0–100, 3rd: 0–1000)/ Estimate position of numbers on a line (0–100,000)	Siegler and Opfer (2003); Booth and Siegler (2006)/Thompson and Opfer (2010)
Fractions (no K equivalent)	Answer comparison and calculation problems with fractions, Estimate numbers on a straight line with labelled endpoints (0–1)	Novel/Hresko et al. (2003)
Proportional reasoning (Wave 2 only)	Choose rectangle matching target in proportion (Distant Foils)/ Choose rectangle matching target in proportion (Close Foils)	Boyer and Levine (2012)
Fraction identification (added to fractions wave 2 only)	Select picture that matches a symbolic traction / No additional 6th grade items	Miura, Okamoto, Vlahovic-Stetic, Kim, and Han (1999); Paik and Mix (2003)

and mathematical domains are separate, but closely related, and this appears to be the case across the entire elementary school age period.

Cross-loading tasks. We also found that a few specific spatial tasks cross-loaded on the mathematical factor beyond the general connection, which changed over the course of the three grades. In kindergarten, mental rotation was significantly related to the mathematical factor, whereas in sixth grade, visuospatial working memory and visuo-motor integration took on a significant relation. These tasks in particular may have special significance in mathematics education at their respective years, should these relationships prove to be causal when tested in training studies. The relative loadings of each task to the general factor in each grade appear remarkably stable across development, which suggests that each task continues to rely on the same resources and processes over the course of development. The few spatial tasks that do show cross-loadings draw at first from dynamic processes (mental rotation) and later from more static processes (visuo-spatial working memory and visuo-motor integration), and in terms of previously identified spatial factors, from spatial visualization to those more associated with perception and working memory.

Open questions. While these factor analyses provided greater certainty of finding shared processing when we examine spatial or mathematical tasks in the same factor analysis, they raised many important questions that remain to be answered. One key question is how spatial skills can influence mathematical reasoning, by what processes or components? While we did find a few cross-loadings between the numeric and spatial factors on spatial and mathematical skills, most spatial skills did not cross-load to mathematical ability, except through the general factor relation. Still, measures like mental rotation in kindergarten, and visuospatial working memory and visuo-motor integration in sixth grade may be the best candidates for training, even taking the instability of the cross-loadings into account. By focusing our training efforts on those specific measures, we can begin to flesh out more causal models that provide a mechanism to the results observed in the correlational factor analysis (see Mix, Levine, Cheng, & Stockton, under review). In the next section, we focus on process models, which can help to identify what components of spatial skills could be important for mathematical reasoning.

Spatial Processes in Mathematics Achievement

The cross-loadings revealed by the factor analyses indicate that there is a link between mental rotation and mathematics in kindergarteners. Other research corroborates this relation as a fruitful connection to explore. For example, Gunderson, Ramirez, Levine, and Beilock (2012) showed that performance on a number line estimation task mediated the relation between performance on a spatial transformation task and an approximate calculation task. However, this finding leads to the questions about the specific processes involved in each of the tasks (number line estimation, spatial transformation, and approximate calculation). Which processes explain the connections seen in these studies? For example, spatial transformation relies on both being able to rotate objects mentally and to match features, either of which might explain its relation to number line estimation, which is itself multifaceted, requiring both an ordinal awareness of numerical magnitude and a mapping from numerical magnitude to spatial extent. Clarifying the specific relations between these measures is important because a well-specified mechanism is key to a well-designed and effective educational intervention.

There is also reason to move beyond the results of both the factor analyses and other previous studies that only analyze summary scores in order to build a theory about spatial-mathematical connections. Each of the measures tested in the previously described factor analyses was comprised of a complicated set of underlying processes and may have relied on integrating multiple orthogonal dimensions of difficulty (e.g., see Cheng, Mix, Reckase, Levine, & Freer, under review, regarding the automatic and deliberate elements of visuospatial working memory). Theorists often tend to assume that specific processes elicit a single, specific type of process, for instance that a mental rotation task reflects dynamic spatial visualization and not static, form perception, but is this actually the case? Moreover, even if it is, which of these processes is also important for young children's mathematical skills? Despite this, we know that we should take care not to reify the processes we measure with tasks. By modeling the underlying processes involved in a task we are better able to understand what proficiency on a task actually indicates. Our ultimate goal when examining spatial tasks for their underlying processes is to answer basic questions asked in developmental research: which spatial processes are causally related to the development of mathematical reasoning (Overton & Reese, 1973)?

Models of cognitive processes break down tasks into interacting components. These models have been used to understand performance and growth on a wide range of tasks, from remembering lists (Henson, 1998) to analogical reasoning (Doumas, Hummel, & Sandhofer, 2008), to simple arithmetic (Ashcraft, 1987). For example, Thompson, Ratcliff, and Mckoon (2016) used a diffusion model to compare children's and adults' symbolic and non-symbolic number discrimination. The model parameters showed differences in how adults and children, beyond simple performance differences between groups. That is, the diffusion model indicated that the reason adults could respond faster than children was because adults acquired enough information to decide whether magnitudes were different more quickly than children, regardless of whether those magnitudes were dots or numerals.

Mental rotation. The question of whether young children connect the dynamic transformation process used in a mental rotation to numerical tasks has been asked since at least since the 1970s. Marmor (1977) asked whether children who are better at mental rotation perform better on number conservation and Davidson (1987) asked whether children who are better at rotational displacement problems perform better on arithmetic problems. While correlations have been observed between tasks requiring mental rotation and a variety of mathematical tasks, it is not clear why this is the case. Below, we will use models of mental rotation to describe potential connections to mathematical concepts and procedures. We also review recent studies that attempt to determine whether training processes that underlie mental rotation performance are beneficial to mathematical reasoning.

A process view of mental rotation. It is useful to first describe the process typically assumed during mental rotation. Imagining the rotation of an object may feel intuitive but it is not so simple to verbally describe how it is done. Even the most general definition of the process involved, e.g., "MR involves transforming a representation held in visual short-term memory" (Provost & Heathcote, 2015) is not wholly uncontroversial, in that the speed, automaticity, and number of transformations are not specified. Mental rotation was hypothesized as a cognitive construct after Shepard and Metzler (1971) reported that the speed with which participants could determine if a three-dimensional block stimulus matched a target was a function of the angular disparity between the presented object and the target. Evidence suggests that participants intuitively rotate the object in the direction that requires the least amount of angle to match its target (Cooper & Shepard, 1973), which shows that participants can quickly identify their target in its typical orientation, and that they are not confused by similar objects (Corballis, 1988). Mental rotation tasks are often treated as if they reflect a "pure" ability to imagine rotations, but cognitive models reveal that mental rotation involves diverse subcomponents, which are relevant to our understanding of why mental rotation is related to mathematics.

Angular disparity. The best-fitting process model of behavioral data suggests participants actually engage in multiple, small but variable rotations in succession, almost as if they were grasping and turning an object until the participant reaches his or her limit of manual flexibility, then repeats (Provost & Heathcote, 2015). The analogy to actual manual rotation is supported by neuroimaging work that shows that the pattern of activation when participants enact mental rotation is similar to activity when participants actually move objects (Thayer & Johnson, 2006; Zacks, 2008). Models also suggest participants need increasing evidence as a function of the angle of disparity in order to make a decision (i.e., when the stimulus is rotated far from the target, participants need to gradually accumulate evidence about the stimuli's angle, causing them to take longer to be certain of their choice, Provost & Heathcote, 2015).

Different processes for complex and simple stimuli. There are also important differences among mental rotation tasks that reveal different processes at work beyond a purely "rotational" process. When comparisons must be made between stimuli that are more complex, reaction times are slower than between simple stimuli (Bethell-Fox & Shepard, 1988; Shepard & Metzler, 1988). Similarly, studies that include MR tasks with distractors find the most frequent incorrect choice in MR tasks is the choice of the mirror image of the correct choice rotated to the same degree as the correct choice (e.g., Kelley, Lee, & Wiley, 2000). Cognitive models account for this with a component that allows for confusability between the target and its mirror (e.g., confusing a "d" for a "b"), particularly when the stimuli are complex. These results and models suggest that all mental rotation stimuli are not equal, and that a separate process of "abstraction" of complex stimuli might need to occur within some mental rotation trials (Lovett & Schultheis, 2014). Thus, it is possible that abstraction, and not angular disparity, could be the source of connection between performance on mental rotation tasks and mathematics rather than the rotation process per se.

Not mental rotation at all. Participants' own descriptions of their strategies in mental rotation more frequently involve description of matching features of targets and choices, rather than mental rotation (Shepard, 1978). It is also clear that for a non-trivial number of trials of a mental rotation task, choices are made without engaging in mental rotation, particularly for 2D objects for which participants may use a fast flipping transformation (Cooper & Shepard, 1973; Kung & Hamm, 2010; Searle & Hamm, 2012). This suggests that the overall performance curve reflects a mixture of slower, rotational trials, and faster, non-rotational trials, overall resulting in the canonical bowed out curve that relates angular disparity to rotation speed (Searle & Hamm, 2017). Perhaps it is the quicker type of transformation, or the ability to pull out relevant feature of a spatial stimuli, which actually relate to mathematics, and not the angle-specific transformation.

Influence of mental rotation on mathematical reasoning. When we consider the form of the mental rotation task that was administered to children in kindergarten through third grade in the factor analytic studies (Mix et al., 2016), a number of potential processing models need to be considered. Participants were presented two-dimensional, scrambled alpha-numeric characters and were instructed to choose two of four stimuli that matched a target. The angles of rotation of the choices included both small and large angular disparities. Thus, it seems likely that non-rotational strategies might be available to children who engaged in the task, but also that both complexity of stimuli and angle might be critical to the relation to mathematical skill. Sixth grade children completed a three-dimensional mental rotation task with cube stimuli, which are potentially less solvable by non-rotational processes, and neither they nor third graders showed any specific cross-loading between mental rotation and mathematical ability.

In thinking about how mental rotation related process might more generally relate to early numeric processes and tasks, it is useful to examine where these connections have been observed. The majority of studies where mental rotation has been connected to mathematics skills were conducted with adults or older children—more comparable to the older children in our studies (third and sixth graders), who performed three-dimensional cube rotations. These other studies reported relations of performance on such mental rotation tasks to performance in broad areas of mathematics such as geometry (Battista, 1990; Delgado & Prieto, 2004; Kyttälä & Lehto, 2008), mental arithmetic (Kyttälä & Lehto, 2008; Reuhkala, 2001), problem solving (Hegarty & Kozhevnikov, 1999), and even to number sense in adults (Thompson et al., 2013). In addition, most of those studies included multiple other spatial measures and many of them were correlated with mathematical and verbal measures. Perhaps these correlations actually represent variation not specific to mental rotation skill or even solely related to spatial skill, rather reflecting a relation of mathematics performance to general intelligence or other domain general cognitive skills.

In the few studies that have focused on relations between mental rotation and mathematics in younger children or that have attempted to train mental rotation in order to improve mathematics, only a few have shown a connection (Kyttälä, Aunio, Lehto, Van Luit, & Hautamäki, 2003; Cheng & Mix, 2014; Lowrie, et al., 2017). It is interesting to note that the study that found a null relation (Carr, Steiner, Kyser, & Biddlecomb, 2008) used a 3-D measure of mental rotation, which models suggest would decrease the incidence of non-rotational responding such as the use feature matching strategies (Hawes, Moss, Caswell, & Poliszczuk, 2015; Xu & LeFevre, 2016). However, it is also possible that a 3D mental rotation task is too difficult for young children, and therefore suppresses individual differences (Neuburger, Jansen, Heil, & Quaiser-Pohl, 2011).

One possible explanation for the kindergarten connection between mental rotation and mathematics is that it reflects the ability to detect form or imagine transformations that could be useful for early mathematics concepts (e.g., better discriminating and encoding of numerical symbols; better imagining of transformations of quantities involved in arithmetic problems; a more easily visualized "mental number line" representation). This last connection was recently substantiated directly; adults who performed better at mental rotation had stronger spatial-numeric associations (Thompson et al., 2013). The subcomponents of mental rotation, both the recognition of parts of objects (affected by the complexity of those objects, as in a computational model of mental rotation, Lovett & Schultheis, 2014) and the process of mentally rotating those objects could be differently engaged throughout mathematical activities. Some processes involved in mental rotation and other spatial visualization type skills, whether they are the processes used in slow rotations or fast transformations, may be more important than others for mathematical reasoning, and this could vary depending on the particular mathematics problems being examined and the ages of the participants—a rich set of questions in need of further study.

Visuospatial working memory. Working memory is a construct that was initially proposed to address gaps in previous theories of memory (Baddeley & Hitch, 1974). Previous theories had suggested that for very recently activated information, short-term memory provided a place to hold in mind a small amount of information without rehearsal, but with considerable loss due to decay of information, before it was consolidated in long-term memory (Broadbent, 2018). In contrast, working memory models proposed buffer areas that come into play as a sort of way-station between perception and processing where information is selectively acted upon. A shared feature of models of working memory was the central executive function, which essentially coordinates all of the different functions that must be carried out (shifting attention, processing, storing, updating, and maintaining information) and some have characterized this model as requiring a sort homunculus rather than providing a real mechanism by which all of these functions are carried out (Wingfield, 2016).

Modality specificity of visuospatial working memory. Competing theories have contested how much of the processing of visuospatial working memory is just representation of sensory modality information that will become important later in processing. They have also questioned when and how information is retained, culled, and transformed, in what is generally described as executive function (Miyake et al., 2001; Smith & Jonides, 1999). Experimental evidence suggests that age has disparate effects on different working memory functions; shifting focus shows age-related improvement, whereas the number of errors due to interference or substitution of information does not change over development (Carriedo, Corral, Montoro, Herrero, & Rucián, 2016; Lendínez, Pelegrina, & Lechuga, 2015). A more detailed understanding of how working memory carries out processing of complex spatial stimuli is needed, particularly in how it handles uniquely spatial information, to fully explain its relation to mathematical processing, and how mathematical tasks impose spatial processing demands.

Insights from visuospatial working memory process models. Working memory is necessary for basic attentional processes, such as keeping the perceptual representation of a recent stimulus activated, and localized processing of different types of stimuli (e.g., verbal vs spatial, comprising movement and location information, vs object, comprising static images; Johnson et al., 2005). Meta-analysis of brain imaging studies showed working memory activation for spatial "where" content was handled by specific brain regions, as was verbal content, while object related activation was not consistently tied to particular brain regions (Nee et al., 2013), suggesting it is handled by many brain regions. Further, areas that are activated for spatial content tend to have differentiated functional roles, with one, the superior frontal sulcus, activating most strongly to refresh a location in memory rather than when perceiving location (Johnson et al., 2005). This suggests that different processing may occur more readily for certain spatial content, which further suggests that how one thinks about mathematical problems may depend on its spatial characteristics (e.g., the size of symbols in a mathematics problem affects an individual's answers and errors to that problem, Landy & Goldstone, 2010).

Visusospatial working memory as a measure of discarding irrelevant information. Models of working memory have shown that the function of visuospatial working memory is more dedicated to active maintenance, particularly removal of unneeded information, than to processing novel information (Ecker, Lewandowsky, & Oberauer, 2014). Further, while the efficiency and overall strength of memory is often measured by the number of items one can keep in mind, the model reveals that how many items we remember is unrelated to how quickly we can remove items from memory. This conforms with other models that suggest the removal of unwanted information, and visuospatial processing more generally, relates directly to "fluid intelligence," not because of shared processing but rather because both systems must update continuously to what are described as "top-down processing goals" (e.g., inferential reasoning, Shipstead, Harrison, & Engle, 2016). It is possible that VSWM is related to mathematics because, particularly for mathematics problems given to older children, information must be changed and relations among numerical variables are fluid (e.g., the identity of "x" in algebra, whether an operation involving a fraction represents an increase or decrease in magnitude).

Influence of visuospatial working memory on mathematics. In contrast to the specific relation of mental rotation and mathematics in kindergarteners, in sixth graders VSWM is specifically related to mathematics. As with mental rotation there are some basic implications that process models might make for the results of the cross-domain factor analysis. First, one limitation of our visuospatial working memory task was that it plausibly involved both spatial storage and processing and central executive functions associated with working memory or fluid intelligence. Our primary measure of working memory was a measure of the location of stimuli, which makes it likely that children were responding with the "where" pathway of working memory rather than the "what," object focused form of working memory. The iterative nature of the task, wherein children were required to respond to many trials of increasingly populated arrays, also suggests that we were not purely measuring the capacity of children's memory, and their ability to maintain items in memory, but also their ability to "actively forget" information from previous trials.

Among the studies that have connected working memory to specific numerical skills, a variety have found general connections to mathematical skills (Reuhkala, 2001; Szűcs, Devine, Soltesz, Nobes, & Gabriel, 2014; Kyttälä & Lehto, 2008; Bull, Espy, & Wiebe, 2008; Casey et al., 1995; Primi, Ferrão, & Almeida, 2010). Fortunately, several of these studies have specifically probed verbal and visuospatial memory, as well as some form of executive functioning, and have shown that VSWM is the construct at work. Studies that provide an account of which function(s) of working memory (e.g., attending, storage capacity, etc.) actually connect to numerical skills are rare (but see Dulaney, Vasilyeva, & O'Dwyer, 2015, showing storage and attention are related to mathematics achievement). One clue to a functional role to analysis and manipulation of specifically spatial information comes from a recent study; fourth graders who specifically fared poorly on mathematical problem-solving had poor spatial working memory, and could not access or envision spatial relations, but performed normally on visual working memory tasks (Passolunghi & Mammarella, 2010). One important venue of future research should look at what about mathematical problem solving in fourth through sixth grades relies on this VSWM.

Several theorists have posited that visuospatial processes facilitate learning numerical skills through specific routes. One suggests that VSWM enables more abstract or conceptual thought (e.g., Nath & Szücs, 2014) while another suggests it provides a resource that allows for more complex and useful numerical strategies (Foley, Vasilyeva, & Laski, 2017). By these accounts the additional storage and visuospatial analysis resources work over time to facilitate learning and improved performance. One possibility is that these resources allow one to first form mental models or to imagine more useful mental models of mathematical problems.

Other longitudinal work examining the relational between several components of working memory and mathematics achievement is consistent with our finding of a significant relation of VSWM to mathematics in sixth graders but not in kindergarteners or third graders. Specifically, Li and Geary (2013) observed no relation between any component of working memory and mathematics ability in first grade, but that those children who increased the most in their visuospatial memory from first to fifth grade scored significantly higher than their peers on measures of numerical operations, while other spatial measures were not significant predictors. These results provide longitudinal evidence of the increasing importance of visuospatial working memory, mirroring the cross-sectional findings that emerged through confirmatory factor analysis (Mix et al., 2017).

What Are the Educational Implications of Relations Between Mathematical and Spatial Skills?

Several studies have investigated whether training spatial skills can improve numeracy or, more generally, mathematical achievement. In this section, we review recent efforts to include spatial skills in educational settings and interventions to improve mathematical outcomes. We then describe general principles to improve mathematical education by incorporating spatial skill instruction.

Spatializing the Mathematics Curriculum

As shown by a meta-analysis carried out by Uttal et al. (2013), spatial skills are malleable. Moreover, training a particular spatial skill leads not only to improvement in that spatial skill but also to spatial skills more generally. However, training of specific spatial skills in order to improve specific mathematical skills sometimes works (e.g., Cheng & Mix, 2014; Nath & Szücs, 2014; Foley et al., 2017; Lowrie, Logan, and Ramful, 2017) and sometimes does not (e.g., Hawes, et al., 2013; Simons, et al., 2016; Xu & LeFevre, 2016). Further, teaching-specific spatial skills to students may present a tall order in actual educational settings, both because these skills are generally outside the bounds of required curricula, and because existing research does not support clear prescriptions about when specific kinds of spatial training would be beneficial.

A more recent effort to improve mathematics education is a more kitchen sink approach, where spatial skills are focused on more generally in their own right. This approach has proven successful at attenuating the effects of low spatial reasoning on mathematics performance in undergraduate students (Sorby, Casey, Veurink, & Dulaney, 2013) as well as providing a core of spatial skills that appear to be generally advantageous for success in the STEM disciplines (Sorby, 2009; Miller & Halpern, 2013). Another effective method of promoting spatial skills quickly and early has been to infuse this kind of thinking into play activities like block play (Casey et al., 2008). Evidence suggests that children's experience with basic spatial play activities like puzzles and blocks has early connections to performance on spatial tasks such as mental rotation (Levine, Ratliff, Huttenlocher, & Cannon, 2012) and to foundational mathematics concepts and practices (Verdine, Irwin, Golinkoff, & Hirsh-Pasek, 2014).

Classroom interventions that incorporate spatial skills training more generally have had some success in early education. Students in one school program that was provided with spatial training activities over the course of the school year showed substantial growth in spatial and mathematical domains (Bruce & Hawes, 2015). Providing 9- to 10-year-old students with weekly lessons that emphasized different aspects of working memory, including VSWM, was also effective in increasing students' visual perception abilities, span, and addition accuracy (Witt, 2011). A more integrated approach to including spatial skills in the classroom focuses on providing teachers with formative assessments, feedback, and professional development geared at making pre-kindergarten teachers aware of spatial skills and their connection to mathematical achievement, with promising preliminary results (Young, Raudenbush, Fraumeni, & Levine, 2017). We believe that these forms of early intervention, which help to get children's spatial and numerical skills on track early, are especially important to closing later gaps in achievement across STEM areas.

General Principles for Leveraging Spatial Skills to Improve Education

A number of studies provide evidence that spatial training is particularly useful when learning new content, as in the case when college students begin visualization intensive organic chemistry (Stieff, 2013). This was born out by the results of regressions analyses conducted after the exploratory factor analysis of space and mathematics. In all three grade levels the authors studied (K, 3rd, and 6th), the results suggested that spatial skills were more closely related to novel mathematical content (Mix et al., 2016). In addition, teaching using spatial tools, such as gesture, rich spatial language, diagrams, and spatial analogies, (Newcombe, 2010), as well as 3D manipulatives (Mix, 2010) has been shown to be helpful to student mathematics learning (e.g., Richland, Stigler, & Holyoak, 2012; Levine, Goldin-Meadow, Carlson & Hemani-Lopez, 2018). Further, those tools appear to be particularly effective in helping students understand difficult concepts and procedures when they are combined (e.g., spatial language and gesture, Congdon, Novack, Brooks, Hemani-Lopez, O'Keefe, Goldin-Meadow, 2017). By providing rich spatial information in multiple ways, educators can help students create a lexicon of spatial relations, terms, and connections to mathematics, which can be utilized again and again as the child encounters novel problems.

Conclusions

In this chapter, we have provided a review of the literature suggesting that spatial skills can be organized into factors and also divided along several meaningful dimensions. We argue that these divisions can help us to understand a set of skills that widely differ and should motivate further exploration of spatial processing. We also believe that more attention should be paid to the way that spatial skills differ in their connection to other cognitive abilities and in how malleable and easily trained they are. We have shown that children's numerical and spatial abilities are related at the level of shared underlying processes across development, yet remain functionally distinct at each time. We have argued for a more fine-grained, process oriented view of spatial numeric relations which does not reify cognitive constructs but breaks them down to search for mechanism. We argue that combining information gained from factor analyses (in this case showing the correlated, overlapping structure of spatial and mathematical skills) with methods and models from cognitive science highlights a way to uncover mechanisms and causal connections between basic processes and achievement. We also believe that these process accounts can be leveraged for educational gains. The research we have reviewed suggests that spatial skills hold promise as pathways by which numerical skills can be improved and mathematics achievement can be maximized.

References

Aiken, L. R., Jr. (1970). Attitudes toward mathematics. *Review of Educational Research, 40*(4), 551–596.

Armstrong, J. S., & Soelberg, P. (1968). On the interpretation of factor analysis. *Psychological Bulletin, 70*(5), 361.

Ashcraft, M. H. (1987). Children's knowledge of simple arithmetic: A developmental model and simulation. In *Formal methods in developmental psychology* (pp. 302–338). New York: Springer.

Atit, K., Shipley, T. F., & Tikoff, B. (2013). Twisting space: Are rigid and non-rigid mental transformations separate spatial skills? *Cognitive Processing, 14*(2), 163–173.

Baddeley, A. D., & Hitch, G. (1974). Working memory. *Psychology of Learning and Motivation, 8*, 47–89.

Baroody, A. J. (2003). *The development of adaptive expertise and flexibility: The integration of conceptual and procedural knowledge*. Mahwah, NJ: Erlbaum.

Barth, H., Beckmann, L., & Spelke, E. S. (2008). Nonsymbolic, approximate arithmetic in children: Abstract addition prior to instruction. *Developmental Psychology, 44*(5), 1466.

Battista, M. T. (1990). Spatial visualization and gender differences in high school geometry. *Journal for Research in Mathematics Education, 21*, 47–60.

Beery, K. E., & Beery, N. A. (2004). *The Beery-Buktenica developmental test of visual-motor integration: Administration, scoring, and teaching manual* (5th ed.). Minneapolis, MN: NCS Pearson.

Bethell-Fox, C. E., & Shepard, R. N. (1988). Mental rotation: Effects of stimulus complexity and familiarity. *Journal of Experimental Psychology: Human Perception and Performance, 14*(1), 12.

Bollen, K. A. (1989). A new incremental fit index for general structural equation models. *Sociological Methods & Research, 17*(3), 303–316.

Booth, J. L., & Siegler, R. S. (2006). Developmental and individual differences in pure numerical estimation. *Developmental Psychology, 42*(1), 189.

Boyer, T. W., & Levine, S. C. (2012). Child proportional scaling: Is 1/3= 2/6= 3/9= 4/12? *Journal of Experimental Child Psychology, 111*(3), 516–533.

Broadbent, D. E. (2018). Successive responses to simultaneous stimuli. Quarterly *Journal of Experimental Psychology, 8*(4), 145–152.

Browne, M. W. (2010). An overview of analytic rotation in exploratory factor analysis. *Multivariate Behavioral Research, 36*(1), 111–150.

Bruce, C. D., & Hawes, Z. (2015). The role of 2D and 3D mental rotation in mathematics for young children: What is it? Why does it matter? And what can we do about it? *ZDM, 47*(3), 331–343.

Bull, R., Espy, K. A., & Wiebe, S. A. (2008). Short-term memory, working memory, and executive functioning in preschoolers: Longitudinal predictors of mathematical achievement at age 7 years. *Developmental Neuropsychology, 33*(3), 205–228.

Carriedo, N., Corral, A., Montoro, P. R., Herrero, L., & Rucián, M. (2016). Development of the updating executive function: From 7-year-olds to young adults. *Developmental Psychology, 52*(4), 666.

Carr, M., Steiner, H. H., Kyser, B., & Biddlecomb, B. (2008). A comparison of predictors of early emerging gender differences in mathematics competency. *Learning and Individual Differences, 18*(1), 61–75.

Carroll, J. B. (1993). *Human cognitive abilities: A survey of factor-analytic studies*. Cambridge University Press.

Casey, M. B., Nuttall, R., Pezaris, E., & Benbow, C. P. (1995). The influence of spatial ability on gender differences in mathematics college entrance test scores across diverse samples. *Developmental Psychology, 31*(4), 697.

Casey, M. B., Nuttall, R. L., & Pezaris, E. (2001). Spatial-mechanical reasoning skills versus mathematics self-confidence as mediators of gender differences on mathematics subtests using cross-national gender-based items. *Journal for Research in Mathematics Education, 32*, 28–57.

Casey, B. M., Andrews, N., Schindler, H., Kersh, J. E., Samper, A., & Copley, J. (2008). The development of spatial skills through interventions involving block building activities. *Cognition and Instruction, 26*(3), 269–309.

Chatterjee, A. (2008, August). The neural organization of spatial thought and language. In *Seminars in speech and language* (Vol. 29, No. 3, pp. 226–238). Thieme Medical Publishers.

Cheng, Y. L., & Mix, K. S. (2014). Spatial training improves children's mathematics ability. *Journal of Cognition and Development, 15*(1), 2–11.

Cheng, Y.-L., Mix, K. S., Reckase, M. D., Levine, S. C., & Freer, D. (under review). The Dimensionality of Visuo- Spatial Working Memory and Arithmetic Computation in Third Grade Students. Cognitive Science.

Clark III, H. T. (1988). Cognitive Style and Competence in Mathematics Problem Solving.

Cliff, N., & Pennell, R. (1967). The influence of communality, factor strength, and loading size on the sampling characteristics of factor loadings. *Psychometrika, 32*(3), 309–326.

Coleman, R.H. (1960) *An analysis of certain components of mathematical ability and an attempt to predict mathematical achievement in a specific situation*. Doctoral dissertation, Indiana University, University Microfilms, Ann Arbor, MI (1956), p. 945 No. 17.

Common Core State Standards Initiative. (2010). *Common Core State Standards for Mathematics (CCSSM)*. Washington, DC: National Governors Association Center for Best Practices and the Council of Chief State School Officers.

Congdon, E. L., Novack, M. A., Brooks, N., Hemani-Lopez, N., O'Keefe, L., & Goldin-Meadow, S. (2017). Better together: Simultaneous presentation of speech and gesture in math instruction supports generalization and retention. *Learning and instruction, 50*, 65–74.

Cooper, L. A., & Shepard, R. N. (1973). Chronometric studies of the rotation of mental images. In Visual information processing (pp. 75–176).

Corballis, M. C. (1988). Recognition of disoriented shapes. *Psychological Review, 95*(1), 115.

Davidson, P. M. (1987). Early function concepts: Their development and relation to certain mathematical and logical abilities. *Child Development, 58*, 1542–1555.

Delgado, A. R., & Prieto, G. (2004). Cognitive mediators and sex-related differences in mathematics. *Intelligence, 32*(1), 25–32.

DeLoache, J. S., Uttal, D. H., & Pierroutsakos, S. L. (1998). The development of early symbolization: Educational implications. *Learning and Instruction, 8*(4), 325–339.

Doumas, L. A., Hummel, J. E., & Sandhofer, C. M. (2008). A theory of the discovery and predication of relational concepts. *Psychological Review, 115*(1), 1.

Dulaney, A., Vasilyeva, M., & O'Dwyer, L. (2015). Individual differences in cognitive resources and elementary school mathematics achievement: Examining the roles of storage and attention. *Learning and Individual Differences, 37*, 55–63.

Ecker, U. K., Lewandowsky, S., & Oberauer, K. (2014). Removal of information from working memory: A specific updating process. *Journal of Memory and Language, 74*, 77–90.

Edens, K. M., & Potter, E. F. (2013). An exploratory look at the relationships among math skills, motivational factors and activity choice. *Early Childhood Education Journal, 41*(3), 235–243.

Feigenson, L., Dehaene, S., & Spelke, E. (2004). Core systems of number. *Trends in Cognitive Sciences, 8*(7), 307–314.

Foley, A. E., Vasilyeva, M., & Laski, E. V. (2017). Children's use of decomposition strategies mediates the visuospatial memory and arithmetic accuracy relation. *British Journal of Developmental Psychology, 35*(2), 303–309.

Frick, A., Möhring, W., & Newcombe, N. S. (2014). Picturing perspectives: Development of perspective-taking abilities in 4-to 8-year-olds. *Frontiers in Psychology, 5*, 386.

Gerbing, D. W., & Hamilton, J. G. (1996). Viability of exploratory factor analysis as a precursor to confirmatory factor analysis. *Structural Equation Modeling: A Multidisciplinary Journal, 3*(1), 62–72.

Ginsburg, H. P., & Baroody, A. J. (2003). *Test of early mathematics ability*. Austin, TX: PRO-ED.

Goldin-Meadow, S., Cook, S. W., & Mitchell, Z. A. (2009). Gesturing gives children new ideas about math. *Psychological Science, 20*(3), 267–272.

Guilford, J. P. (1967). The nature of human intelligence. New York, NY, US: McGraw-Hill.

Gunderson, E. A., Ramirez, G., Levine, S. C., & Beilock, S. L. (2012). The role of parents and teachers in the development of gender-related math attitudes. *Sex Roles, 66*(3–4), 153–166.

Hair, J. F., Jr., Anderson, R. E., Tatham, R. L., & Black, W. C. (1995). *Multivariate data analysis with readings*. Upper Saddle River, NJ: Prentice Hall.

Hawes, Z., Moss, J., Caswell, B., & Poliszczuk, D. (2015). Effects of mental rotation training on children's spatial and mathematics performance: A randomized controlled study. *Trends in Neuroscience and Education, 4*(3), 60–68.
Hegarty, M., & Kozhevnikov, M. (1999). Types of visual–spatial representations and mathematical problem solving. *Journal of Educational Psychology, 91*(4), 684.
Hegarty, M., Montello, D. R., Richardson, A. E., Ishikawa, T., & Lovelace, K. (2006). Spatial abilities at different scales: Individual differences in aptitude-test performance and spatial-layout learning. *Intelligence, 34*(2), 151–176.
Hegarty, M., & Waller, D. (2004). A dissociation between mental rotation and perspective-taking spatial abilities. *Intelligence, 32*(2), 175–191.
Hegarty, M., & Waller, D. (2005). Individual differences in spatial abilities. In P. Shah & A. Miyake (Eds.), *The Cambridge handbook of visuospatial thinking* (pp. 121–169). Cambridge: Cambridge University Press.
Henson, R. N. (1998). Short-term memory for serial order: The start-end model. *Cognitive Psychology, 36*(2), 73–137.
Hresko, W. P., Schlieve, P. L., Herron, S. R., Swain, C., & Sherbenou, R. J. (2003). *Comprehensive mathematical abilities test*. Austin, TX: PRO-ED.
Huttenlocher, J., Newcombe, N., & Vasilyeva, M. (1999). Spatial scaling in young children. *Psychological Science, 10*(5), 393–398.
Huttenlocher, J., & Presson, C. C. (1973). Mental rotation and the perspective problem. *Cognitive Psychology, 4*(2), 277–299.
Johnson, M. K., Raye, C. L., Mitchell, K. J., Greene, E. J., Cunningham, W. A., & Sanislow, C. A. (2005). Using fMRI to investigate. *Cognitive, Affective, & Behavioral Neuroscience, 5*(3), 339–361.
Kaiser, H. F. (2016). The application of electronic computers to factor analysis. *Educational and Psychological Measurement, 20*(1), 141–151.
Kaminski, J. A., Sloutsky, V. M., & Heckler, A. F. (2009). Concrete instantiations of mathematics: A double-edged sword. *Journal for Research in Mathematics Education, 40*(2), 90–93.
Kaufman, A. S., & Kaufman, N. L. (1983). *Kaufman Assessment Battery for Children (K-ABC) administration and scoring manual*. Circle Pines, MN: American Guidance Service.
Kelley, T. D., Lee, F. J., & Wiley, P. (2000). *Developing an ACT-R model of spatial manipulation* (ARL Tech. Rep. No. 2179). Aberdeen, MD: Aberdeen Proving Ground.
Kline, W. E. (1960). *A synthesis of two factor analyses of intermediate algebra*. Doctoral dissertation, Princeton University, University Microfilms, Ann Arbor, MI.
Kozhevnikov, M., & Hegarty, M. (2001). A dissociation between object manipulation spatial ability and spatial orientation ability. *Memory & Cognition, 29*(5), 745–756.
Kozhevnikov, M., Kozhevnikov, M., Yu, C. J., & Blazhenkova, O. (2013). Creativity, visualization abilities, and visual cognitive style. *British Journal of Educational Psychology, 83*(2), 196–209.
Krakowski, M., Ratliff, K., Gomez, L. M., & Levine, S. (2010, June). Spatial intelligence and the research: Practice challenge. In *Proceedings of the 9th International Conference of the Learning Sciences* (Vol. 1, pp. 556–563). International Society of the Learning Sciences.
Kung, E., & Hamm, J. P. (2010). A model of rotated mirror/normal letter discriminations. *Memory & Cognition, 38*(2), 206–220.
Kyttälä, M., Aunio, P., Lehto, J. E., Van Luit, J., & Hautamäki, J. (2003). Visuospatial working memory and early numeracy. *Educational and Child Psychology, 20*(3), 65–76.
Kyttälä, M., & Lehto, J. E. (2008). Some factors underlying mathematical performance: The role of visuospatial working memory and non-verbal intelligence. *European Journal of Psychology of Education, 23*(1), 77–94.
Landy, D., & Goldstone, R. L. (2010). Proximity and precedence in arithmetic. *The Quarterly Journal of Experimental Psychology, 63*(10), 1953–1968.

Landy, D., Brookes, D., & Smout, R. (2011, January). Modeling abstract numeric relations using concrete notations. In Proceedings of the Annual Meeting of the Cognitive Science Society, (Vol. 33, No. 33).

Lendínez, C., Pelegrina, S., & Lechuga, M. T. (2015). Age differences in working memory updating: The role of interference, focus switching and substituting information. *Acta Psychologica, 157*, 106–113.

Levine, S. C., Ratliff, K. R., Huttenlocher, J., & Cannon, J. (2012). Early puzzle play: a predictor of preschoolers' spatial transformation skill. *Developmental Psychology, 48*(2), 530.

Levine, S. C., Foley, A., Lourenco, S., Ehrlich, S., & Ratliff, K. (2016). Sex differences in spatial cognition: Advancing the conversation. *Wiley Interdisciplinary Reviews: Cognitive Science, 7*(2), 127–155.

Levine, S. C., Goldin-Meadow, S., Carlson, M. T., & Hemani-Lopez, N. (2018). Mental transformation skill in young children: the role of concrete and abstract motor training. *Cognitive Science, 42*(4), 1207–1228.

Libertus, M. E., Feigenson, L., & Halberda, J. (2013). Numerical approximation abilities correlate with and predict informal but not formal mathematics abilities. *Journal of Experimental Child Psychology, 116*(4), 829–838.

Li, Y., & Geary, D. C. (2013). Developmental gains in visuospatial memory predict gains in mathematics achievement. *PLoS One, 8*(7), e70160.

Liben, L. S., & Downs, R. M. (1989). Understanding maps as symbols: The development of map concepts in children. *Advances in Child Development and Behavior, 22*, 145–201.

Linn, M. C., & Petersen, A. C. (1985). Emergence and characterization of sex differences in spatial ability: A meta-analysis. *Child Development, 56*, 1479–1498.

Little, R. J., & Rubin, D. B. (1989). The analysis of social science data with missing values. *Sociological Methods & Research, 18*(2–3), 292–326.

Lohman, D. F. (1988). Spatial abilities as traits, processes, and knowledge. In R. J. Sternberg (Ed.), Advances in the psychology of human intelligence (Vol. 4, pp. 181–248). Hillsdale, NJ: Erlbaum.

Lohman, D. F. (1994). Spatially gifted, verbally inconvenienced. In Talent development (Vol. 2, pp. 251–264).

Lovett, A., & Schultheis, H. (2014, January). Modeling spatial abstraction during mental rotation. In Proceedings of the Annual Meeting of the Cognitive Science Society (Vol. 36, No. 36).

Lowrie, T., Logan, T., & Ramful, A. (2017). Visuospatial training improves elementary students' mathematics performance. *British Journal of Educational Psychology, 87*(2), 170–186.

Lyons, I. M., Price, G. R., Vaessen, A., Blomert, L., & Ansari, D. (2014). Numerical predictors of arithmetic success in grades 1–6. *Developmental Science, 17*(5), 714–726.

MacCallum, R. C., Widaman, K. F., Zhang, S., & Hong, S. (1999). Sample size in factor analysis. *Psychological Methods, 4*, 84–99.

Markey, S. M. (2009). *The relationship between visual-spatial reasoning ability and math and geometry problem-solving*. American International College.

Marmor, G. S. (1977). Mental rotation and number conservation: Are they related? *Developmental Psychology, 13*(4), 320.

Matthews, P. G., Lewis, M. R., & Hubbard, E. M. (2016). Individual differences in nonsymbolic ratio processing predict symbolic math performance. *Psychological Science, 27*(2), 191–202.

Maul, A. (2013). Method effects and the meaning of measurement. *Frontiers in Psychology, 4*, 169.

Mazzocco, M. M., Feigenson, L., & Halberda, J. (2011). Preschoolers' precision of the approximate number system predicts later school mathematics performance. *PLoS One, 6*(9), e23749.

McGee, M. G. (1979). Human spatial abilities: Psychometric studies and environmental, genetic, hormonal, and neurological influences. *Psychological Bulletin, 86*(5), 889.

Michael, W. B., Guilford, J. P., Fruchter, B., & Zimmerman, W. S. (1957). The description of spatial-visualization abilities. *Educational and Psychological Measurement, 17*(2), 185–199.

Miller, D. I., & Halpern, D. F. (2013). Can spatial training improve long-term outcomes for gifted STEM undergraduates? *Learning and Individual Differences, 26*, 141–152.

Miura, I. T., Okamoto, Y., Vlahovic-Stetic, V., Kim, C. C., & Han, J. H. (1999). Language supports for children's understanding of numerical fractions: Cross-national comparisons. *Journal of Experimental Child Psychology, 74*(4), 356–365.
Mix, K. S. (2010). Spatial tools for mathematical thought. *Space and Language*, 41–66.
Mix, K. S., & Cheng, Y. L. (2011). The relation between space and math: Developmental and educational implications. *Advances in Child Development and Behavior, 42*, 197–243.
Mix, K. S., & Cheng, Y. L. (2012). The relation between space and math: Developmental and educational implications. In Advances in child development and behavior (Vol. 42, pp. 197–243). JAI.
Mix, K. S., Levine, S. C., Cheng, Y. L., Young, C., Hambrick, D. Z., Ping, R., & Konstantopoulos, S. (2016). Separate but correlated: The latent structure of space and mathematics across development. *Journal of Experimental Psychology: General, 145*(9), 1206.
Mix, K. S., Levine, S. C., Cheng, Y. L., Young, C. J., Hambrick, D. Z., & Konstantopoulos, S. (2017). The latent structure of spatial skills and mathematics: A replication of the two-factor model. *Journal of Cognition and Development, 18*(4), 465–492.
Mix, K. S., Hambrick, D. Z., Satyam, V. R., Burgoyne, A. P., & Levine, S. C. (2018). The latent structure of spatial skill: A test of the 2× 2 typology. *Cognition, 180*, 268–278.
Mix, K. S., Levine, S. C., Cheng, Y.-L. & Stockton, J. D. (under review). Does Spatial Training Improve Mathematics Performance? A Comparison of Training Type, Age, and Mathematics Outcome. Child Development
Miyake, A., Friedman, N. P., Rettinger, D. A., Shah, P., & Hegarty, M. (2001). How are visuospatial working memory, executive functioning, and spatial abilities related? A latent-variable analysis. *Journal of Experimental Psychology: General, 130*(4), 621.
Möhring, W., Newcombe, N. S., & Frick, A. (2015). The relation between spatial thinking and proportional reasoning in preschoolers. *Journal of experimental child psychology, 132*, 213–220.
Mundy, E., & Gilmore, C. K. (2009). Children's mapping between symbolic and nonsymbolic representations of number. *Journal of Experimental Child Psychology, 103*(4), 490–502.
Nath, S., & Szücs, D. (2014). Construction play and cognitive skills associated with the development of mathematical abilities in 7-year-old children. *Learning and Instruction, 32*, 73–80.
National Council of Teachers of Mathematics. (2000). *Principles and standards for school mathematics*. Reston, VA: National Council of Teachers of Mathematics.
National Research Council, and Geographical Sciences Committee. (2005). *Learning to think spatially*. National Academies Press.
Nee, D. E., Brown, J. W., Askren, M. K., Berman, M. G., Demiralp, E., Krawitz, A., & Jonides, J. (2013). A meta-analysis of executive components of working memory. *Cerebral Cortex, 23*(2), 264–282.
Neubauer, A. C., Bergner, S., & Schatz, M. (2010). Two-vs. three-dimensional presentation of mental rotation tasks: Sex differences and effects of training on performance and brain activation. *Intelligence, 38*(5), 529–539.
Neuburger, S., Jansen, P., Heil, M., & Quaiser-Pohl, C. (2011). Gender differences in pre-adolescents' mental-rotation performance: Do they depend on grade and stimulus type? *Personality and Individual Differences, 50*(8), 1238–1242.
Newcombe, N. S. (2010). Picture this: Increasing math and science learning by improving spatial thinking. *American Educator, 34*(2), 29.
Newcombe, N. S., & Shipley, T. F. (2015). Thinking about spatial thinking: New typology, new assessments. In J. S. Gero (Ed.), *Studying visual and spatial reasoning for design creativity* (pp. 179–192). Amsterdam, The Netherlands: Springer.
Overton, W. F., & Reese, H. W. (1973). Models of development: Methodological implications. In J. R. Nesselroade & H. W. Reese (Eds.), *Life-span developmental psychology: Methodological issues* (pp. 65–86). New York: Academic Press.
Paik, J. H., & Mix, K. S. (2003). US and Korean Children's Comprehension of Fraction Names: A reexamination of cross–national differences. *Child Development, 74*(1), 144–154.

Passolunghi, M. C., & Mammarella, I. C. (2010). Spatial and visual working memory ability in children with difficulties in arithmetic word problem solving. *European Journal of Cognitive Psychology, 22*(6), 944–963.

Peters, M., Laeng, B., Latham, K., Jackson, M., Zaiyouna, R., & Richardson, C. (1995). A redrawn Vandenberg and Kuse mental rotations test-different versions and factors that affect performance. *Brain and Cognition, 28*(1), 39–58.

Preacher, K. J., & MacCallum, R. C. (2002). Exploratory factor analysis in behavior genetics research: Factor recovery with small sample sizes. *Behavior Genetics, 32*, 153–161.

Primi, R., Ferrão, M. E., & Almeida, L. S. (2010). Fluid intelligence as a predictor of learning: A longitudinal multilevel approach applied to math. *Learning and Individual Differences, 20*(5), 446–451.

Provost, A., & Heathcote, A. (2015). Titrating decision processes in the mental rotation task. *Psychological Review, 122*(4), 735.

Raghubar, K. P., Barnes, M. A., & Hecht, S. A. (2010). Working memory and mathematics: A review of developmental, individual difference, and cognitive approaches. *Learning and Individual Differences, 20*(2), 110–122.

Reuhkala, M. (2001). Mathematical skills in ninth-graders: Relationship with visuo-spatial abilities and working memory. *Educational Psychology, 21*(4), 387–399.

Richland, L. E., Stigler, J. W., & Holyoak, K. J. (2012). Teaching the conceptual structure of mathematics. *Educational Psychologist, 47*(3), 189–203.

Rips, L. J., Bloomfield, A., & Asmuth, J. (2008). From numerical concepts to concepts of number. *Behavioral and brain sciences, 31*(6), 623–642.

Rittle-Johnson, B., Siegler, R. S., & Alibali, M. W. (2001). Developing conceptual understanding and procedural skill in mathematics: An iterative process. *Journal of Educational Psychology, 93*, 346–362. https://doi.org/10.1037//0022-0663.93.2.346

Rittle-Johnson, B., Fyfe, E. R., Hofer, K. G., & Farran, D. C. (2016). Early Math Trajectories: Low-Income Children's Mathematics Knowledge From Ages 4 to 11. *Child Development, 88*(5), 1727–1742.

Rousselle, L., & Noël, M. P. (2007). Basic numerical skills in children with mathematics learning disabilities: A comparison of symbolic vs non-symbolic number magnitude processing. *Cognition, 102*(3), 361–395.

Rummel, R. J. (1970). Applied factor Analysis. Evanston, IL: Northwestern University Press.

Rusch, C. E. (1957). *An analysis of arithmetic achievement in grades four, six, and eight*. Doctoral dissertation, University of Wisconsin, University Microfilms, Ann Arbor, MI.

Sarnecka, B. W., & Carey, S. (2008). How counting represents number: What children must learn and when they learn it. *Cognition, 108*(3), 662–674.

Searle, J. A., & Hamm, J. P. (2012). Individual differences in the mixture ratio of rotation and nonrotation trials during rotated mirror/normal letter discriminations. *Memory & Cognition, 40*(4), 594–613.

Searle, J. A., & Hamm, J. P. (2017). Mental rotation: An examination of assumptions. *Wiley Interdisciplinary Reviews: Cognitive Science, 8*(6). https://doi.org/10.1002/wcs.1443

Schneider, M., Beeres, K., Coban, L., Merz, S., Susan Schmidt, S., Stricker, J., & De Smedt, B. (2016). Associations of non-symbolic and symbolic numerical magnitude processing with mathematical competence: A meta-analysis. *Developmental Science, 20*(3).

Schoenfeld, A. H. (2014). *Mathematical problem solving*. Elsevier.

Shepard, R. N. (1978). The mental image. *American Psychologist, 33*(2), 125.

Shepard, R. N., & Metzler, J. (1971). Mental rotation of three-dimensional objects. *Science, 171*(3972), 701–703.

Shepard, S., & Metzler, D. (1988). Mental rotation: Effects of dimensionality of objects and type of task. *Journal of Experimental Psychology: Human Perception and Performance, 14*(1), 3.

Shipstead, Z., Harrison, T. L., & Engle, R. W. (2016). Working memory capacity and fluid intelligence maintenance and disengagement. *Perspectives on Psychological Science, 11*(6), 771–799.

Siegler, R. S. (1999). Strategic development. *Trends in Cognitive Sciences, 3*(11), 430–435.

Siegler, R. S., & Opfer, J. E. (2003). The development of numerical estimation: Evidence for multiple representations of numerical quantity. *Psychological Science, 14*(3), 237–250.

Simons, D. J., Boot, W. R., Charness, N., Gathercole, S. E., Chabris, C. F., Hambrick, D. Z., & Stine-Morrow, E. A. (2016). Do "brain-training" programs work? *Psychological Science in the Public Interest, 17*(3), 103–186.

Skemp, R. R. (1961). Reflective intelligence and mathematics. *British Journal of Educational Psychology, 31*(P1), 45–55.

Slusser, E. B., Santiago, R. T., & Barth, H. C. (2013). Developmental change in numerical estimation. *Journal of Experimental Psychology: General, 142*(1), 193.

Smith, E. E., & Jonides, J. (1999). Storage and executive processes in the frontal lobes. *Science, 283*(5408), 1657–1661.

Sorby, S. A. (2009). Educational research in developing 3-D spatial skills for engineering students. *International Journal of Science Education, 31*(3), 459–480.

Sorby, S., Casey, B., Veurink, N., & Dulaney, A. (2013). The role of spatial training in improving spatial and calculus performance in engineering students. *Learning and Individual Differences, 26*, 20–29.

Spearman, C. (1927). *The abilities of man*. Oxford: Macmillan.

Star, J. R. (2005). Reconceptualizing procedural knowledge. *Journal for Research in Mathematics Education, 36*(5), 404–411.

Stieff, M. (2012). Sex differences in the mental rotation of chemistry representations. *Journal of Chemical Education, 90*(2), 165–170.

Szűcs, D., Devine, A., Soltesz, F., Nobes, A., & Gabriel, F. (2014). Cognitive components of a mathematical processing network in 9-year-old children. *Developmental Science, 17*(4), 506–524.

Thayer, Z. C., & Johnson, B. W. (2006). Cerebral processes during visuo-motor imagery of hands. *Psychophysiology, 43*(4), 401–412.

Thompson, J. M., Nuerk, H. C., Moeller, K., & Kadosh, R. C. (2013). The link between mental rotation ability and basic numerical representations. *Acta Psychologica, 144*(2), 324–331.

Thompson, C. A., & Opfer, J. E. (2010). How 15 hundred is like 15 cherries: Effect of progressive alignment on representational changes in numerical cognition. *Child Development, 81*(6), 1768–1786.

Thompson, C. A., Ratcliff, R., & Mckoon, G. (2016). Individual differences in the components of children's and adults' information processing for simple symbolic and non-symbolic numeric decisions. *Journal of Experimental Child Psychology, 150*, 48–71.

Thurstone, L. L. (1933). *A simplified multiple factor method and an outline of the computations*. Chicago: University of Chicago Bookstore.

Thurstone, L. L. (1938). *Primary mental abilities*. Chicago: University of Chicago Press.

Tomarken, A. J., & Waller, N. G. (2005). Structural equation modeling: Strengths, limitations, and misconceptions. *Annual Review of Clinical Psychology, 1*(1), 31–65.

Tucker, L. R., & Lewis, C. (1973). A reliability coefficient for maximum likelihood factor analysis. *Psychometrika, 38*(1), 1–10.

Uttal, D. H., Meadow, N. G., Tipton, E., Hand, L. L., Alden, A. R., Warren, C., & Newcombe, N. S. (2013). The malleability of spatial skills: A meta-analysis of training studies. *Psychological Bulletin, 139*(2), 352–402.

Verdine, B. N., Irwin, C. M., Golinkoff, R. M., & Hirsh-Pasek, K. (2014). Contributions of executive function and spatial skills to preschool mathematics achievement. *Journal of Experimental Child Psychology, 126*, 37–51.

Voyer, D., Voyer, S., & Bryden, M. P. (1995). Magnitude of sex differences in spatial abilities: a meta-analysis and consideration of critical variables. *Psychological bulletin, 117*(2), 250.

Wai, J., Lubinski, D., & Benbow, C. P. (2009). Spatial ability for STEM domains: Aligning over 50 years of cumulative psychological knowledge solidifies its importance. *Journal of Educational Psychology, 101*(4), 817.

Wechsler D, Kaplan E, Fein D, Kramer J, Morris R, Delis D, Maerlender A. (2004). WISC-IV Technical and Interpretative Manual. San Antonio, TX: NCS Pearson.

Werdelin, I. (1966). A synthesis of two factor analyses of problem solving in mathematics. *Didakometry, 8*, 14.

Wingfield, A. (2016). Evolution of models of working memory and cognitive resources. *Ear and Hearing, 37*, 35S–43S.

Witt, M. (2011). School based working memory training: Preliminary finding of improvement in children's mathematical performance. *Advances in Cognitive Psychology, 7*, 7–15.

Xu, C., & LeFevre, J. A. (2016). Training young children on sequential relations among numbers and spatial decomposition: Differential transfer to number line and mental transformation tasks. *Developmental Psychology, 52*(6), 854.

Young, C., Raudenbush, S., Fraumeni, B., & Levine, S. (August, 2017). The Structure of Young Children's Numerical and Spatial Abilities. Poster presented at the Annual Meeting of the Cognitive Science Society, London, UK.

Zacks, J. M. (2008). Neuroimaging studies of mental rotation: A meta-analysis and review. *Journal of Cognitive Neuroscience, 20*(1), 1–19.

Part II
Commentaries

Chapter 6
Part I Commentary 1: Deepening the Analysis of Students' Reasoning About Length

Michael T. Battista, Leah M. Frazee, and Michael L. Winer

To highlight mutually beneficial intersections between research in psychology and mathematics education, in this commentary, we connect our measurement research to that of Congdon, Vasilyeva, Mix, and Levine (2018). We illustrate how our qualitative investigation of measurement reasoning can elaborate, deepen, and introduce additional perspectives and insights into the research. We discuss three points of intersection: non-measurement reasoning as elaboration of intuition and a bridge to measurement reasoning; understanding and misunderstanding of rulers; and what students actually count in their attempts at length iterations. Our research also extends some of the ideas from early childhood to elementary school.

Non-measurement Reasoning as Elaboration of Intuition and Bridge to Measurement Reasoning

Congdon, et al. (2018) state, "One reason that measurement may prove difficult for young children is that it requires them to integrate their preexisting imprecise intuitions about quantity and continuous extent with conventional, number-based measurement tools such as rulers" (p. 31). Battista's (2012) Cognition Based

M. T. Battista (✉)
Department of Teaching and Learning, The Ohio State University, Columbus, OH, USA
e-mail: battista.23@osu.edu

L. M. Frazee
Department of Mathematical Sciences, Central Connecticut State University, New Britain, CT, USA
e-mail: frazee@ccsu.edu

M. L. Winer
Department of Mathematics, Eastern Washington University, Cheney, WA, USA
e-mail: mwiner@ewu.edu

K. S. S. Mix, M. T. Battista (eds.), *Visualizing Mathematics*, Research in Mathematics Education, https://doi.org/10.1007/978-3-319-98767-5_6

Assessment (CBA) measurement research is consistent with Congdon et al.'s hypothesis and indicates that (a) imprecise intuitive reasoning about measurement can become more sophisticated, (b) there are several paths from intuitive to formal measurement reasoning, and (c) there are difficulties with measurement reasoning beyond using rulers.

To illustrate, Battista's (2012) learning progression for geometric measurement distinguishes non-measurement (visual-intuitive) from measurement reasoning (Table 6.1). *Non-measurement reasoning does not use numbers.* It uses visual, holistic, intuitive, or vague judgments; direct and indirect comparisons; imagined motions; decomposition; or deductive inferences based on geometric properties. *Measurement reasoning involves using numbers to indicate how many measurement-units are contained in an object.* Measurement reasoning involves measurement-unit enumeration and numerical operation on measurement numbers. As will be illustrated by the episodes below, the same student often uses both non-measurement and measurement reasoning.

One component of CBA length research (59 students, grades 1–5) included five tasks in which non-measurement and measurement reasoning were both clearly

Table 6.1 Outline of Battista's Length Learning Progression

Level	Non-measurement reasoning about length: Students	Level	Measurement reasoning about length: Students
NM0	compare objects' lengths in vague visual ways	**M0**	use numbers in ways unconnected to appropriate length-unit iteration
NM1	correctly compare objects' lengths directly or indirectly	**M1**	incorrectly iterate length-units
NM1.1	*compare objects' lengths directly*	*M1.1*	*iterate non-length units to get incorrect length*
NM1.2	*use third object to compare objects' lengths indirectly*	*M1.2*	*iterate length-units but get incorrect length*
NM2	compare objects' lengths by systematically manipulating or matching their parts	**M2**	correctly iterate length-units
NM2.1	*rearrange parts to compare objects' lengths*	*M2.1*	*iterate non-length units to get correct length for straight paths*
NM2.2	*match parts one-to-one to compare objects' lengths*	*M2.2*	*iterate non-length units to get correct length for non-straight paths*
		M2.3	*iterate length-units to get correct length for straight and non-straight paths*
		M3	correctly operate on composites of length-units
		M4	correctly and meaningfully operate on length using only numbers (no visible units or iteration)
NM3	compare objects' lengths using geometric properties or transformations	**M5**	understand and use procedures and formulas for the perimeter of polygons

implementable, neither was suggested by the task statement, and using measurement reasoning was the best strategy. Although there was some shift, primary to intermediate grades, from non-measurement to measurement reasoning, measurement was used only 38% of the time, and appropriate use of measurement reasoning (Level M2) was very low (14%).

Connections Between Non-measurement and Measurement Reasoning

There were several ways that students connected non-measurement and measurement reasoning. RC, like many students, used Level NM2.1 reasoning to straighten then directly compare path lengths.

I: Tony the turtle has to walk on one of these paths to get his food. He wants to take the shortest path. Which path is shorter, or are they the same [Fig. 6.1]?

[RC (Grade 5) counts 5 segments for each path then 6 dots for each path, but concludes that the bottom path is shorter, "because [top path]—you've got the little triangle—if you pull that out, it'll be like right there [motioning horizontally outwards along top path]. You can't pull this [bottom path] out anymore."

RC: [Places a line of cylindrical rods held together by a wire horizontally straight over the top path, and marks a rod past its end—point X in figure.] This one [top] is longer. This one [bottom] is shorter.

RC, like numerous students on numerous problems, compared lengths by counting unequal line segments. However, despite the fact that RC counted the same number of segments in each path, to judge the path lengths, he relied on intuitive Level NM2.1 straightening reasoning to correctly decide that the bottom path is shorter.

For several previous problems, RC wrote sequences of numbers on each segment, starting at zero, then either added the end numbers for each segment, or

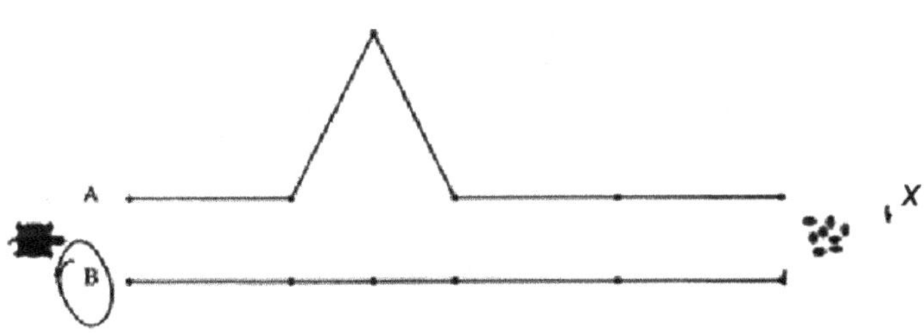

Fig. 6.1 Student RC Problem 1

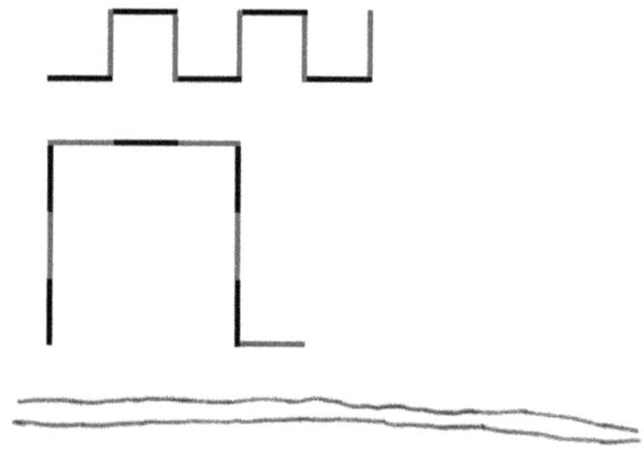

Fig. 6.2 Student RC Problem 2

added all the numbers. So the interviewer tried to help RC move beyond this Level M0 incorrect numerical reasoning by connecting measurement to non-measurement reasoning.

I: [Showing two sets of cylindrical inch rods with holes in them with wire threaded through] These are pictures of these wires [Fig. 6.2]. Which wire is longer, or are they the same length?...

[RC draws horizontal segments one at a time for both wires, and concludes "They're the same length."]

I: Is there any way that counting... might help you? Like how many lines were there?

RC: [After counting 10 segments on each path 1–10.] They're the same length. [After straightening the wires and laying them side-by-side] They're the same length.

RC's final statement suggests that he needed to re-justify his "same-length" counting conclusion by using the non-measurement strategy of straightening and directly comparing (Level NM 2.1). Subsequently, RC did several similar problems in which he first counted unit segments, then verified his correct conclusions using inch rods to straighten the wires. Later, RC used only counting. However, on a different kind of problem, RC counted unequal segments, and his connection between non-measurement straightening and measurement counting seemed to break.

The following examples illustrate other ways in which students' non-measurement and measurement reasoning interact. In Fig. 6.3, SL combined counting and straightening but made a visualization error in straightening that led to counting errors.

SL(Grade 5): This one [bottom wire] I counted like 1–8. This one [top wire] has 1–7. ... So this one's [bottom] bigger. It has 8.

In Fig. 6.4, AW used unsophisticated non-measurement reasoning to overrule his correct measurement reasoning.

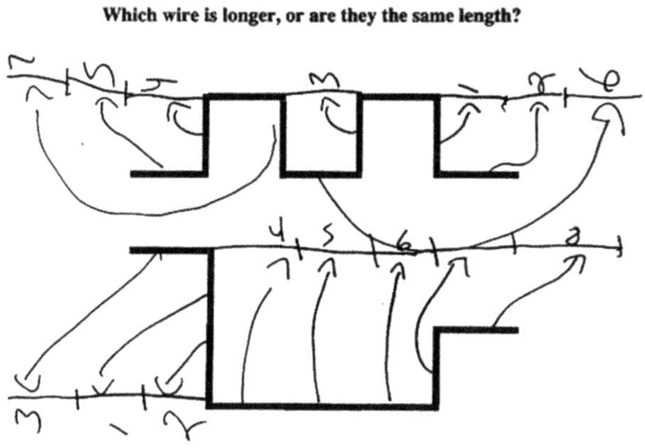

Fig. 6.3 Student SL Problem 1 [alternating length-units were black and gray]

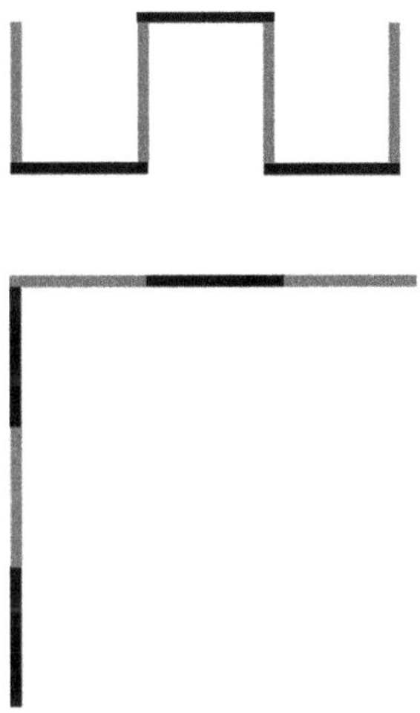

Fig. 6.4 Student AW Problem 1

I: *Suppose I pull the wires so they are straight. Which wire would be longer, or would they be the same?*

[AW(Grade 1) says the bottom wire is longer. After AW and the interviewer place inch rods on the two wires, AW correctly counts 7 rods on the top and 6 rods on the bottom and concludes that the top wire is longer, but then changes his mind.]

AW: *No, this one [bottom]... since it goes down...*

[AW and the interviewer rearrange the rods from each wire into straight lines.]

AW: *... It's actually the first [top] one! ... Because it goes [pointing to the right end of the top rearranged rods] kind of off the paper. And this one doesn't [pointing to the right end of the bottom row of rearranged rods].*

Importantly, AW did not abandon his unsophisticated non-measurement reasoning (Level NM 0) until he implemented the straightening process (Level NM 2.1).

For AW, non-measurement reasoning was primary, so it was only after straightening and comparing wires directly that he believed the top 7-rod wire was longer than the bottom 6-rod wire. In the next problem (Fig. 6.5), however, AW continued to struggle to reconcile non-measurement and measurement reasoning.

I: Which of these two wires would be longer if you straighten them? Or are they the same length? ...

[AW says they are "about the same" because the left and right endpoints are vertically aligned., "the same." ... AW and the interviewer place inch rods on top of each wire picture.]

I: When we straighten those wires out, which one will be longer?

AW: [Counts 9 rods on the top wire and 8 on the bottom wire] But I think they're still the same.

[AW and the interviewer straighten the rods as shown in Fig. 6.6, and AW concludes that the top wire is longer, indicating surprise.]

I: So what do you think?

AW: It's still that one [pointing to the top wire in Fig. 6.6], because that one, well, ... if you add one more of those [pointing to the missing rod in the bottom wire], then they would be the same. But this one just has (counting 1–8) 8. This one has (counting 1–9) 9.

I: Which one's longer?

AW: [Pointing to the top wire in Fig. 6.6] This one.

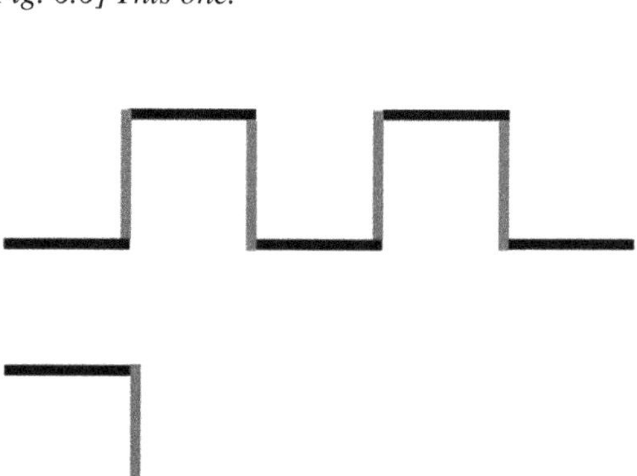

Fig. 6.5 Student AW Problem 2

Fig. 6.6 Student AW use of inch rods to straighten the two wires for Problem 2

AW reconciled his non-measurement and measurement reasoning only when he recognized how the count of one more rod for the top wire was connected to its extra extent in the straightened wires.

Congdon et al. (2018) propose that formal understanding of geometric measurement requires children to integrate their intuitive understandings of continuous spatial extent with counting discrete, countable entities. CBA research suggests one way this can happen. Children's initial measurement concepts seem based on abstraction of the physical processes of direct comparison and straightening, which can help them create a valid concept of length of non-straight paths, and with appropriate guidance, can help them understand the validity of comparing lengths using unit-iteration. Before students understand the connection between straightening and unit-iteration, they seem to have difficulty conceptually connect unit-counting and judging length (and subsequently developing an understanding of the properties of measurement).

Understanding and Misunderstanding Rulers

Congdon et al. (2018) suggest that even when students measure correctly, they may not grasp unit lengths' linear extents, and they may believe that they are being asked to find the right endpoint of an object on a ruler, irrespective of where the object begins. They also argue that students are much better at counting discrete units than segmenting continuous quantities into discrete units. We examine several CBA students' reasoning on ruler tasks to gain additional insight into these issues. And we note that even though CBA questions were asked in various ways, we observed similar types of reasoning throughout students' work. So it may not be how the questions are asked, but perhaps how previous instruction provided answers to the questions.

Task. Use the ruler to find the length of the black stick (Fig. 6.7).

TMM (Grade 2): It's 7, because it stops right at the 7. ... [To check, places inch rods from left to right, one at a time, on the black stick.]...
I: How many?
TMM: [Counting rods by ones] 1, 2, 3, 4, 5, 6.
I: Does that change your mind about how long the black stick is?
TMM: No.
I: No, it's still 7?

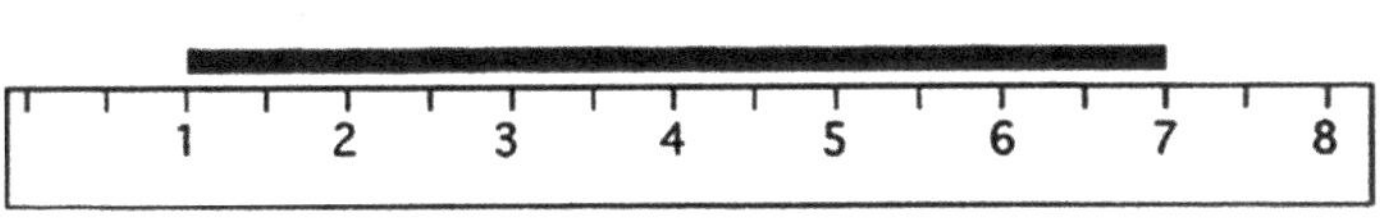

Fig. 6.7 Student TMM Problem 1

TMM: It's 6 of these long [pointing to the rods]. And it stopped at the 7.
I: So it's 7 in. long?
TMM: Yeah.
I: But it takes 6 of these inch rods to make it?
TMM: Yeah.

This problem indicates that TMM's reading of ruler lengths does not signify to him unit-length iteration. After counting the number of inch rods in the black rod, which differs from his length reading, he still maintained that the length was 7. For TMM, length was something you determine by reading a numeral on a ruler. So absent was his connection between numbers on the ruler and unit-length iteration that TMM did not see the incompatibility of his two responses of 6 and 7. The inconsistent answers to the question of "how long?", as measured by the ruler, and the question of "how many unit-lengths to make the stick?" was quite common among students.

Additional insight into students' understanding of rulers is provided by tasks in which they had to judge the validity of other students' reasoning.

AW: Right. ... Because he said it was 7. And that starts out at 1 and stops at 7 (Fig. 6.8).

AW: Wrong. ... Because he didn't start where 1 was (Fig. 6.9).

AW: Wrong. Because it goes to 8 [pointing at the 8- and the 7-in. marks as he names them] and it doesn't go to 7 (Fig. 6.10).

Because students count sets of objects starting at 1, many believe that if a stick starts at 1, the right endpoint number indicates its length.

The next episode delves more deeply into AW's ruler reasoning, exposing its inconsistency.

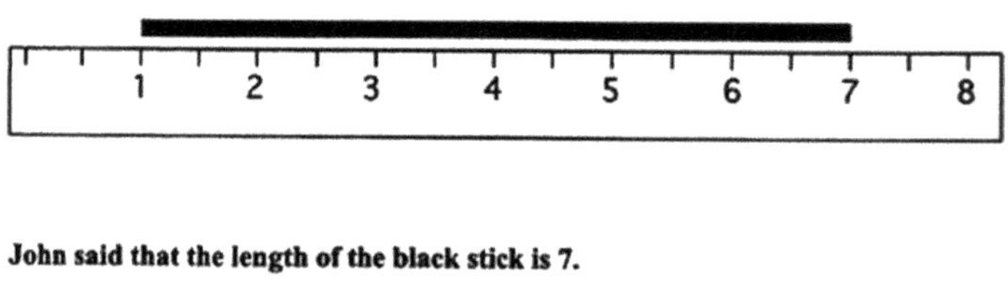

Fig. 6.8 Student AW Problem 4

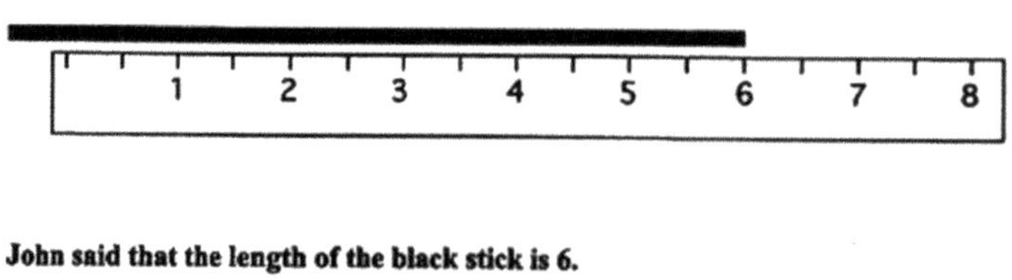

Fig. 6.9 Student AW Problem 5

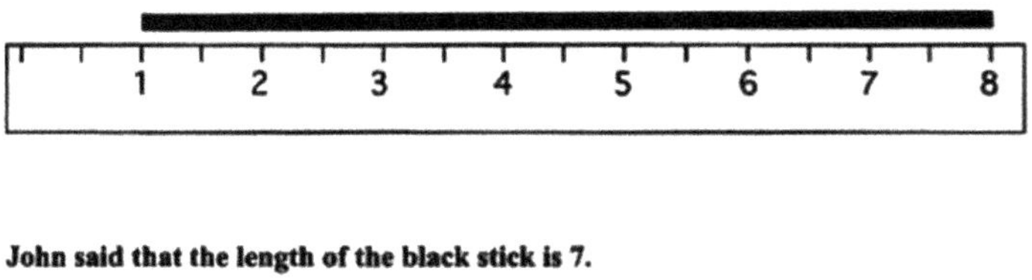

Fig. 6.10 Student AW Problem 3

Fig. 6.11 Student AW Problem 3

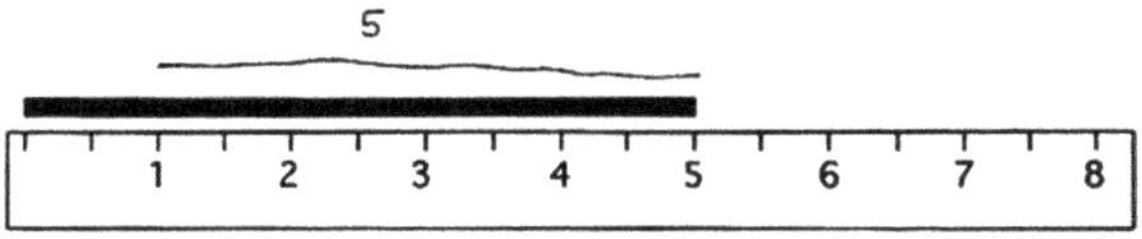

I: Use the ruler to find the length of the black stick (Fig. 6.11).

AW: [Traces stick, left to right, stopping at its right endpoint] 5. ...
I: How did you decide it was 5?
AW: Because it started right here [ruler left end] and it went to 5. ...
I: Suppose I drew a stick from 1 out to 5 [drawing a segment that starts above the 1 and ends above the 5]. How long would that stick be? What would we say the length of that stick is?
AW: 5.
I: Okay, so this one would be 5, too? And how did you figure out that one's 5?
AW: Because you started right here [left endpoint of interviewer's segment, then tracing to its right endpoint]. And it went to 5.

This episode shows even more profoundly the disconnection that can exist between ruler reading and linear extent. Even though AW evidenced emerging understanding between unit-iteration and linear extent in Figs. 6.4 and 6.5, he seemed to abandon this valid reasoning in the context of ruler reading. Reading right-endpoint numbers on rulers seems to overpower students, probably because of the authority of the curriculum. Although teachers and curricula correctly show students how to measure, many students misinterpret what they see.

What Students Count

Congdon et al. (2018) suggest that children are biased to "count discrete physical entities that are readily labeled" and "to estimate continuous quantities based on perceptual spatial cues alone even when salient, helpful discrete cues are present."

We wonder if it is what is labeled or what is visually and discretely salient (or a combination) that attracts students' attention. Below, we discuss data that provides additional insight on this issue.

CBA research extends our understanding of how students' counting for length measurement is abstracted and conceptualized. For instance, many students determine lengths by counting dots or hash marks. But students' conceptualizations of dot counting are varied. Many students count dots because they believe they should count something, so they count the most visually salient visual items, or they mimic, without conceptually understanding, what others count. Some students count dots because they think the dots are somehow connected to unit segments, but they do not know how, or they once understood but have forgotten the connection. Indeed, this dot-as-unit-length-indicator representation can be very tenuous for students, as shown in Fig. 6.12.

[KG(Grade 5) recognizes immediately that one dot-to-dot segment in the top leg of Fig. 6.12b is equivalent to 2 dot-to-dot segments in the top leg in Fig. 6.12a.]

KG: So that'd be [tracing and counting 2 dot-to-dot segments at a time on the top leg in Fig. 6.12c] 1, 2, 3, 4, 5, 6, 7, 8, 9. And that's [sweeping along the first segment, then touching dots on the top leg in Fig. 6.12d] 1, 2, 3, 4, 5, 6, 7, 8, 9.

[For the bottom leg in Fig. 6.12a KG said (see Fig. 6.12e), "1" as he swept to the right-side dot of the first segment; swept across the second segment, saying "2"

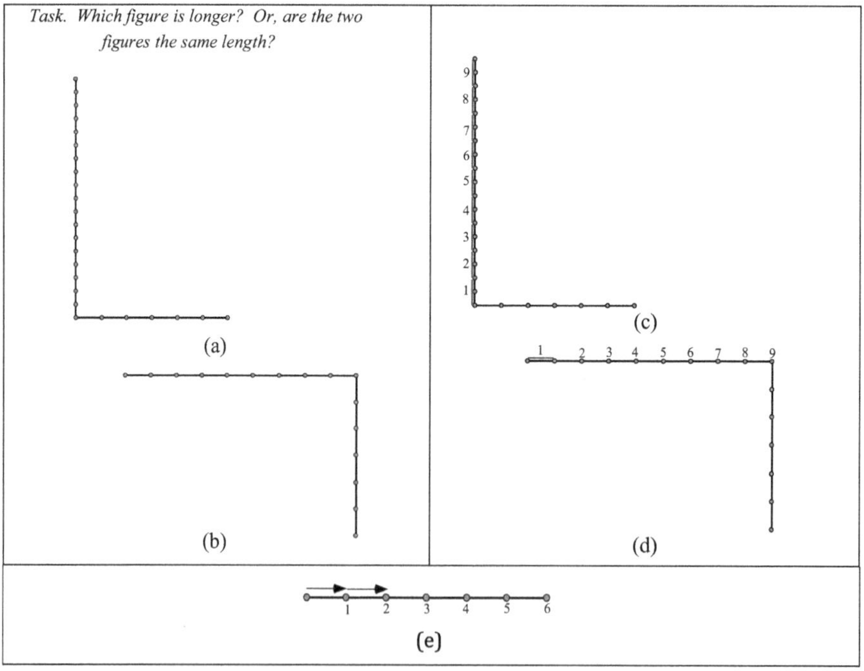

Fig. 6.12 Student KG Problem 1

as he got to the dot on the right side of this segment; then said "3" as he got to the right dot for the next segment (with no sweeping motion), and then continued to count dots.]

KG: And [pointing to each dot but the top one on the lower leg in Fig. 6.12b, going downward] 1, 2, 3, 4, 5, 6...

Wait. [Counts dots on the lower leg in Fig. 6.12a 1–7, then the lower leg in Fig. 6.12b 1–7] So they're equal. ... [Counting 2-segment spans on the top leg in Fig. 6.12c] 1, 2, 3, 4, 5, 6, 7, 8, 9. And there's [counting dots on top leg in Fig. 6.12b] 1, 2, 3, 4, 5, 6, 7, 8, 9, 10 here. So this one [top leg in Fig. 6.12b] is longer.

Several times, KG swept across a segment between dots, left-to-right, then said a counting word as he reached the right-endpoint. The switch from sweeping motions to just pointing at dots is a strong indication that KG initially curtailed unit-length sweeps to dot counting, thereafter using dots as representations of unit-lengths. However, immediately after KG shifted from sweeping-motions to dot-counting, on other segments, he counted just dots, but included the first dot in his count, losing the connection to correct unit-length iteration.

KG's episode is fascinating. First, KG's sweeping is consistent with Steffe's contention that the length property of segments arises from abstracting scanning motions over segments, their endpoints, and the duration of the motions (Battista, 2006; Steffe, 2010). Second, it seems plausible that initially KG had abstracted and connected three things—the sweeping motions over the length segments, the stopping actions on the visually salient segment-ending dots, and the count-words he uttered when he reached end-dots. But when he curtailed the sweeping motions to dot-counting only, perhaps because he did many more dot-counts than sweeps, the abstraction that he retained was of dot-counting, with only vague connections to unit-length sweeps. As this episode indicates, students who initially understood dots (or hash-marks) as indicators of unit lengths often lose the fragile conceptual attachment to unit-length iteration.

The following task provides additional evidence that students sometimes lose track of what they are counting when they use unit-length indicators instead of segments or sweeping motions.

How many black rods does it take to make a line segment as long as the gray rod [Fig. 6.13]?

SA(Grade 2) said that she knew that the black rod "takes 3 hash marks" on the gray segment. Moving from left to right, SA counted the fourth, fifth, and sixth hash marks, "1, 2, 3," [Fig. 6.14] marked the sixth hash mark, and said, "have one."

Fig. 6.13 Student SA Problem 1

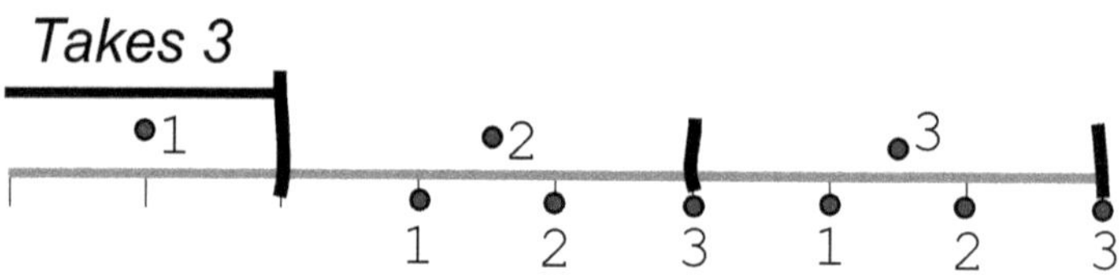

Fig. 6.14 Student SA Problem 1

Which sidewalk from home to school is longer or are they the same? [No rulers were available to students.]

TM(Grade 3): [As shown, labels and counts 15 squares along the gray/solid path and 15 along the dotted path] I think they're the same.

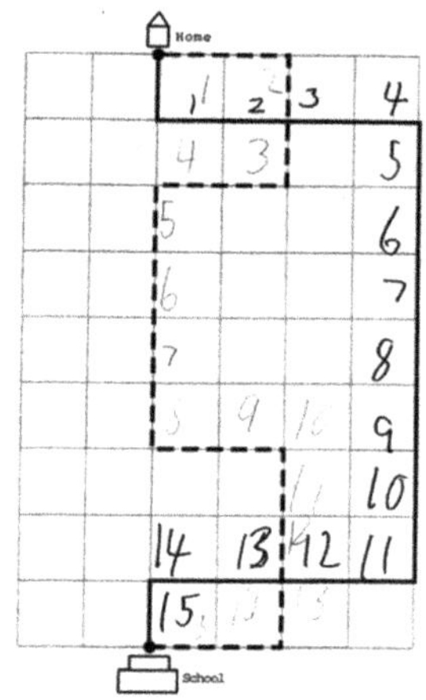

Fig. 6.15 Student TM Problem 1

She counted, "1, 2, 3" on the seventh, eighth, and ninth hash marks and said "have one." She returned to the beginning of the gray rod, pointed to each section she created, and counted "1, 2, 3."

SA focused on the set of three hash marks delimiting the black rod, but she lost track of the original length unit as she used these hash marks as representations of the unit iterations.

Another task that we used to investigate students' understanding of unit-length iteration is the home-to-school problem (Fig. 6.15).

TM counted squares that appeared along the paths, seemingly unaware that his square-counting did not correspond to length-unit iteration. CBA research indicates that this kind of mistaken counting was common. No first or second grade students, and only 6% of third graders, 12% of fourth graders, and 21% of fifth graders correctly iterated a length unit on this problem. Note that this square-counting procedure might be similar to Congdon et al.'s (2018) circle counting, indicating that circle counting may not be as conceptually valid length-reasoning as it first appears.

Conclusion

We have illustrated how quantitative research in psychology can be integrated with qualitative research in mathematics education to the benefit of both research programs. We believe there are additional productive research intersections, perhaps the next best example, being area measurement.

References

Battista, M. T. (2006). *Levels of sophistication in elementary students' reasoning about length*. Presentation given at the annual meeting of the American Educational Research Association, San Francisco.

Battista, M. T. (2012). *Cognition based assessment and teaching of geometric measurement (length, area, and volume): Building on students' reasoning*. Heinemann.

Congdon, E. L., Vasilyeva, M., Mix, K. S., & Levine, S. C. (2018). From intuitive spatial measurement to understanding of units. In K. S. Mix & M. T. Battista (Eds.), *Visualizing mathematics: The role of spatial reasoning in mathematical thought*. Springer.

Steffe, L. (2010). Articulation of the reorganization hypothesis. In L. P. Steffe & J. Olive (Eds.), *Children's fractional knowledge*. New York: Springer.

Chapter 7
Part I Commentary 2: Visualization in School Mathematics Analyzed from Two Points of View

A. Gutiérrez

In this book, terms like spatial reasoning, spatial thinking, or spatial visualization are used by the authors to refer, maybe with some subtle differences, to the use of elements, abilities or skills, vocabulary, and gestures having to do with characteristics of mathematical concepts which are perceived through sight. In this text, I will use the term *visualization* to refer to all them.

The interest of researchers in the role of visualization in school mathematics did not begin in mathematics education, but in educational psychology. As I commented in Gutiérrez (1996), several relevant psychologists, such as Denis, Kosslyn, Krutetskii, Paivio, Shepard, Yakimanskaya, and others, made seminal works to characterize visualization that influenced the emergence of the mathematics education approach to visualization. However, mathematics educators' interest in focusing specifically on the teaching and learning of mathematics made them explore their own way, and open new approaches that became specific theoretical constructs, like those proposed by Bishop, K. Clements, Gutiérrez, Mitchelmore, Presmeg, and Wheatley in the 1980s and 1990s. Nowadays, the mathematics education research on visualization is not a part of the educational psychology research on this topic, but there continue to be relevant links. The chapters in this book illustrate some of those links.

The use of visualization and visual strategies in school mathematics is usually associated with the teaching and learning of geometry, as shown in several chapters of this book by mathematics educators and many other publications (Presmeg, 2006). However, visualization is useful also to understand and learn many other content areas in school mathematics (arithmetic, algebra, functions, statistics, etc.), since all them may benefit from the use of some kind of visual representations like graphs, diagrams, drawings, dynamic representations of calculations, and so on, as shown by several papers in Hitt (2002). Both in educational psychology and mathematics education, there have been researchers interested in exploring the role of visualization in other areas of school mathematics, from kindergarten to future

A. Gutiérrez (✉)
Departamento de Didáctica de la Matemática, University of Valencia, Valencia, Spain
e-mail: angel.gutierrez@uv.es

K. S. S. Mix, M. T. Battista (eds.), *Visualizing Mathematics*, Research in Mathematics Education, https://doi.org/10.1007/978-3-319-98767-5_7

teachers and other undergraduate students. In this book, there is wide interest in elementary arithmetic: Casey and Fell reflect on the relation between visualization and different aspects of elementary arithmetic, namely the choice of strategies by young children when acquiring numeracy and early addition and subtraction, and the use of more or less complex strategies when doing higher level arithmetic in the upper primary grades, like decomposition of numbers or the choice of calculation strategies of counting-all, counting-on, decomposition, and retrieval. These results agree with well-known results by mathematics educators, like Fuson, describing procedures used by children that combine visual images, motor actions with fingers and verbal recitations to count and calculate additions or subtractions. These strategies are eventually internalized as dynamic mental images (Presmeg, 1986b) and used by children to progress to the learning of more automatic and algorithmic calculation procedures.

Looking at middle primary school grades, Cipora, Schroeder, Soltanlou, and Nuerk's chapter summarizes research results, based on statistical correlation techniques, on the kinds of relationships between space and arithmetic. They have found narrow correlations between visualization and the use of multi-digit numbers. Jirout and Newcombe present results about the relation between arithmetic proportionality and a specific kind of visualization, named *spatial scaling*, based on the ability to reason about contexts where some spatial relationships are identified and then applied to a different sized context. They mention, as a difficulty in teaching and learning numbers in primary school, that it is not always clear what numeracy or number means. However, if the teaching of numbers and arithmetic is approached from a phenomenological point of view (Freudenthal, 1983), the changes in the meaning of numeracy and number along the primary and secondary grades (from natural to complex numbers) may be seen as a continuum of increasingly complex mathematical objects created to solve increasingly complex real problems, each new kind of numbers including the previous ones. Numbers are different because they solve different problems.

An important aspect of learning numbers raised by Jirout and Newcombe is the need to make explicit the differential characteristics of the visual representations of each new set of numbers. Most of their chapter is devoted to analyzing spatial scaling and proportional relationships in the context of relative magnitudes. They compile results demonstrating that visual representations of numbers and operations are necessary for a good understanding of early arithmetic and a basis for later understanding of mathematics. Although the most common context for spatial scaling is that of distances in maps and the real world, the authors present other contexts where visualization plays a relevant role in making the concepts accessible to primary school children. These kinds of results, conclusions, and proposals are also present in mathematics education research publications, like those synthesized in some handbooks for arithmetic in general (Verschaffel, Greer, & Torbeyns, 2006), rational number in particular (Lamon, 2007), and other areas of primary school mathematics (Mulligan & Vergnaud, 2006).

A particularity of the educational psychology chapters in the book, unlike much of mathematics education research, is that all of them focus on young children, including most of their references. Also, some of them not only deal with geometry, but with other curricular topics and mathematical concepts. Casey and Fell, besides thinking about the context of elementary arithmetic, discuss the issue of the relation between visualization and measurement sense. For instance, visualization is very helpful to develop the concept of array and apply it to calculate or estimate measurement of length, area, or volume with the help of mental representations of the number line, and tiled surfaces or volumes. Their research is also related to the learning of fractions conceptualized as parts of the unit of measurement and the graphical representation of calculations with fractions.

Congdon, Vasilyeva, Mix, and Levine analyze the transition in the primary grades from an intuitive perception to a metric understanding of space and the usefulness of visualization in this transition. They pay attention to the understanding of the unit of measurement because this concept is recognized as central in the process of acquisition of measurement. Congdon and colleagues review the well-known results of Piaget on this topic and relate them to the difficulties students show in the international assessment like TIMSS or PISA. A main reason for such failure is that teaching of measurement in schools tends to be algorithmic, based on memorizing formulas and applying them to calculate perimeters, areas, or volumes of figures, but teachers do not pay enough attention to the meaning of units of measurement and their manipulation. Congdon and colleagues' chapter also presents a detailed review of literature, from both educational psychology and mathematics education, related to teaching and understanding measurement. They show the evolution of the learning of length, area and volume, and angles within the primary grades and the role that visualization should play in such learning processes, by describing the different procedures and successes of children using rote procedures or procedures where visual representations are part of a scaffolding for their learning.

These results are aligned with results from mathematics education, like David and Tomaz (2012), who showed that drawings and manipulatives helped students to gain an understanding of the concepts of area and area measurement deeper than their pairs receiving a more algorithmic teaching. Although the statistical comparison of pre and post-tests of experimental and control groups did not show significant differences, a qualitative analysis of students' procedures of solution showed clear differences.

It would have been interesting to see data from the educational psychology research about higher educational levels, to see whether they support that visual images and visualization are not just accessory elements for mathematicians, teachers, and students, but they play an important role, since images may help us understand a new concept or suggest a way to prove a new conjecture (Giaquinto, 2007).

Another question analyzed by both educational psychologists and mathematics educators is the relation between students' use of visualization and their achievement in mathematics. Casey and Fell discuss literature showing a relation between the development of visualization skills and arithmetic skills in early grades (K-2) and, as a consequence, a relation between good visualization skills and mathematical achievement. Their conclusion is that there is evidence for a relation between the use of visualization abilities and the development of addition and subtraction skills in kindergarten and grade 1. This agrees with Young, Levine, and Mix, who conclude that teachers' support of visual reasoning is an effective way to promote students' achievement. In the same vein, the chapters by Lowrie and Logan, and Gutiérrez, Ramírez, Benedicto, Beltrán, and Jaime analyze the relation, confirmed by many studies, between visualization and performance or mathematical talent; likewise, the chapter by Sinclair, Moss, Hawes, and Stephenson focus on children's drawings, as a vehicle to show their visual reasoning, and mathematical achievement.

In spite of much data reported by the different chapters in this book in favor of the relation between achievement in mathematics and spatial reasoning, there is also literature concluding the opposite. Krutetskii (1976) described the components of the *structure of mathematical giftedness* and also discussed some elements of mental mathematical activity that he considered to not be obligatory components of the structure, such as computational ability; memory for symbols, numbers, and formulas; ability for spatial concepts; and ability to visualize abstract mathematical relationships and dependencies. Lean and Clements' (1981) analysis of literature concluded that there is not clear support for the relation between visualization and mathematical performance. Presmeg (1986a) stated that most talented students prefer non-visual procedures due to several factors like the nature of mathematics they study, economy of time, preferences of their teachers, and so on. However, more recent authors, like Rivera (2011), Gruessing (2011), Ramírez (2012), and Paz-Baruch, Leikin, and Leikin (2016), concluded that there is a positive relation between expertise in the use of visualization abilities and mathematical talent.

As a closing synthesis, the chapters in this book show that educational psychology and mathematics education share an interest in analyzing the role of visualization in teaching and learning mathematics. There is also agreement in some results and conclusions, but there are clear differences in specific research objectives; namely, educational psychology seems to be mostly interested in the elementary school level, while mathematics education explores also secondary school, undergraduate, and graduate levels and even professional mathematicians' reasoning. For instance, Giaquinto (2007) and Alcock and Inglis (2010) analyze the role of visualization in highly formalized mathematics areas, like algebra or calculus, and the activity of writing formal proofs. They show that this kind of mathematical activity, purely textual and symbolic, is based on the application of axioms, definitions, theorems, etc., but that visualization plays an important role to help give sense to such symbol manipulations.

There are also differences in research methodologies since educational psychology prefers psychometric methods, showing panoramic pictures of broad questions, while mathematics education prefers qualitative methods, producing fine grained results answering specific questions. Those commonalities and differences are a good basis for productive interactions and exchange of ideas between educational psychology and mathematics education.

References

Alcock, L., & Inglis, M. (2010). Visual considerations in the presentation of mathematical proofs. *International Journal of Media, Technology and Lifelong Learning, 6*(1), 43–59.

David, M. M., & Tomaz, V. S. (2012). The role of visual representations for structuring classroom mathematical activity. *Educational Studies in Mathematics, 80*(3), 413–431.

Freudenthal, H. (1983). *Didactical phenomenology of mathematical structures*. Dordrecht, The Netherlands: D. Reidel.

Giaquinto, M. (2007). *Visual thinking in mathematics. An epistemological study*. Oxford, UK: Oxford University Press.

Gruessing, M. (2011). Spatial abilities and mathematics achievement among elementary school children. In B. Ubuz (Ed.), *Proceedings of the 35th Conference of the International Group for the Psychology of Mathematics Education* (Vol. 1, p. 306). Ankara, Turkey: PME.

Gutiérrez, A. (1996). Visualization in 3-dimensional geometry: In search of a framework. In L. Puig, & A. Gutiérrez (Eds.), *Proceedings of the 20th Conference of the International Group for the Psychology of Mathematics Education* (Vol. 1, pp. 3–19). Valencia, Spain: Universidad de Valencia.

Hitt, F. (Ed.) (2002). *Representations and mathematics visualization. Papers from the PME-NA Working Group on Representations and Mathematics*. México DF: Cinvestav-IPN.

Krutetskii, V. A. (1976). *The psychology of mathematical abilities in schoolchildren*. Chicago, IL: The University of Chicago Press.

Lamon, S. J. (2007). Rational numbers and proportional reasoning. In F. K. Lester (Ed.), *Second handbook of research on mathematics teaching and learning* (pp. 629–667). Reston, VA: NCTM.

Lean, G., & Clements, M. A. (1981). Spatial ability, visual imagery, and mathematical performance. *Educational Studies in Mathematics, 12*(3), 267–299.

Mulligan, J., & Vergnaud, G. (2006). Research on children's early mathematical development. Towards integrated perspectives. In A. Gutiérrez & P. Boero (Eds.), *Handbook of research on the psychology of mathematics education. Past, present and future* (pp. 117–146). Rotterdam, The Netherlands: Sense Publishers.

Paz-Baruch, N., Leikin, R., & Leikin, M. (2016). Visual processing in generally gifted and mathematically excelling adolescents. *Journal for the Education of the Gifted, 39*(3), 237–258.

Presmeg, N. C. (1986a). Visualization and mathematical giftedness. *Educational Studies in Mathematics, 17*(3), 297–311.

Presmeg, N. C. (1986b). Visualization in high school mathematics. *For the Learning of Mathematics, 6*(3), 42–46.

Presmeg, N. (2006). Research on visualization in learning and teaching mathematics. In A. Gutiérrez & P. Boero (Eds.), *Handbook of research on the psychology of mathematics education. Past, present and future* (pp. 205–235). Rotterdam, The Netherlands: Sense Publishers.

Ramírez, R. (2012). *Habilidades de visualización de los alumnos con talento matemático*. Ph.D. dissertation, University of Granada, Granada, Spain. Retrieved from http://fqm193.ugr.es/produccion-cientifica/tesis/ver_detalles/7461/descargar

Rivera, F. D. (2011). *Towards a visually oriented school mathematics curriculum*. New York: Springer.

Verschaffel, L., Greer, B., & Torbeyns, J. (2006). Numerical thinking. In A. Gutiérrez & P. Boero (Eds.), *Handbook of research on the psychology of mathematics education. Past, present and future* (pp. 51–82). Rotterdam, The Netherlands: Sense Publishers.

Chapter 8
Part I Commentary 3: Proposing a Pedagogical Framework for the Teaching and Learning of Spatial Skills: A Commentary on Three Chapters

Tom Lowrie and Tracy Logan

Education, generally, and mathematics education specifically, have long-held associations with the field of psychology. Schoenfeld (1987) and Mayer (1992) both described the connections between the two fields and indeed, many educational theories of development evolved from psychology. To this point, one of the longest running groups in mathematics education derived from the field of cognitive psychology, namely, The International Group for the Psychology of Mathematics Education (IGPME). IGPME was established in 1976 under the guidance of Efraim Fischbein, a cognitive psychologist. Initially, the focus was, as the name suggested, on the developmental and psychological complexities of learning various mathematical concepts and processes. However, over the years, the organization has broadened to include new ways of thinking about mathematics learning that go beyond the purely cognitive aspect. In fact, very few cognitive psychologists attend the annual conference these days. Although the direct insights and engagement of cognitive psychology researchers are not commonplace, some overlap remains.

In an article published recently in the journal *Educational Studies in Mathematics*, Bruce and colleagues (Bruce et al., 2017) outlined influences and pathways that cognitive psychologists and mathematics educators have followed, many independent seemingly from one another. Their network analysis revealed a number of factors inhibiting transdisciplinary connections including discipline-based validity and outcome expectations and unawareness of work outside researchers own fields. Nevertheless, they advocate that the two fields have reconnected through the work being undertaken in spatial reasoning. Across mathematics education and cognitive psychology, there has been a focus on identifying the ways in which spatial reasoning is linked to mathematics learning and how spatial thinking provides the foundational support for mathematics reasoning. Three of the chapters drawn from the cognitive psychology section of this book provide great insight into the work being

T. Lowrie (✉) · T. Logan
Faculty of Education, University of Canberra, Bruce, ACT, Australia
e-mail: thomas.lowrie@canberra.edu.au; tracy.logan@canberra.edu.au

K. S. S. Mix, M. T. Battista (eds.), *Visualizing Mathematics*, Research in Mathematics Education, https://doi.org/10.1007/978-3-319-98767-5_8

done to understand the relation between spatial thinking and mathematics reasoning. Collectively, Casey and Fell, Jirout and Newcombe, and Young, Levine and Mix considered the effects of training spatial skills and how such skills can be predictive of and influence later mathematics performance. The three chapters focus on young children through to middle school and begin to draw out causal relationships between spatial thinking and mathematical reasoning, as studies move away from correlational data. As a collective, these chapters identify and define the underlying cognitive processes and skills that are common across spatial thinking and mathematics reasoning. This body of work, and related studies within the cognitive psychology domain, provide insights into learning seldom addressed by the mathematics education community. By way of example, studies that map curriculum content to the specific structures of spatial skills, provide opportunities to align skill development to school curricula experiences (Mix et al., 2017).

Young, Levine, and Mix point to the fact that both spatial ability and mathematics ability are multifaceted and complex, yet little research has "examined mathematical measures broadly enough to reveal separate skills. Yet this examination is vital because mathematics is frequently divided by differences in content rather than skills" (p. 121). Their studies have found that spatial processing and mathematics processing are separate latent factors that are highly correlated, through the early to late elementary school. A few factors indicated cross-loadings, with specific links found between mathematics and mental rotation at Kindergarten—possibly because it helps with encoding, imagery, and the transformations of quantities important for arithmetic problems—and visuospatial working memory and visuo-motor integration in sixth grade. They also highlight literature that indicates that the teaching of spatial constructs affords opportunities for spatial tools to be used more purposefully in teaching—with tools such as gesture and the use of spatial language helpful in mathematics learning, especial when students are exposed to novel situations.

Casey and Fell consider students' strategy use across spatial and mathematical processing. They argue that spatial reasoning can be improved through embedding a variety of spatial strategies within different mathematics content areas, namely: fractions; word problems; and geometry. They highlight the links between being able to visualize and generate mental images and the use of spatial strategies on mathematics problems. Research has suggested that "children with higher spatial skills have greater ability to draw on spatial as well as analytical strategies when solving mathematical problems" (p. 55). This approach provides greater flexibility in mental processing strategies that may be beneficial as students move into more complex mathematics.

Jirout and Newcombe describe research associated with relative magnitude and spatial-relational reasoning when solving number line, fraction, and proportional reasoning problems using external representations, such as diagrams and models. They argue that correct interpretation and use of external representations relies on spatial scaling and understanding the relative magnitude of the representation. They identify that "interventions that promote relational reasoning and use spatial representations seem to have positive impacts on relative magnitude understanding" (p. 19), where the interventions have occurred across several different areas such as linear spatial relations; approximate number system acuity; and number lines. Jirout and Newcombe conclude that relative magnitude may rely more heavily on spatial-relational reasoning than exact number magnitude and that "education should consider ways of explicitly prioritizing relative magnitude learning along with more traditional whole-number knowledge and arithmetic processes" (p. 21).

From an education perspective, all three chapters present a range of suggestions on how to improve children's spatial reasoning skills. There is a tension here though, which warrants further investigation. Young, Levine, and Mix highlight the success training spatial skills separately has at an undergraduate level and the play-based approach to spatial thinking in the early years. It seems they advocate for a segregated approach, where training spatial skills are taught separately from mathematics. Elsewhere, in a well-cited and highly influential article, Cheng and Mix (2014) provided evidence that spatial training could improve children's mathematics performance. This worked had been replicated, although with somewhat moderate effect sizes (Mix & Cheng, 2018). Alternatively, Casey and Fell, and Jirout and Newcombe suggest that spatial thinking should be taught explicitly for the type of mathematics skills and thinking it relates to and promotes. In fact, Casey and Fell suggest that different types of spatial thinking need to be aligned to the most appropriate mathematics content via focused interventions. It could be argued there is a need for a synergy between the two approaches. There is great potential for cognitive psychology to have a large and meaningful impact in classrooms, beyond its influence to date. To do so, there needs to be more of a focus on context and pedagogy, with connections between curriculum and classroom practices.

Many of the intervention programs coming out of cognitive psychology have been implemented by the members of the respective author's research teams. By way of example, Uttal et al.' (2013) meta-analysis described training programs delivered via video games; course training, usually at undergraduate level; and spatial task training, predominantly undertaken in laboratory settings. Fewer programs are implemented by classroom teachers in situ (e.g., Bruce & Hawes, 2014; Casey et al., 2008). Perhaps this is understandable given the nature of experimental design and the associated fidelity measures required in the field. However, as we investigate how spatial training programs relate to, and improve, mathematics understanding and skills, closer attention needs to be paid to the classroom settings where most mathematics learning takes place and to those charged with educating children, teachers, and educators.

Building on these ideas, this commentary proposes a way of moving forward in the spatial reasoning literature by connecting the cognitive psychology training to the mathematics education practices through a pedagogical framework that provides a structure for classroom-based interventions.

ELPSA Framework

Our classroom-based intervention research to date has tried to incorporate high levels of fidelity (where at all possible) into intervention programs that promote spatial training through classroom activities that were both connected to curriculum and distinctively skill based (Lowrie, Logan, & Ramful, 2017). In parallel, we have developed a pedagogical tool that could be embraced by classroom teachers, one which utilized a framework that drew on well-regarded sociological and psychological understandings of learning (Adler, 1998; Cobb, 1988; Lerman, 2003)—the Experience-Language-Pictorial-Symbolic-Application (ELPSA) learning framework (Lowrie & Patahuddin, 2015). ELPSA was used to design the lessons for the spatial reasoning intervention program and explain how students developmentally understood concepts within the respective spatial reasoning constructs. The framework promotes learning as an active process where individuals construct their own ways of knowing (developing understanding) through discrete, scaffolded activities and social interactions. Each step of the framework is critical for establishing sense making, and the sequence provides a logical structure to scaffold, reinforce, and apply knowledge and concepts.

The first element of the learning framework (Experience) draws on the knowledge that students possess. In this stage, the teacher should determine what the students know and what new information needs to be introduced to scaffold their understanding. In this first phase, students are encouraged to make connections between their own spatial practices and specific spatial forms (e.g., how they orientate a map to determine which direction they should navigate). Pedagogically, this phase also provides opportunities for the classroom teachers to understand "what individuals know." The second component of the framework (Language) outlines how specific terminology is used to promote understanding—that is, being explicit about spatial features and intrinsic connections. This stage of the process is also associated with particular pedagogy practices, since it is important for teachers to model appropriate terminology and encourage students to use this language to describe their understandings in ways that reinforce their knowledge and promote discourse with others. The third component of the learning framework (Pictorial) is characterized by the use of spatial and concrete representations to exemplify ideas and concepts (Burte, Gardony, Hutton, & Taylor, 2017; Pillay, 1998). Such representations could be constructed by the teacher (including shared resources and artifacts) or students (including drawing diagrams or visualizing). The fourth component (Symbolic) is aligned to the formalization of ideas or concepts. This stage draws on students' capacity to represent, construct, and manipulate analytic information with flexibility and a degree of fluency (Stieff, 2007). In this phase, capable spatial thinkers are encouraged to go beyond visual forms of reasoning, particularly when automation is possible. The final component of the learning framework (Application) highlights how symbolic understanding can be applied to new situations. This is evident in students' ability to transfer their knowledge to novel situations.

An example of the ELPSA framework in action is described below, accompanied by student work samples and anecdotes aligned to the teachers' pedagogy. The lesson focused on lines of symmetry and visualizing symmetry. Symmetry is part of the Australian Curriculum Mathematics. The concept is introduced around Grades 2 and 3 and elaborated on through all grades to Grade 7. The lesson was designed to encourage the students to visualize horizontal, vertical, and diagonal lines of symmetry (or reflection) and attempt to discover a pattern in the way images were represented when reflected. Symmetry and reflection are integral aspects of mathematical and scientific thinking (e.g., Hargittai & Hargittai, 2009; Livio, 2006) and as such, children require a solid foundation in understanding the spatial concept. Throughout the lesson, the teachers reinforced the need to visualize by engaging students in a cyclic process of Visualize, Predict, Experiment, Check. This cycle encouraged students to undertake the mental process of imagining what the reflection or symmetrical image would be, then attempting to describe or represent that prediction, experiment through undertaking the task, then compare their predictions to their results.

Experience

What is symmetry? The teachers began the lesson from the viewpoint of what students knew about the topic and encouraged active engagement through contextualized whole-class discussions. Students were asked to design a symmetrical design using geometric pattern blocks or on a geometric pattern block app (see Fig. 8.1). This gave students the opportunity to illustrate to teachers their understanding of symmetry. Students also completed a task where they were required to draw the other side of a symmetrical image, in this case, a leaf (see Fig. 8.2). This was completed with varying degrees of success.

Fig. 8.1 Student representing symmetry on a digital device

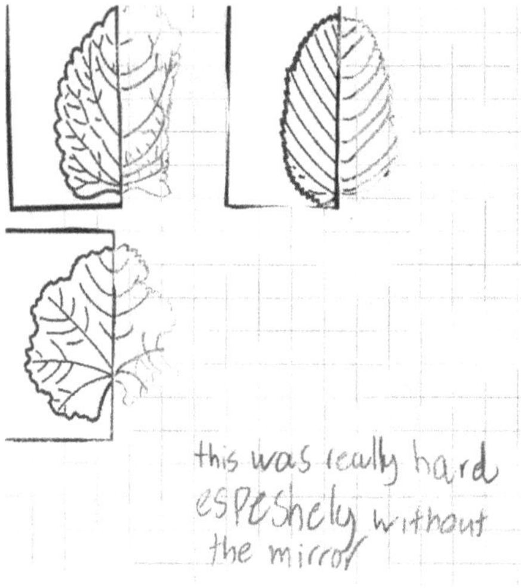

Fig. 8.2 Student completing symmetry tasks by completing templates

Language

What are the language conventions associated with symmetry? The teachers were explicit about the terminology used, increasing the complexity of the language conventions throughout the topic and encouraged students to reflect upon the relevance of this language at the completion of the lessons. Figure 8.3 shows a poster from one of the classrooms that students could refer to during their lessons to help with language. Below are some of the key terminology identified within the lesson.

visualize→predict→experiment→check; reflection, reflective symmetry, line symmetry, reflection line, horizontal, vertical, diagonal, inclined, reflect, translate, upside down, sideways.

Pictorial Reasoning

In the Pictorial phase, the teachers modeled symmetry concepts through diagrams, and encouraged students to do the same, aiding the transition from concrete and diagrammatical representations to more sophisticated visualization strategies. Students began with reflections of more familiar letters and symbols along the y and x axes. They were then asked to consider reflections of similar letters and symbols along the diagonal axis. During the pictorial phase, the visualize, predict, experiment, check process was used to help students from the concrete to the visual, encouraging the students to rely less on the materials.

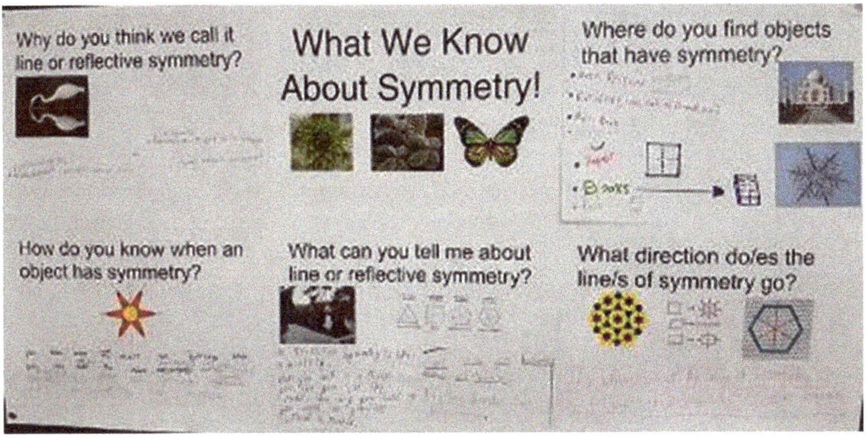

Fig. 8.3 Teacher generated scaffold of symmetrical ideas, which include students work samples

Fig. 8.4 A pictorial representation of vertical lines of reflection

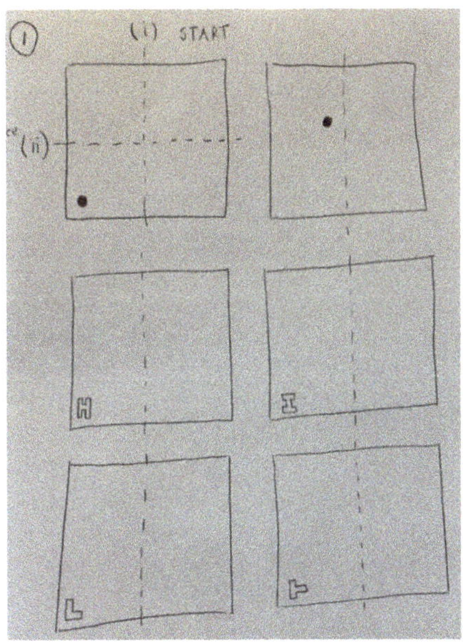

Figure 8.4 shows a teacher's example, where they started with the vertical line of reflection. The representation was used to provide students with a mental model of the process. In this example, the teacher has encouraged students to consider reflections on the same fold (vertical fold) from objects in the same corner (bottom left). Thus, the only difference is the orientation (the letter H on a different rotation) or shape of the figure (and L and T). Thus, the actual is building pattern noticing.

Symbolic Reasoning

The symbolic stage of the cycle requires analytic thinking. Typically, this form of reasoning involves the appropriate use of symbolic tools and representations. When content is spatial in nature, symbolic reasoning is associated with pattern noticing and a capacity to interpret spatial demands in a more automatic manner, often without the concrete or visual demands typically required to decode novel spatial tasks. In this phase, the classroom teachers encouraged students to reason analytically, as a transition beyond representing information "in the mind's eye" or concretely. This symbolic reasoning was evident in the development of rules such as the orientation of objects after a diagonal reflection.

Here students needed to recognize conventions associated with lines of reflection on vertical, horizontal, and diagonal axes. Students begin to reason that for reflections on the x and y axes, horizontal stays horizontal and vertical stays vertical. However, with diagonal reflections, horizontal moves to vertical and vertical moves to horizontal. See Figs. 8.5, 8.6, and 8.7 for the symbolic concept of perpendicularity.

These diagrammatical representations are more than concrete representations or drawings of spatial information, since the pattern noticing affords opportunities for analytic reasoning. Thus, the representations become analytic thinking (see Fig. 8.7).

Applications

The final stage involves the application of ideas to related symmetry and problem-solving tasks. In this stage, the teachers presented open-ended activities that required students to apply concepts to other situations (see Fig. 8.8). Geogebra was also used as a way for students to explore the diagonal line of symmetry as an application (see Fig. 8.9).

Conclusion

In Chap. 5 of this manuscript, Young, Levine, and Mix maintain that a range of spatial tools should be used to promote spatial thinking, beyond the specific content and skills typically used in intervention programs. They acknowledge that spatial tools are especially effective when students encounter novel problems, advocating that "by providing rich spatial information in multiple ways, educators can help students create a lexicon of spatial relations, terms and connections to mathematics" (p. 140). Jirout and Newcombe describe the importance of providing students with a variety of representations and advocate that certain types of spatial representations may provide explicit mathematics concepts in ways that extrapolate information more purposefully (e.g., continuous proportional representations). We argue that the ELPSA framework encourages mathematics ideas to be represented in different

Horizontal Line of Reflection	Vertical Line of Reflection	Diagonal (Incline) Line of Reflection
Vertical → Vertical	Vertical → Vertical	Horizontal → Vertical
Horizontal → Horizontal	Horizontal → Horizontal	Vertical → Horizontal

Fig. 8.5 Symbolic thinking can be used to move beyond the traditional mental processes required to visualize across lines

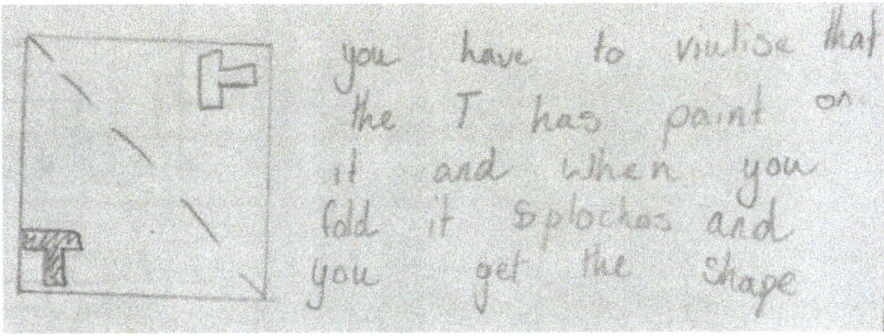

Fig. 8.6 A student moving toward symbolic reasoning, while still evoking visual approaches

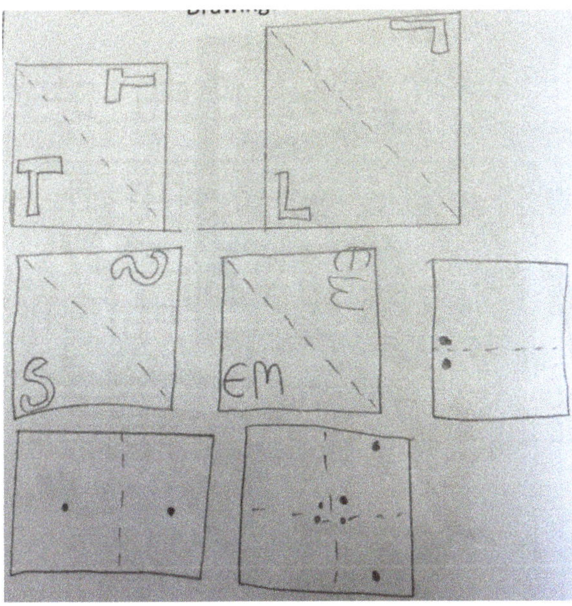

Fig. 8.7 A student "symbolizing" horizontal, vertical, and diagonal representations of symmetry

ways, encouraging classroom teachers to re-represent spatial ideas to consolidate student's understanding. Each component of the ELPSA framework provides a distinct pedagogical approach to foster a repertoire of representations and encourages teachers to present information across embodied, verbal, pictorial, and symbolic representations, with the pictorial aspect open-ended with respect to the types of

Fig. 8.8 A students representation of a "real-life" symmetical experience

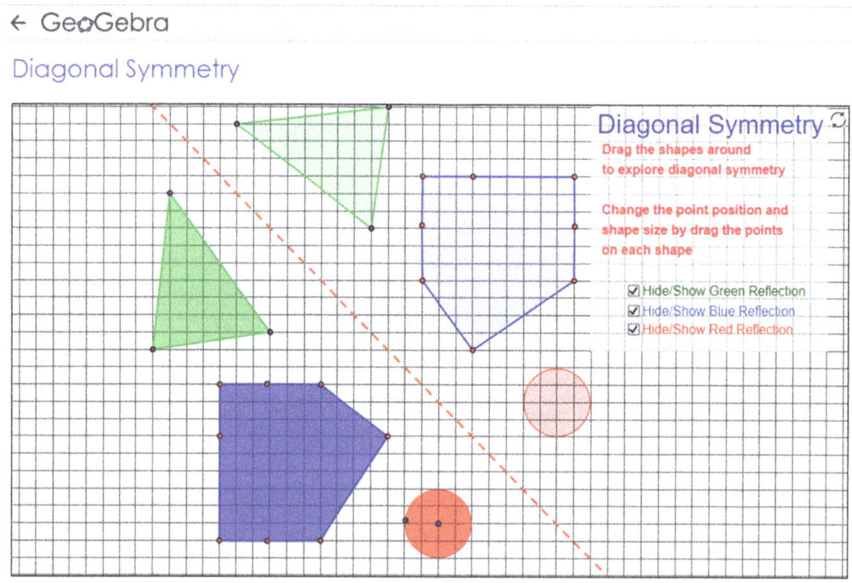

Fig. 8.9 Students utilize other tools to explore symmetry (https://www.geogebra.org/m/cwYhQmMU)

visual and diagrammatic tools utilized. For example, the Experience component encourages gesture and tacit thinking, the Language component specific use of rich spatial terminology, the Pictorial component the use of 2D and 3D manipulatives—both concrete and mental representations. The framework also goes some way to ensuring that an over emphasis on symbolic representations does not occur frequently or too early in concept formation (as described by Jirout and Newcombe).

The ELPSA framework advocates for the use of concrete manipulatives. Research outlined by Jirout and Newcombe suggested that use of concrete manipulatives can assist students to think spatially while acting as a scaffold for more abstract mathematical processing. The embodied nature of engaging with manipulatives in both the Experience and Pictorial phases, assists with language development as students discover explicit language associated with spatial concepts and undertake tasks with a focus on lived experiences. In a similar vein, Casey and Fell described methods for teachers to help students generate images, which they suggest is a critical aspect of children thinking spatially and essential for utilizing spatial strategies across mathematics content areas. This aligns with the pictorial phase of ELPSA. As students proceed through the visualize, predict, experiment, and check cycle, explicit opportunities for generating images, both mentally and concretely, are created.

The transition of the ELPSA framework from the pictorial to symbolic aligns with Casey and Fell's conclusions since students with stronger spatial skills are better equipped to utilize both spatial and analytic strategies when solving arithmetic problems. ELPSA allows students the flexibility to move between pictorial and symbolic/analytic processing, and as they gain content knowledge and confidence with the analytic strategies, they will fold back less and less to representing their thinking spatially or pictorially. When students are faced with a novel or complex task, they should be encouraged to revert to spatial/pictorial strategies until they are more fluent with the analytic approaches (Lowrie & Kay, 2001; Martin, 2008). The iterative nature of the framework offers a solid pedagogical foundation for students' spatial concept development, since the framework encourages three phases of representation to be considered before symbolic representations are introduced or applied.

Our colleagues in cognitive psychological tend to be more focused in their research designs, than those typically framed in mathematics education—providing opportunities for aspects of learning and concept developed to be quarantined. The ELPSA framework allows for the blending of the two approaches in ways that provide synergies between cognitive psychology and mathematics education.

References

Adler, J. (1998). A Language of teaching dilemmas: Unlocking the complex multilingual secondary mathematics classroom. *For the Learning of Mathematics, 18*(1), 24–33.

Bruce, C. D., Davis, B., Sinclair, N., McGarvey, L., Hallowell, D., Drefs, M., ... Woolcott, G. (2017). Understanding gaps in research networks: Using "spatial reasoning" as a window into the importance of networked educational research. *Educational Studies in Mathematics, 95*(2), 143–161. https://doi.org/10.1007/s10649-016-9743-2

Bruce, C. D., & Hawes, Z. (2014). The role of 2D and 3D mental rotation in mathematics for young children: What is it? Why does it matter? And what can we do about it? *ZDM—The International Journal on Mathematics Education, 47*(3), 331–343. https://doi.org/10.1007/s11858-014-0637-4

Burte, H., Gardony, A. L., Hutton, A., & Taylor, H. A. (2017). Think3d!: Improving mathematics learning through embodied spatial training. *Cognitive Research: Principles and Implications, 2*(13), 1–18. https://doi.org/10.1186/s41235-017-0052-9

Casey, B. M., Andrews, N., Schindler, H., Kersh, J. E., Samper, A., & Copley, J. (2008). The development of spatial skills through interventions involving block building activities. *Cognition and Instruction, 26*(3), 269–309.

Cheng, Y. L., & Mix, K. S. (2014). Spatial training improves children's mathematics ability. *Journal of Cognition and Development, 15*(1), 2–11.

Cobb, P. (1988). The tension between theories of learning and instruction in mathematics education. *Educational Psychologist, 23*(2), 87.

Hargittai, M., & Hargittai, I. (2009). *Symmetry through the eyes of a chemist* (3rd ed.). Dordrecht, The Netherlands: Springer.

Lerman, S. (2003). Cultural, discursive psychology: A sociocultural approach to studying the teaching and learning of mathematics learning discourse. In C. Kieran, E. Forman, & A. Sfard (Eds.), *Learning discourse: Sociocultural approaches to research in mathematics education* (pp. 87–113). Dordrecht, The Netherlands: Springer.

Livio, M. (2006). *The equation that couldn't be solved: How mathematical genius discovered the language of symmetry*. New York: Simon and Schuster.

Lowrie, T., & Kay, R. (2001). Task representation: The relationship between visual and nonvisual solution methods and problem difficulty in elementary school mathematics. *Journal of Educational Research, 94*(4), 248–253.

Lowrie, T., & Patahuddin, S. M. (2015). ELPSA as a lesson design framework. *Journal of Mathematics Education, 6*(2), 1–15.

Lowrie, T., Logan, T., & Ramful, A. (2017). Visuospatial training improves elementary students' mathematics performance. *British Journal of Educational Psychology, 87*(2), 170–186. https://doi.org/10.1111/bjep.12142

Mayer, R. E. (1992). *A series of books in psychology. Thinking, problem solving, cognition* (2nd ed.). New York, NY: W H Freeman/Times Books/Henry Holt and Company.

Martin, L. C. (2008). Folding back and the dynamical growth of mathematical understanding: Elaborating on the Pirie-Kieren Theory. *Journal of Mathematical Behavior, 27*, 64–85.

Mix, K., & Cheng, Y. (2018, April). *More than just numbers: Varied predictors of mathematical knowledge*. Paper presented at the American Education Research Association Conference, New York, NY.

Mix, K., Levine, S., Cheng, Y., Young, C., Hambrick, D., & Konstantopoulos, S. (2017). The latent structure of spatial skills and mathematics: A replication of the two-factor model. *Journal of Cognition and Development, 18*(4), 465–492. https://doi.org/10.1080/15248372.2017.1346658

Pillay, H. (1998). Cognitive processes and strategies employed by children to learn spatial representations. *Learning and Instruction, 8*(1), 1–18.

Schoenfeld, A. H. (Ed.). (1987). *Cognitive science and mathematics education*. Hillsdale, NJ: Lawrence Erlbaum Associates.

Stieff, M. (2007). Mental rotation and diagrammatic reasoning in science. *Learning and Instruction, 17*, 219–234.

Uttal, D. H., Meadow, N. G., Tipton, E., Hand, L. L., Alden, A. R., Warren, C., & Newcombe, N. S. (2013). The malleability of spatial skills: A meta-analysis of training studies. *Psychological Bulletin, 139*(2), 352–402.

Chapter 9
Part I Commentary 4: Turning to Temporality in Research on Spatial Reasoning

Nathalie Sinclair

The chapters in this book make it clear that the long-standing attempts to reclaim the visual register[1]—spatial reasoning—in mathematics thinking have reached a new intensity. Although most of the chapters written by the cognitive psychologists are concerned with the potential for spatial reasoning to significantly support children's success in the quantitative domain of arithmetic, their work raises the possibility for also engaging children in more qualitative forms of mathematical thinking that are important in and of themselves. While evidently being a complex construct involving several different kinds of capabilities and tendencies, one of the dimensions of spatial reasoning that was evoked in several of the chapters centrally involves motion—that is, some kind of temporal change. Given the importance of kinetic thinking in mathematics (see Núñez, 2006; Sinclair & Gol Tabaghi, 2010; Whiteley, 2002) and its growing significance not only in contemporary mathematics (Zalamea, 2012) but also in STEM careers (medical imaging, animation, protein modeling, etc.), I am interested in exploring ways in which cognitive psychologists and mathematics educators can contribute to better understanding the more mobile aspects of spatial reasoning.

I begin by describing the temporal dimension of spatial reasoning that has been developed in the cognitive psychology literature and examining how it has been taken up in the mathematics education literature. I then consider some research emerging from the study of dynamic geometry environments (DGEs), which have provided a rich context for studying students' geometric thinking about continuously transforming geometric objects, and which might be a catalyst for investigating the importance of the temporal aspects spatial reasoning more broadly in the mathematics curriculum. Finally, I show how temporality could be at play in the mathematical concepts investigated in the cognitive psychology chapters.

[1] In the eighteenth century it was considered an advantage to be blind, and therefore not distracted by the empirical world; much later, the Bourbaki mathematicians banned the use of diagrams, producing an entire geometry textbook with not one single image.

N. Sinclair (✉)
Faculty of Education, Simon Fraser University, Burnaby, BC, Canada
e-mail: nathsinc@sfu.ca

K. S. S. Mix, M. T. Battista (eds.), *Visualizing Mathematics*, Research in Mathematics Education, https://doi.org/10.1007/978-3-319-98767-5_9

The Quadrant Model and Its Relation to Mathematics Education Research

As Young and Levine write in their chapter, spatial reasoning has been categorized along two major dimensions: an intrinsic-extrinsic dimension and a static-dynamic dimension. While some disagreements persist over whether this model is sufficiently complete or adequately differentiates different forms of spatial reasoning, researchers have found that mathematicians excel at tasks involving dynamic spatial reasoning such as mentally transforming objects (folding, bending, rotation, scaling, cross sectioning, or comparing 2-D and 3-D views) (Newcombe & Shipley, 2012; Uttal et al., 2013) and that student success on tasks involving working with moving objects correlates with mathematical achievement.

While intrinsic tasks focus on the properties of objects, and how sub-parts relate to each other, extrinsic tasks examine the relationships between objects to an external referent (such as horizontality or verticality). Much more research has been done with intrinsic-dynamic tasks than with extrinsic-dynamic tasks. Tasks that have been used in the cognitive psychology literature to assess extrinsic-dynamic capability include perspective taking and navigating. These tasks frame the extrinsic-dynamic category as being about how one's perception of the relations among objects changes as one moves through the environment. However, in mathematics, extrinsic-dynamic thinking is at play in a large number of situations where co-variation is involved. Indeed, functions are fundamentally extrinsic-dynamic as they involve thinking about how a dependent variable might change in relation to the changing value of an independent variable: for example, how the volume of a cylinder changes as its height increases. In the next section, I will be examining potential connections between the concepts and tasks presented in the cognitive psychology chapters and a more mathematically focussed notion of extrinsic-dynamic spatial reasoning.

The second aspect of temporal spatial reasoning that I will be examining relates to invariance. In intrinsic-dynamic mental rotation tasks, the shape itself is not changing—it remains rigid.[2] This is different from a situation in which one imagines stretching a cube or shearing a rectangle, that is, when the transformation is not an isometry. Indeed, Atit, Shipley, and Tikoff (2013) suggest that rigid/nonrigid might be an additional dimension of spatial reasoning. In mathematics, these kinds of nonrigid transformations are used not only to generate different examples of particular shapes (in addition to the prototypical rectangle, there might be a squashed rectangle or a square-like rectangle), but also to identify what does not change across all these shapes (in each particular case of a rectangle, the diagonals will always bisect). In exploring spatial reasoning in mathematics, we might focus attention on how both isometric and non-isometric transformation of objects can be used

[2]When students engage in mental rotation of a cube, they are transforming the cube through the isometric transformations of rotation, translation and reflection, though not usually doing so intentionally.

in order *to see invariance*. In the next section I will discuss how new digital technologies have changed the way teachers and students can work with variation and invariance.

New Opportunities for Spatial Reasoning: The Case of Dynamic Geometry Software

Developed in the early 1990s, (DGEs) such as *The Geometer's Sketchpad* and *Cabri-géomètre* enabled users to continuously transform geometric objects on the screen. If at first many mathematicians and teachers found this new computer-based innovation anathema to the traditional static environments of geometry, dynamic mathematics software has now been adopted broadly by the research community, by teachers and also by textbook writers and curriculum designers. These dynamic transformations enable users to generate whole families of shapes while maintaining their defining mathematical features. The transformations include isometries (rotations, reflection, translation) but also non-isometries such as dilations and affine transformations. For example, in Fig. 9.1, the quadrilateral ABCD on the left undergoes several transformations: dragging A transforms it into something that looks like a rectangle; dragging C turns it into something that looks like a square; and dragging C in the different direction turns it into something that looks like a rhombus. In all of these transformations, the lengths of the sides have change, the angles have changed, the orientation of the quadrilateral has changed. However, what has not changed is the fact that the opposite sides are always parallel. In this case, the software is performing the transformation, but the spatial reasoning at stake in a geometry classroom might involve noticing this invariance, noticing what does *not* change during the transformation. In geometry, and mathematics more broadly, identifying invariance is a necessary step in being able to name properties, create definitions and state theorems.

In a study of students investigating quadrilaterals with *Sketchpad*, Battista (2007) formulated the *transformational-saliency hypothesis*, which posits that humans are good at noticing invariance in change. In the example described above, you might not notice that the opposite sides of the quadrilateral are parallel at first,

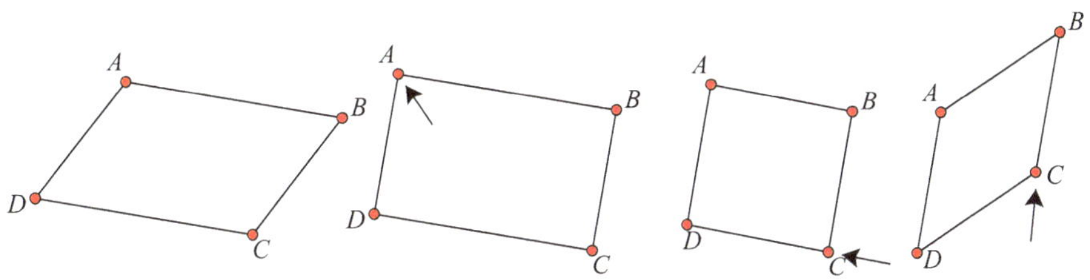

Fig. 9.1 Starting with a quadrilateral ABCD, point A is dragged to form a rectangle, then point C is dragged to form a square and finally point C is dragged to form a rhombus

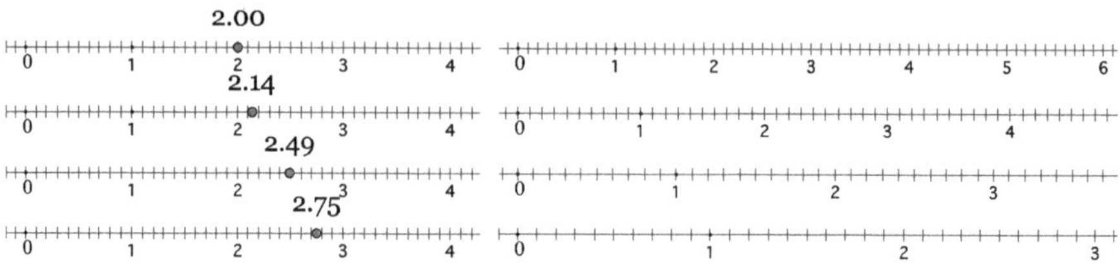

Fig. 9.2 (**a**) The point is varying along the number line between 2 and 3, which leaves the unit value invariant. (**b**) The whole number line is being stretched out

when you look at the left-most quadrilateral only, but as the quadrilateral is varied continuously you may indeed notice what is *not* changing.[3] Any geometric shape can thus be defined by its invariances, something that is much easier to do in an environment where it is possible to vary the shape and continuously generate a large number of particular cases. As the past two decades of research with DGEs has shown, it turns out that it is not just in geometry where this kind of variation is relevant. For example, in the domain of number, a dynamic number line (Fig. 9.2) enables two forms of variation. By dragging a point along the line in the interval between 2 and 3, for example, the value of the tenths and hundredths places changes, but the value of the unit does not. The transformational-saliency hypothesis asserts that learners would notice what remains invariant. The number line itself can also be stretched by moving the unit point. What invariances arise as the unit point gets further from the origin (Fig. 9.2b)? For example, the distance between 0 and 1 and between 1 and 2 is always the same, even though they increase as one moves from the top line to the bottom one.

In this quantitative context, the focus is less on rigid or nonrigid transformations; however, like the parallelogram example, it involves noticing invariance in a situation of controlled variation. In addition to working with variance and invariance, DGEs have provided a strong impetus for re-temporalizing a broad range of mathematical concepts in number, algebra, measurement, calculus, linear algebra, etc. This has also happened in non-technology situations. For example, Lakoff and Núñez (2000) describe the example of addition as motion along a path, where the very idea of adding something is metaphorically understood as walking forward a certain distance, which is a temporalized conception of addition. Number itself can be thought of in more temporal terms when conceived in ordinal rather than cardinal terms (Coles & Sinclair, 2017).

The above discussion raises the general question of whether a more dynamic or temporal approach to mathematical concepts might be helpful in student learning and whether it might also help students develop spatial reasoning (which is itself helpful in student learning). In the next section, I first consider how noticing invariance relates to the mathematical concepts discussed in the chapters. The goal will be to show how some of these concepts can be reframed in terms of variance and

[3] The static medium of this book limits such noticings, because the continuity of the transformation is lost. Dear reader, please drag for yourself: www.desmos.com/geometry/geaqgl0vc2

invariance and how this relates to spatial reasoning. I then propose some questions for future research that relate to how continuous variation and invariance can be studied and developed in research on spatial reasoning.

Moving Time into the Research on Spatial Reasoning

In their chapter, Young and Levine note that symbolic number sense is a strong predictor of mathematical achievement across the elementary years and ask whether "spatial stills training can be used to train symbolic and/or non-symbolic number sense" (p. 127). Symbolic fluency includes ordinal awareness, that is, a sense of the sequence of numbers and their relations (knowing that 23 comes after 22), which is very different from a cardinal number sense. Might the transformational-saliency hypothesis be relevant to the kind of spatial reasoning involved in symbolic number sense? Knowing that 23 comes after 22 is the same as knowing that 33 comes after 32 (and that 3 comes after 2 and 13 comes after 12) when one notices that the tens digit is invariant. Interestingly, the kind of variation required to notice such an invariance is usually not available to kindergarten and grade 1 students, who are typically limited to the 1–20 range of numbers. In order to notice any kind of invariance, children would need to be able to have encounters with a broader set of numbers (see Coles & Sinclair, 2017). The questions that arise from this shift to temporality relate both to whether students notice invariance in the number symbols and whether there might be tasks (perhaps involving the hundreds chart or the Gattegno chart) that could promote an attention to variation and invariance.

Still in the realm of the quantitative, Jirout and Newcombe's chapter focusses on relative magnitudes, which are involved in comparing fractions and calculating proportions. They discuss the role of both spatial scaling and spatial representation in students' understanding of relative magnitudes and consider the number line as a useful task environment for inquiring about and also supporting students' thinking about relative magnitude. On the number line, ½ can be conceptualized as part of a whole, and therefore in a proportional way, which is different from ½ as an absolute quantity. Indeed, as a relative magnitude, ½ is the class of fractions x/y such as y is twice as big as x. This way of thinking of ½ invites thinking in terms of variation—the numerator and the denominator can change, but what stays the same is the fact that the denominator is twice as big. A more visual way of working with ½ is exemplified in continuous, intensive quadrant of Fig. 1.1 of Jirout and Newcombe's chapter, but the two instances of ½ do not provide sufficient variation for the kind of transformation-saliency that Battista hypothesizes. Imagine an environment where the height of the rectangle can be changed continuously, then what would remain invariant is the relative quantity of liquid in the container. In a number line context, we might also imagine a stretchy ruler (Fig. 9.3a) whose endpoint can be dragged so that the length of the ruler varies, but the relative quantities remain invariant. Such a stretchy ruler could be placed on an image such as the one

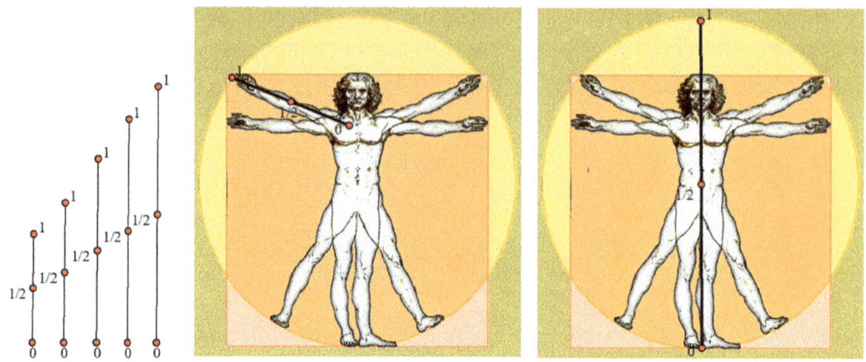

Fig. 9.3 A "stretchy ruler" that focuses on relations between lengths

Fig. 9.4 Varying the position of the tangram pieces changes the shape, but not the area

shown in Fig. 9.3a , b to show how the elbow is halfway down the arm and the belly button is halfway up the Vitruvian man.

Casey and Fell's chapter discusses the importance of decomposition especially in relation to solving addition and subtraction problems. For example, in solving 8 + 5, Casey and Fell argue that a higher-level mental strategy would involve breaking the numbers into simpler facts: this might involve starting with 5 + 5 to get 10 and then adding the remaining 3 to get 13. A more temporal strategy might explicitly involve moving 2 from the 5 to get 10 and then putting 3 on to get 13. In this way, the decomposition is tethered to a mental motion that might occur on the number line or even a tens frame. This might be coupled with a more geometric set of actions using tangrams, for example, where students would focus on how moving a tangram piece (from Fig. 9.4a to Fig. 9.4b, where the bottom green triangle moves in a rigid way upward)—decomposing and recomposing—doesn't change the area of the whole shape. This brings invariance into focus as well, highlighting the possible variations that would maintain the same area.

An even more dynamic strategy would involve shearing (also known as Cavalieri's principle), which is a transformation that moves each "slice" of a shape along a vector parallel to its base. Imagine a stack of books or cards all lined up,

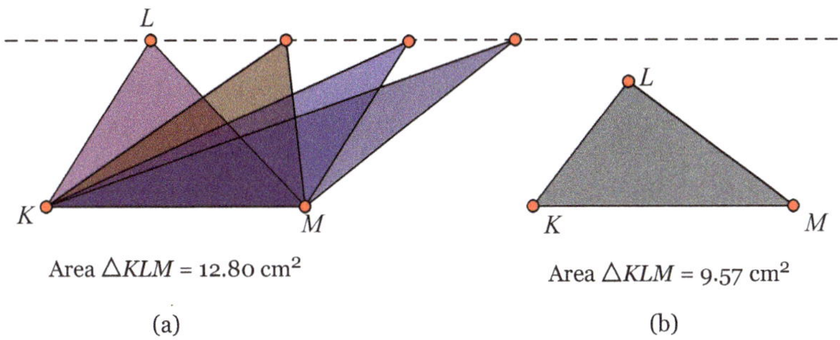

Fig. 9.5 (**a**) Shearing a triangle in order to preserve area; (**b**) changing the area of the triangle by splitting L from the parallel line

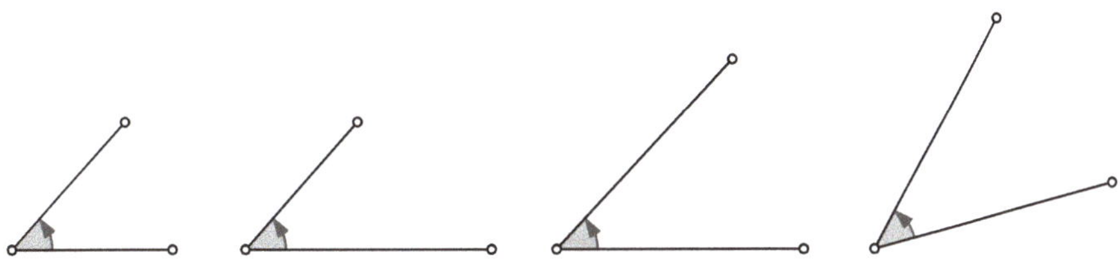

Fig. 9.6 A family of angles in which the amount of turn remains invariant

then slide each one to the right a little. The rectangle has been transformed into a parallelogram, without changing its area—since the thickness and lengths of each book/card have not changed). In Fig. 9.5a, the vertex L of triangle KLM can be dragged along a line parallel to KM, which will change its shape (make it more obtuse, for example) but not its area. Each infinitely small slice of the triangle is being pushed further to the right. In Fig. 9.5b, vertex L is no longer on the parallel line and even though KM is the same length as in Fig. 9.5a, the area of KLM on the right has changed. The formula for the area of a triangle encapsulates this shearing idea as it says that the height and the base determine the area of a triangle.

Finally, the chapter by Congdon et al. focusses on spatial measurement and includes a discussion of children's understanding of angle measure. Angles are very complex concepts that can be defined in many different ways. [Henderson and Taimina (2005) have written that depending on your definition of an angle, a triangle might have 3, 6, 9, or 12 angles!] A dynamic approach to angle involves defining angle as the amount of turn, an approach to early angle learning that does not depend on the use of degrees and that does not confuse the size of the angle with the length of its arms (Kaur, 2017). An angle can be defined as the class of straightedge (segments, ray) pairs that are produced by the same amount of turn, as shown in Fig. 9.6 (the arrows show the direction and amount of turn). By varying the lengths of the arms, and by varying the orientation, the transformation-saliency hypothesis would assert that children would notice the invariance of this turn.

I have been attempting to describe some more motion-based approaches to mathematical concepts (area, angle, arithmetic, number). Since motion involves time, and since most conceptions of time are very spatially structured (Núñez & Cooperrider, 2013), putting mathematical objects in motion (be they triangles or numerals) may inevitably call upon spatial reasoning. These approaches are clearly powerful ways of thinking mathematically, but their relevance to mathematics education will depend both on figuring out first, whether the transformational-saliency hypothesis can be validated—and if so, in what contexts, with what supports—and second, if the hypothesis does not hold, how to support learners' development of this aspect of spatial reasoning. The former research endeavour strikes me as an especially fruitful area of collaboration for cognitive psychologists and mathematics education researchers, where the methods of cognitive psychologists could help identify how children are seeing situations involving change. In the case of the parallelogram, for example, are they seeing continuous change (one quadrilateral being transformed) or a set of discrete changes (many different instances of a quadrilateral)? Are there particular aspects of quadrilaterals that they notice more than others? For example, is angle a more salient invariance or is side length or area? Might colour be used to help focus learners on what is changing and what is not changing? Are certain constraints on change helpful? For example, instead of dragging the vertex of the parallelogram arbitrarily, would a more controlled variation (say, along a certain, invisible line) make it easier to notice invariance?

If the transformation-saliency hypothesis does not hold, what kinds of tasks and activities might help students become more proficient at noticing invariance? The chapters have provided many examples of tasks that have been used to test and also improve students' spatial reasoning. In the same vein, we can ask what kinds of tasks might help support students' noticing of invariance. Casey and Fell rightly underscore the importance of "the ability to *generate images*" (p. 48, *emphasis in original*). What kinds of tasks could help children develop the ability to generate *dynamic* images, to be able to imagine the vertex of a triangle moving around and generate examples of obtuse, acute, and right triangles (and maybe even degenerate ones)? Does watching a dynamic visualisation and discussing it in a classroom setting improve this ability? Might drawing the dynamic images help support students' ability to generate them mentally later on?

On a final note, I point to the important connections that might arise in relation to the temporal forms of reasoning that are intertwined with sensorimotor activities, particularly those of the hands. In relation to relative quantity, for example, Abrahamson, Shayan, Bakker, and van der Schaaf's (2016) Mathematics Imagery Trainer enables children to use their hands to explore proportions. As they move their hands, they get direct feedback about whether the relative height of their hands stays the same. This provides not only a visual but also motor and kinaesthetic sense of invariance that seems highly relevant to relative magnitude. Since gestures are temporal in nature, how might their controlled or designed use of them help support the dynamic dimensions of spatial reasoning (cf. Battista, 2002)?

References

Abrahamson, D., Shayan, S., Bakker, A., & van der Schaaf, M. (2016). Eye-tracking Piaget: Capturing the emergence of attentional anchors in the coordination of proportional motor action. *Human Development, 58*(4–5), 218–244.

Atit, K., Shipley, T. F., & Tikoff, B. (2013). Twisting space: Are rigid and non-rigid mental transformations separate spatial skills? *Cognitive Processing, 14*(2), 163–173.

Battista, M. T. (2002). Learning geometry in a dynamic computer environment. *Teaching Children Mathematics, 8*(1), 333–339.

Battista, M. T. (2007). The development of geometric and spatial thinking. In F. Lester (Ed.), *Second handbook of research on mathematics teaching and learning* (pp. 843–908). Reston, VA: National Council of Teachers of Mathematics.

Coles, A., & Sinclair, N. (2017). Re-thinking place value: From metaphor to metonym. *For the Learning of Mathematics, 37*(1), 3–8.

Henderson, D., & Taimina, D. (2005). *Experiencing geometry. Euclidean and non-Euclidean with history*. Ithaca, NY: Cornell University.

Kaur, H. (2017). *Young children's understanding of angle in a dynamic geometry environment.* Unpublished doctoral dissertation. Simon Fraser University.

Lakoff, G., & Núñez, R. (2000). *Where mathematics come from: How embodied mind brings mathematics into being*. New York: Basic Books.

Newcombe, N., & Shipley, T. F. (2012). Thinking about spatial thinking: New typology, new assessments. In J. S. Gero (Ed.), *Studying visual and spatial reasoning for design creativity*. New York, NY: Springer.

Núñez, R. (2006). Do real numbers really move? Language, thought, and gesture: The embodied cognitive foundations of mathematics. In R. Hersh (Ed.), *18 unconventional essays on the nature of mathematics* (pp. 160–181). New York, NY: Springer.

Núñez, R., & Cooperrider, K. (2013). The tangle of time and space in human cognition. *Trends in Cognitive Sciences, 17*(5), 220–229.

Sinclair, N., & Gol Tabaghi, S. (2010). Drawing space: Mathematicians' kinetic conceptions of eigenvectors. *Education Studies in Mathematics, 74*(3), 223–240.

Uttal, D. H., Meadow, N. G., Tipton, E., Hand, L. L., Alden, A. R., Warren, C., & Newcombe, N. (2013). The malleability of spatial skills: A meta-analysis of training studies. *Psychological Bulletin, 139*(2), 352–402. https://doi.org/10.1037/a0028446

Whiteley, W. (2002). *Teaching to see like a mathematician*. Retrieved from http://www.math.yorku.ca/~whiteley/Teaching_to_see.pdf.

Zalamea, F. (2012). *Synthetic philosophy of contemporary mathematics*. New York: Sequence Press.

Part III
Educational Perspectives

Chapter 10
Analyzing the Relation Between Spatial and Geometric Reasoning for Elementary and Middle School Students

Michael T. Battista, Leah M. Frazee, and Michael L. Winer

Abstract Numerous studies have found that spatial ability and mathematical ability are positively correlated. But specifying the exact nature of the relation between these types of reasoning has been elusive, with much research focused on understanding correlations between mathematical performance and specific spatial skills as measured by spatial tests. We attempt to deepen understanding of the relationship between spatial and mathematical reasoning by precisely describing the spatial processes involved in reasoning about specific topics in geometry. We focus on two major components of spatial reasoning. *Spatial visualization* involves mentally creating and manipulating images of objects in space, from fixed or changing perspectives on the objects, so that one can reason about the objects and actions on them, both when the objects are and are not visible. *Property-based spatial analytic reasoning* decomposes objects into their parts using geometric properties to specify how the parts or shapes are related, and, using these relationships, operates on the parts. Spatial analytic reasoning generally employs concepts such as measurement, congruence, parallelism, and isometries to conceptualize spatial relationships.

Keywords Mathematics · Geometry · Spatial visualization · Analytic · Reasoning · Mental model · Learning progression · Structure · Shape · Isometries · Reflections · Rotations · Measurement · Length · Angle · Area · Volume

---

Much of the research and development referenced in this chapter was supported in part by the National Science Foundation under Grant Numbers 0099047, 0352898, 554470, 838137, and 1119034 to the first author. The opinions, findings, conclusions, and recommendations, however, are the authors' and do not necessarily reflect the views of the National Science Foundation.

M. T. Battista (✉)
Department of Teaching and Learning, The Ohio State University, Columbus, OH, USA
e-mail: battista.23@osu.edu

L. M. Frazee
Department of Mathematical Sciences, Central Connecticut State University, New Britain, CT, USA
e-mail: frazee@ccsu.edu

M. L. Winer
Department of Mathematics, Eastern Washington University, Cheney, WA, USA
e-mail: mwiner@ewu.edu

K. S. S. Mix, M. T. Battista (eds.), *Visualizing Mathematics*, Research in Mathematics Education, https://doi.org/10.1007/978-3-319-98767-5_10

Numerous studies have found that spatial ability and mathematical ability are positively correlated (Mix, Levine, Cheng, Young, & Hambrick, 2016). But specifying the exact nature of the relation between these types of reasoning has been elusive, with much research focused on understanding correlations between mathematical performance and specific spatial skills as measured by spatial tests (Mix & Cheng, 2012; Mix et al., 2016). *The research described in this chapter attempts to deepen our understanding of this relation by precisely describing the spatial processes involved in reasoning about specific topics in geometry.* The chapter focuses on two topics, spatial reasoning in geometric measurement, and spatial reasoning about geometric shapes and isometries. Before discussing the empirical findings, the theoretical frameworks used in the research are described.

Theoretical Frameworks

Organizational Theme

Both spatial reasoning and geometric reasoning are accomplished by creating and operating on mental models (Battista, 2007). For spatial reasoning to properly support geometric reasoning, these mental models must incorporate operational knowledge of relevant geometric properties and concepts. Thus, to investigate the nature of spatial reasoning in geometry, the research in this chapter focuses on how students use both spatial visualization and geometric properties, and the degree to which students' spatial reasoning does or does not incorporate these properties. The components of this theory will be elaborated, then data and analysis related to the theory will be presented.

Spatial Reasoning

Underlying most geometric thought is *spatial reasoning*, which is the ability to recognize, generate, inspect, operate on, and reflect on spatial objects, images, relationships, movements, and transformations (Battista, 1990; Battista, 2007; Clements & Battista, 1992; Mix et al., 2016). The two major components of spatial reasoning that we focus on in this chapter—and that are critical in geometric reasoning—are visualization and analytic reasoning (Hegarty, 2010).

Spatial visualization[1] involves mentally creating and manipulating images of objects in space, from fixed or changing perspectives on the objects, so that one can reason about the objects and actions on them, both when the objects are and are not visible (Battista, 2017a; Mix et al., 2016). Spatial visualization enables one to predict what

[1] There are many definitions of spatial visualization in the research literature. Our integrated definition seems most useful in thinking about geometric reasoning.

will happen when spatial operations are performed on objects. For our purposes, spatial visualization includes imagining rotations of objects in 2D and 3D space, imagining how objects look from different perspectives, how 2D patterns (nets) can be folded to make 3D boxes and vice versa, imagining a 3D object from its orthogonal projections and vice versa, and visually decomposing and recomposing shapes into parts.

Property-based spatial analytic reasoning decomposes an object (or set of objects) into its constituent parts using geometric properties to specify how the parts or shapes are related, and, using these relationships, operates on the parts in a way to answer questions about the whole. In geometry, analytic reasoning generally employs formal concepts such as measuring length, angle, area, and volume; congruence, parallelism, and isometries to conceptualize spatial relationships within and among shapes. We include in analytic reasoning the ability to represent spatial extent on a number line and to specify locations of objects in 2D and 3D coordinate systems.

In geometric problem solving, spatial visualization and property-based spatial analytic reasoning generally are used together either simultaneously or sequentially in a back-and-forth manner. Generally, in spatial analytic reasoning, students visualize (a) parts and operations on parts being analyzed, and (b) the big picture to integrate and monitor the parts-based analysis. And property-based analysis can affect holistic spatial visualization (e.g., visualizing the rotation of a rectangle to be a congruent rectangle).

Taking a complementary perspective on the visualization/analysis framework, Newcombe and Shipley (2015) and Uttal et al. (2013) organized the set of skills measured by spatial tests in cognitive psychology into a 2-by-2 classification system based on the four factors of intrinsic/extrinsic spatial information and static/dynamic tasks. They identified intrinsic spatial information as "the specification of the parts and relation between the parts, that defines a particular object," whereas extrinsic information "refers to the relation among objects in a group, relative to one another or to an overall framework" (Uttal et al., 2013, p. 353). Static versus dynamic tasks distinguish between objects that are fixed in comparison with objects that can move or change. Applying this classification system to the tasks we discuss in this chapter, determining the spatial relations between the sides of a parallelogram deals with the intrinsic spatial structure of the shape, while determining how a set of unit-cubes can be arranged into a 3D rectangular array deals with the extrinsic spatial structure of the set of cubes. Students' reasoning about 2D rotations by examining how a shape's pre-image and image are related can be a static analysis, while examining, visualizing, and reflecting on an animation of the motion of a rotating object is a dynamic analysis, especially in the special computer environment we describe.

Geometric Reasoning

Formal, pre-proof geometric reasoning consists of using formal conceptual systems to investigate shape and space (Battista, 2007). For instance, mathematicians employ a property-based conceptual system to analyze and define various types of quadrilaterals and triangles. This system uses concepts such as angle measure, length, congruence, and parallelism to conceptualize spatial relationships within and among the shapes. So, defining a square to be a four-sided polygon that has four right angles and all sides the same length and parallel creates an idealized property-based concept that precisely describes the critical spatial relationships that exist between the sides and angles in the class of shapes we label as squares.

Geometric Properties

Properties play an essential role in all geometric reasoning. *Geometric properties* are precise, formal specifications of critical attributes of geometric objects and transformations. Learning geometric properties is an essential component of developing geometric reasoning (Battista, 2007; Gorgorió, 1998; van Hiele, 1986). As will be illustrated later in this chapter, spatial reasoning that supports valid geometric reasoning is accomplished using mental models that have incorporated appropriate geometric properties into their structure and operation. In contrast, students generally make reasoning errors when (a) they perform spatial operations on geometric objects that violate the objects' geometric properties, or (b) they apply property-based knowledge without the appropriate support of spatial visualization.

Some researchers (e.g., Sinclair et al. and Lowrie et al., this book) suggest that the traditional instructional focus on geometric properties may somehow limit the integration of spatial reasoning in geometry. Certainly, that may be true if the focus is on students memorizing verbal statements of properties. However, as we demonstrate in this chapter, property reasoning and spatial reasoning can and should be deeply and inextricably related. In fact, one cannot really "do geometry" without understanding the properties of geometric objects; for doing geometry, as opposed to spatial reasoning in general, means reasoning about spatial objects using formal geometric concepts. In subsequent sections, we specify in detail how spatial reasoning is related to the properties of shapes, isometries, and measurement.

Differentiating Geometric Properties. A critical issue in understanding students' development of geometric reasoning is specifying which properties of geometric objects are most important for them to learn. Initially, the properties that are critically important to students' learning about geometric objects are properties that express prototypical, defining characteristics of those objects, which we call "*prototypical defining properties*." As an example, the prototypical defining properties of rectangles are: opposite sides congruent and parallel, and four right angles. These properties express in formal geometric terms the most visually salient spatial characteristics that students use in identifying rectangles. Of course, there are other, less visually salient properties of rectangles. For instance, in rectangles, the congruent diagonals bisect each other. This property, that "the diagonals are congruent and bisect each other," could be used to define rectangles. However, it is the prototypical defining properties that students typically derive from visual examples of rectangles and that students use to determine if a shape is a rectangle or not. Geometric properties are so important that they play a central role in learning progressions for geometric shape. For example, the progression outlined in Table 10.1 (Battista, 2007, 2012b) describes how students progress from thinking about shapes holistically, to decomposing shapes into their basic components and specifying their properties by spatially interrelating these components, to logically interrelating properties, to understanding and creating geometric proofs within an axiomatic system.

Mental Models

An abundant amount of research indicates that many forms of reasoning are accomplished with mental models (Battista, 1994; 2007; Calvin, 1996; English & Halford, 1995; Greeno, 1991; Hegarty, 2004; Johnson-Laird, 1983; Markovits, 1993). According to mental model theories, individuals understand and make sense of a situation when they activate or construct, then operate on, a mental model to represent the situation (Johnson-Laird, 1983). Mental models are nonverbal recall-of-experience-like mental versions of situations that have structures isomorphic to the perceived structures of the situations they represent (Battista, 1994; Greeno, 1991; Johnson-Laird, 1983, 1998). That is, "parts of the model correspond to the relevant parts of what it represents, and the structural relations between the parts of the model are analogous to the structural relations in the world" (Johnson-Laird, 1998, p. 447). Individuals can reason by operating on a mental model because "the behavior of objects in the

Table 10.1 Learning progression for geometric shapes

Level	Sub-level	Description
1		**Student identifies shapes as visual-wholes**
	1.1	Student incorrectly identifies shapes as visual-wholes
	1.2	Student correctly identifies shapes as visual-wholes
2		**Student describes parts and properties of shapes**
	2.1	Student informally describes parts and properties of shapes
	2.2	Student uses informal and insufficient formal descriptions of shapes' properties
	2.3	Student formally describes shapes' properties completely and correctly
3		**Student interrelates properties and categories of shapes**
	3.1	Student uses empirical evidence to interrelate properties and categories of shapes
	3.2	Student analyzes shape construction to interrelate properties and categories of shapes
	3.3	Student uses logical inference to relate properties and understand minimal definitions
	3.4	Student understands and adopts hierarchical classifications of shape classes
4		**Student understands and creates formal deductive proofs**

model is similar to the behavior of objects that they represent" (Greeno, 1991, p. 178). Indeed, much of what happens when we form and manipulate a mental model reflects our underlying knowledge and beliefs about what would happen if we were dealing with the objects they represent. So, the properties and behavior of objects in a mental model simulate the *properties* and behavior we believe the objects they represent possess. When using a mental model to reason about a situation, a person can mentally view; mentally move around, on, or into; decompose and combine; and transform objects, as well as perform other operations like those that can be performed on objects in the physical world.

Johnson-Laird and colleagues argued that it is important to distinguish between visual images and spatial mental models. Visual images are derived from perception or generated from mental models, with images being static views of objects or mental models from particular viewpoints. Images are mentally manipulable (operable) only when they are embedded in appropriate mental models (Johnson-Laird, 1998). Thus, in spatial reasoning, sometimes people merely construct and scan visual images, and other times they operate on spatial mental models (Knauff, Fangmeier, Ruff, & Johnson-Laird, 2003). Furthermore, sometimes, when aspects of visual images are irrelevant to a task, the evocation of the images can actually interfere with construction and use of an appropriate mental model and thus impede reasoning (Knauff et al., 2003). Additionally, according to Johnson-Laird, "The operations that are carried out in reasoning with models … are conceptual and semantic" (1998, p. 457). Thus, to reason geometrically about a spatial entity (object, diagram, or concept), one must construct a proper mental model of the entity, one that captures its relevant spatial structure and geometric properties.

As an example of the nature of mental models used in geometry, consider visualizing ±90° rotations in a coordinate grid. We know from our current research with students that, in reasoning about ±90° rotations of triangles, for instance, many students visualize/imagine an L-shaped figure attached to the triangle, with one side of the L horizontal and the other side vertical. The L-strategy is important because it seems to make accessible to students the integration of visualization and analytic measurement-based reasoning (e.g., it is easy to determine the lengths of the vertical-horizontal legs on a grid). We now look closely at this situation because it illuminates the integration of visualization and analytic strategies that is so critical in geometry.

Consider the process of visualizing a +90° rotation of Triangle A and the connected "3:9 L" indicated in Fig. 10.1. A person can visualize the rotation of the "3:9 L" and triangle in Fig. 10.1 when they have constructed, through abstraction, a mental model that preserves the configuration's static and dynamic properties. Such a mental model could be structured to incorporate the following properties of a +90° rotation:

1. Rotations preserve the lengths of line segments (e.g., the image of a line segment that is three units long is a line segment that is three units long).
2. Rotations preserve angles (so the right angle between the 3-side and 9-side is preserved, as is the right angle connecting the 9-side to the right triangle).
3. Rotations map shapes to congruent shapes (so the image of the right triangle is a congruent right triangle, and the image of a 3:9 L is a congruent 3:9 L).
4. 90° rotations map vertical segments to horizontal segments, and horizontal segments to vertical segments.

Fig. 10.1 +90° rotation of triangle and connected 3:9 L

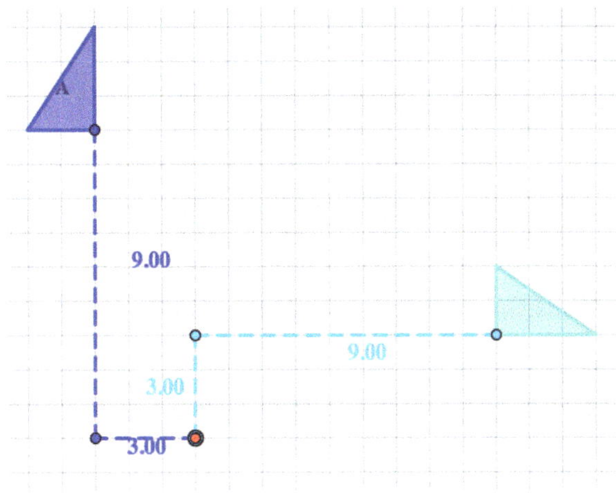

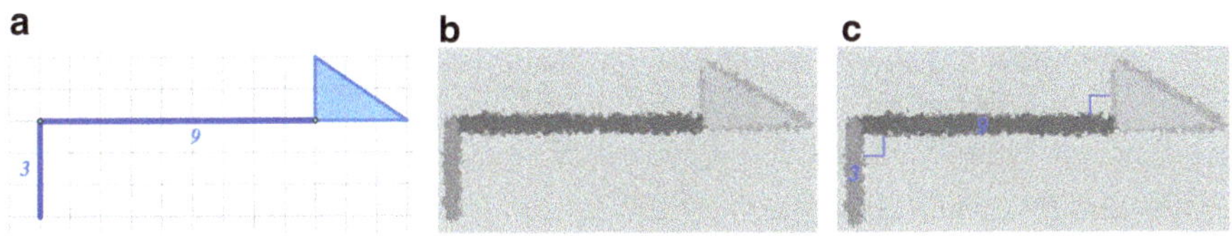

Fig. 10.2 What shape-rotation visualization models look like—tokens, not detailed pictures (**a**) Screen object. (**b**) Visualized spatial token for screen object. (**c**) Visualized spatial token for screen object with connected properties

It is highly likely that people do not visualize rotating exact L side-lengths as detailed pictures, but instead use mental models consisting of structure-preserving spatial tokens. For example, consider visualizing a +90° rotation of the 3:9 L and triangle configuration shown in Fig. 10.2a. In Fig. 10.2, we can see (a) a shape that gets rotated on the computer screen; (b) the visualized spatial token for this screen object that gets rotated in a mental model; and (c) the conceptual properties (in blue) that the mental model connects to the image components (i.e., specific length and angle specifications). No unit-lengths appear on the spatial tokens for the 3- and 9-sides; there are just mental connections to the measurements (almost like hot text, the measurements are not pictured, but can be retrieved when needed).

Mentally connected to the sparse-visual-detail spatial token in Fig. 10.2b is conceptual knowledge that the image of the vertical 3-side is a horizontal 3-side and the image of the horizontal 9-side is a vertical 9-side for ±90° rotations. We visualize rotating tokens without visual detail to reduce the cognitive load of the visualization, since the process of visualizing a rotation can be easily overwhelmed if too much detail is preserved throughout the duration of the rotation. After the token rotation is completed, so that we know where the component images must be placed, we can use the connected conceptual knowledge to place the images with correct lengths on the grid. Visualizing where the 3-side rotates, for instance, we can correctly place its 3-side image on the grid. This example illustrates well how spatial visualization can be connected to spatial analytic reasoning.

Structuring

Spatial and geometric structuring are types of spatial and geometric reasoning, respectively, that play vital roles in the construction of appropriate mental models for geometric reasoning (Battista, 1999, 2007, 2008).

Spatial structuring. Spatial structuring is the mental process of constructing a spatial organization or form for an object or set of objects. Spatial structuring determines a person's perception/conception of an object's spatial nature by identifying its parts, combining parts into spatial composites, and establishing spatial interrelationships between and among parts and composites. For example, different spatial structurings of a quadrilateral might cause it to be perceived as a closed path consisting of sequence of four straight line segments, as a composite of four line segments connected at their endpoints, or even as a composite of four connected but partially overlapping angles. As a spatial structuring is incorporated into a person's mental model for a spatial object, it determines the person's mental representation of the object's spatial essence, and it enables the person to mentally manipulate, reflect on, analyze, and understand it. For instance, one way we distinguish different types of quadrilaterals is by decomposing them into sides, then establishing spatial relationships between their sides. Two spatial structurings that are surprisingly difficult for many elementary school children to distinguish are the opposite versus adjacent side equality relationships shown in Fig. 10.3a (Battista, 2012a, 2017b). Once the distinction has been made, students might spatially structure and reason about them as "sides across from each other are even" for the first arrangement and "sides next to each other are even" for the second arrangement.

In a similar manner, we can spatially structure a right rectangular prism (a rectangular box) by decomposing it into its edges and faces and noting that "sides" (faces) across from each other (opposite) are "the same" (congruent), opposite faces are "lined up" (parallel), and that all intersecting edges form "square corners" (are perpendicular) (Fig. 10.3b).

Geometric structuring. Geometric structuring specifies spatial structuring using formal geometric concepts. That is, to geometrically structure a spatial entity, a person uses formal geometric concepts such as angle measure, slope, parallelism, length, coordinate systems, and geometric transformations to conceptualize and operate on the entity. So, for example, and as will be explained in more detail later, a parallelogram might be spatially structured as a *visual configuration* that might be intuitively described as having "two pairs of even sides that are across from each other"—the spatial structuring is a visual mental model (not the verbal description). A geometric structuring of this same characteristic is an explicit, verbally stated conceptualization of this imagistic relation in terms of formal geometric concepts—a parallelogram is a four-sided polygon with opposite sides parallel and congruent. As another example, the two spatial structurings shown in Fig. 10.3a can be geometrically structured by stating that parallelograms have "two pairs of opposite sides congruent" and kites have "two non-intersecting pairs of adjacent sides congruent."

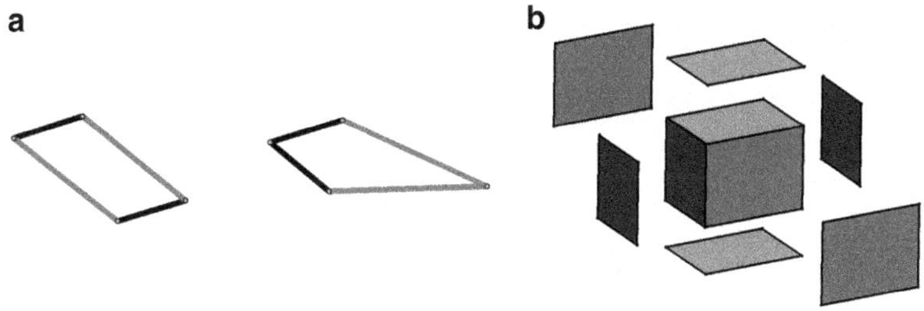

Fig. 10.3 Quadrilaterals and 3-D rectangular box

Critical relation between spatial and geometric structuring. For a geometric structuring of an object to make sense to a person, it must evoke an appropriate spatial structuring of the mental model for that object. Conversely, to geometrically structure an object, a person must have abstracted its spatial structure to a sufficient degree that it is accessible to decomposition, analysis, and formal specification (Battista, 2007).

Method

The research studies reported in this chapter were conducted in several NSF-funded projects. In each project, researchers conducted one-on-one interviews, one-on-one teaching experiments (Steffe & Thompson, 2000), or case-studies of pairs of students in classroom teaching experiments, all of which were video-recorded and transcribed. For all individual interviews and teaching experiments, elementary or middle school students worked problems while sitting with researchers who asked students to think-aloud and asked questions about students' thinking.

The examples on isometries, length, and angle are taken from an online, computer-based, individualized, interactive Dynamic Geometry learning system. The *Individualized Dynamic Geometry Instruction* (*iDGi*)[2] project focuses on the Common Core Standards for School Mathematics (*CCSSM*) topics in 2D geometry and measurement for grades 3–8, with the curriculum sequenced to be consistent with research-based learning progressions (Battista, 2012a, 2012b). The iDGi computer modules provide students with manipulable screen objects, like a parallelogram maker that has draggable vertices, and animations, like a triangle rotating 90° about a given point.

[2] All iDGi software and screens, Copyright 2012–2018.

Part 1: Geometric Reasoning About Shapes and Isometries

Example from Study 1: Property Based Visualization and Reasoning About the Structure of Shapes

Consider a student's reasoning about the following task.

Task. A shape and its measurements are shown below.

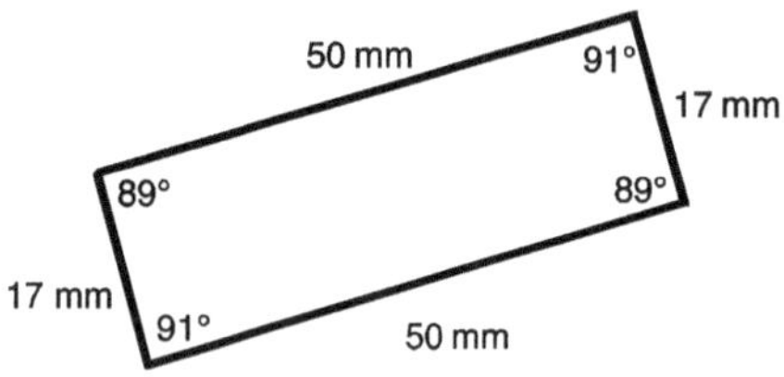

Tell whether the statements below are true or false. Describe how you would convince someone that your answers are correct.

(a) *The shape is a rectangle.*
(b) *The shape is a parallelogram.*

The dialogue below shows how one middle school student reasoned about this problem.

JO: (a) False. No, because a rectangle has to have 90°; this shape does not have all 90° angles, here and here and here and here [pointing to each vertex], it has 91, 91, 89, and 89.
Teacher: Why do you think that rectangles have all 90° angles?
JO: Because they wouldn't be rectangles if they didn't have all 90° angles?
Teacher: Why wouldn't they be rectangles if they didn't have all 90° angles?
JO: Because if the angles were too far [points to the vertices of the figure as he is explaining], it would be a parallelogram. *If they were too far inward or too far out here, it wouldn't be a rectangle, it would be a parallelogram.*
JO: (b) Yes. Because the opposite sides are equal. If you draw a quadrilateral that has both pairs of opposite sides equal, then the opposite sides are parallel too. *If the top and bottom sides are equal but not parallel, it makes the other pair of sides not equal* [student illustrates by making a drawing step-by-step].

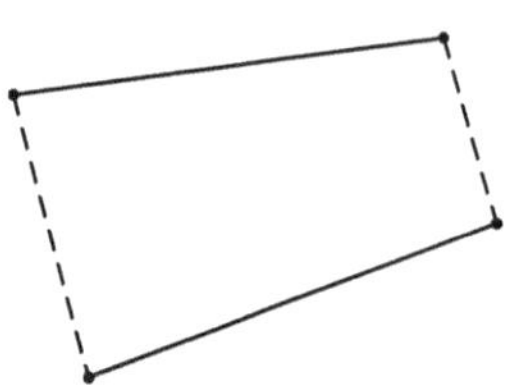

This student not only specified the prototypical defining geometric property that rectangles have four 90° angles, he explained why this property must be true using spatial reasoning; he explicitly related his geometric structuring to his spatial structuring. In saying that if the angles were not 90° they would be "too far inward or too far out," he made explicit his understanding of the spatial relation between adjacent sides in a rectangle, which is captured by the 90° angle measure. He related the formal concept of 90° angles to the spatial relation between pairs of adjacent sides of the shape. Furthermore, in his explanation for part (b), he spatially reasoned and visualized that if two equal, opposite sides of a quadrilateral are not parallel, the other two opposite sides cannot be equal.

While spatial visualization is sometimes viewed as a less sophisticated method of reasoning in comparison with formal mathematical reasoning, the example provided here illustrates that spatial visualization can be linked to geometric property knowledge in a way that allows the student to make sense of the properties of shapes. Appropriate spatial reasoning facilitates movement toward more formal mathematical thinking.

Example from Study 2: Property-Based Visualization and Reasoning About Isometries

For many spatial tasks, such as the Vandenberg Mental 3D Rotation Test, "most students reported using a mental imagery strategy (either imagining the rotation of the objects or imagining changing their perspective with respect to the objects), but there were also a variety of analytic strategies used, including spatial analytic strategies that abstract the relative directions of the different segments of the object, and more abstract analytic strategies in which participants counted the number of cubes in the different segments of the object" (Hegarty, 2010, p. 273).[3] Similarly, in students' learning about isometries—distance- and congruence-preserving transformations—both spatial visualization and spatial analytic reasoning are especially relevant. In Table 10.2, we adapt Hegarty's spatial strategy conceptualizations to apply to students' reasoning about 2D rotation isometries of a right triangle in the context of a grid. Importantly, in geometric reasoning, valid analytic strategies require knowledge and application of relevant geometric properties.

In this section we discuss students' understanding of the properties of rotations and their visualization of turning motions in ways that are consistent with the prototypical defining properties of rotations listed in Table 10.3 (Battista, 2007; Battista & Frazee, 2018; Battista, Frazee, & Winer, 2017). Battista et al. (2017) found that middle school students were successful in learning prototypical defining properties required to solve problems with translations, rotations, and reflections in the iDGi dynamic geometry environment.

[3] Many spatial analytic strategies decompose a problem into parts and operate on the parts so that less visualization is necessary at one time. Counting, measuring, and applying rules (like the volume formula) are common operations included in spatial analytic strategies (Hegarty, 2010).

Table 10.2 Adaptation of Hegarty's strategies for 2D rotation task

Mental imagery strategies
1.1. I imagined the triangle turning in my mind.
1.2. I looked at the turn center and imagined the triangle turning about it in my mind.
Spatial analytic strategies
2.1. I noted the directions of corresponding sides of the triangles and decided if they had been turned through the correct angle.
2.2. I counted the number of units up/down/right/left between the turn center and corresponding vertices.
2.3. I visualized a vertical-horizontal "L" connected to the turn center then I counted to find the lengths of corresponding sides of the L and examined the angles between these sides.
2.4. I visualized rotating a triangle side, then counted how long the preimage was to know how long the image is.

Table 10.3 Prototypical defining properties of rotations

Prototypical defining properties of rotations
P1. Rotations are determined by a turn center and an amount of turn specified as a signed amount of degrees.
P2. Preimage and image polygons have corresponding points (preimage and image point pairs).
P3. The angle between the turn center and any pair of corresponding points equals the rotation angle.
P4. Pairs of corresponding points are the same distance from the turn center.

In the problems discussed below, students are given a right triangle pre-image (the original triangle) and its rotation image on a square grid, and they must find the turn center, first, selecting from five possibilities, then, knowing only that the turn center occurs at a grid intersection point.[4]

Consider seventh grader MR, who on previous problems had evidenced knowledge of Properties P1-P4. As MR started problems in which no possible turn centers were shown, she elaborated her earlier developed vertical-horizontal counting strategy (shown in Fig. 10.4) to include an adjustment scheme for when her turn-center locating counting scheme failed.

MR: [Fig. 10.4.] I'm guessing this is another 90° problem so if I match these two [points X and Y]. This is 1-2-3-4-5-6-7-8-9-10-11-12, so 1-2-3-4-5-6-7-8-9-10.… So if I move it [turn center C] down 1 and across 1. Cause if I want to move it [C] across 1 to reduce this [distance from C to triangle A], I have to move this [C] down 1 so that it doesn't match up with this [triangle A] but not that [triangle B]. Because when it does that [not "match up"], it forms an angle like this [gesturing off-screen], which doesn't work, cause that's definitely not going to be 90°. So I'm going to move it down 1 and [across 1]… again [to C; Fig. 10.5.], so

[4] Students did this second type of problem after numerous problems in which they had to find the image of a point or right triangle, given the turn center and amount of turn, or choosing the turn center from five possibilities.

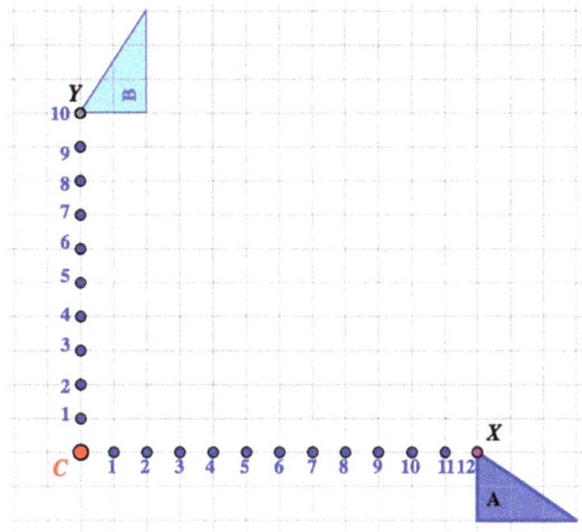

Fig. 10.4 MR's counting from first hypothesized turn center

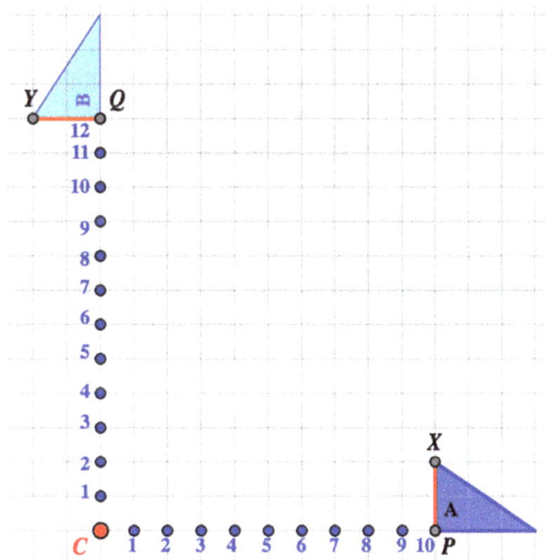

Fig. 10.5 MR's counting from second hypothesized turn center

> 2 again 2 again [indicating segments XP, YQ; Fig. 10.5.]. So that's 1-2-3-4-5-6-7-8-9-10-11-12 and that's 1-2-3-4-5-6-7-8-9-10 [Fig. 10.5.]. Ok, so up 1, across 1 [Fig. 10.6.]; so that should work [which she verifies by clicking on the appropriate angle rotation button].

In this problem, MR's spatial reasoning and knowledge of rotation properties were sufficient for her to implement an analytic strategy to locate the correct turn center for the rotation. She correctly found horizontal and vertical distances from corresponding points X and Y on the preimage and image triangles to the first point she thought might be the turn center (Fig. 10.4). Next, because the distance between corresponding points and the turn center must be equal, she made a reasonable spatial movement of the possible turn center to reduce the

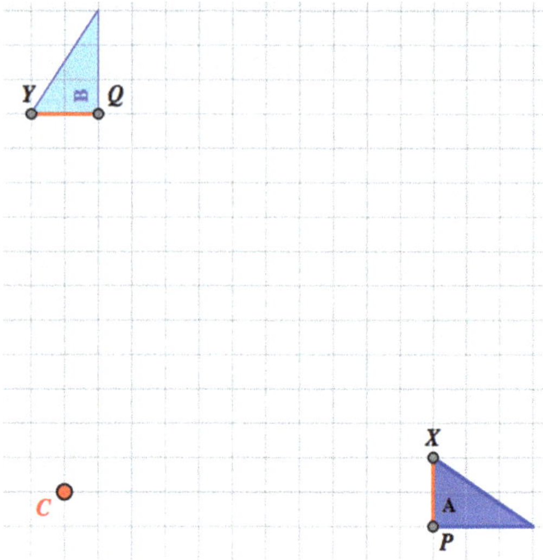

Fig. 10.6 MR's final hypothesized turn center

imbalance in distances to corresponding points—it reduced the distance that was too large (12) and increased the distance that was too small (10). But the distances were now unbalanced in the opposite way (Fig. 10.5). Finally (Fig. 10.6), she made an adjustment that equalized the distances from the possible turn center to midpoints on corresponding sides of the triangles. Each move, from placing her original estimate for the turn center in Fig. 10.4, to adjusting her turn-center estimates with her horizontal and vertical counting, required spatial reasoning and analytic reasoning-based operational knowledge of the property that the distance of the turn center to corresponding points on the pre-image and image must be equal.

However, on several other problems, MR's reasoning was hindered by visualization difficulties. For instance, as shown below, sometimes she failed to recognize correct rotation angles while implementing her analytic counting strategy.

MR: This is going to be 90° so it's going to be here or there [indicates circular regions in Fig. 10.7]. … You know I think I'm going to actually put it [turn center C] up here [in the upper left circular region in Fig. 10.7]. So that's 1-2-3-4-5-6-7 and 2 across [counts up from Triangle B and left from C; Fig. 10.7]. So that has to be 1-2-3-4-5-6-7 and 2 up [counts left and up from Triangle A; Fig. 10.7]. Which doesn't work. Or 2 across and 7 up, which doesn't work. [Moves turn center C; Fig. 10.8] So then this is 1-2-3-4-5-6-7-8 across and 1-2-3-4-5-6-7 up [counts from Triangle B]....Ok, so then this is 1-2-3-4-5-6-7-8 ac—[counts from Triangle A]. Wait 8 across and 7 up [from Triangle B], so this would be 7 across and 8 up [from Triangle A—moves cursor along segments indicated in Fig. 10.8], right?

In this problem, MR repeatedly tried to use her analytic vertical-horizontal counting strategy, but failed to see that rotation was 180°, not 90°, until later when

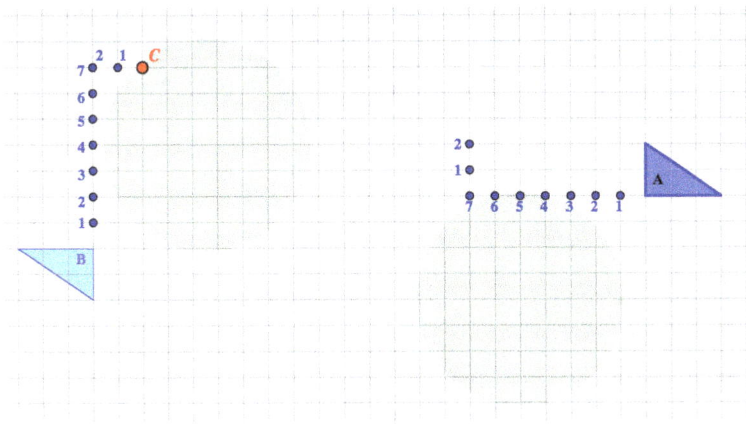

Fig. 10.7 MR's hypothesized turn center location regions for 90° rotation

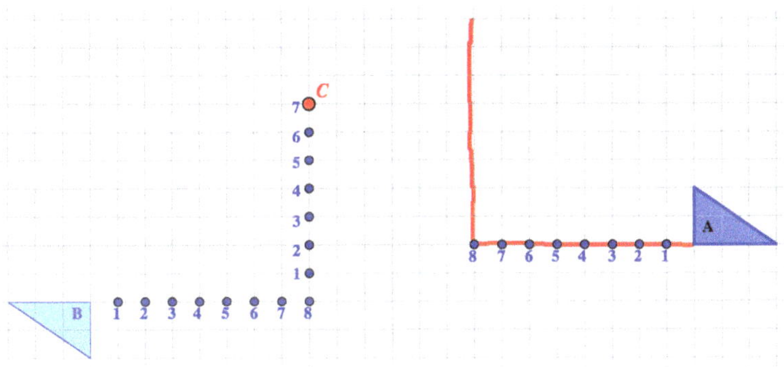

Fig. 10.8 MR's hypothesized turn center for 90° rotation

her interviewer specifically asked her about the rotation angle. Her spatial visualization, which should have indicated to her that this was not a 90° rotation, was not properly guiding her analytic strategy. Also, MR's visualization difficulties often caused her to make errors in determining which way to move the turn center to correct failed up-down/across guided placements. Note also that MR understood that for some rotations (±90°), across counting moves from one triangle to the turn center turned into up/down moves for the other triangle, consistent with the L strategy described in an earlier section, which of course did not work here because the rotation was 180°.

In summary, MR used her knowledge of the properties of rotations to develop an analytic strategy for finding turn centers that was quite effective when she simply had to choose the turn center from five possibilities and problems in which she could visually approximate the location of the turn center when no turn centers were provided. But in some more difficult turn center location problems in which no turn center alternatives were provided, her strategy broke down because she was not able to use spatial visualization to properly guide her analytic strategy. Consequently, *we hypothesize that in many situations, spatial reasoning is needed to monitor an analytic strategy, joining together the components of the analytic reasoning into a coherent whole.*

Part 2: Reasoning About Geometric Measurement

Measurement Properties

For spatial reasoning and numerical reasoning to be properly connected in geometric measurement, certain basic properties of measurement functions must be followed, as described by Krantz, Luce, Suppes, and Tversky: "When measuring some attribute of a class of objects or events, we associate numbers … with objects in such a way that properties of the attribute are faithfully represented as numerical properties" (1971, p. 1). That is, if M is the function that assigns measurement values to objects—so $M(a)$ is the measure of object a—then, consistent with Krantz et al.'s properties and basic axioms for geometric measurement (Moise, 1963), M satisfies the following properties:

1. If object a and object b are congruent, then $M(a) = M(b)$.
2. Object a is spatially larger than object b if and only if $M(a) > M(b)$
3. If object a and object b are disjoint, and we join object a and object b (object a union object b), then:

$$M(a \text{ joined with } b) = M(a \cup b) = M(a) + M(b)$$

4. Given n copies of congruent and non-overlapping unit-measure objects $a_1 \ldots a_n$:

$$\text{If } \bigcup_{i=1}^{n} a_i = b, \text{ then } nM(a_1) = M(b)$$

These properties justify the measurement iteration process in which we determine a measurement by iterating a unit-measure to cover or fill the object being measured with no gaps or overlaps. If there are gaps in a unit-measure covering so that it is a proper subset of the object being measured, then Property 2 implies that the measure of the covering will be less than the measure of the object. If there are overlaps, then Properties 3 and 4 are not satisfied, so we cannot count/add the unit measures to find the measure of the object.

Spatial-Numerical Linked Structuring

Spatial analytic reasoning in geometric measurement requires not only knowledge of measurement properties, but what we call *spatial-numerical linked structuring* (SNLS). As was already described, spatial structuring is the mental act of constructing a spatial organization or form for an object or set of objects. *Numerical structuring* is the mental act of constructing an organization or form for a set of computations. *Spatial-numerical linked structuring in measurement* is a coordinated process in which numerical operations on measurement numbers are performed based on a linked spatial structuring of the measured objects in a way that is consistent with properties of numbers and measurement. Incorrect measurement is generally based on a SNLS that violates at least one of the measurement properties. Note that each measurement property expresses a generalized spatial-numerical linked structuring. In this part of the chapter, illustrations of spatial-numerical linked structuring (some correct, some incorrect) are provided for angle, length, area, and volume. The examples describe and discuss student actions and interpret them using the spatial-numerical linked structuring conceptual framework.

Order of Structuring. The normal order of activation of SNLS for meaningful geometric reasoning is (1) spatial structuring, then (2) numerical structuring (even though it may appear that both types of structuring are happening simultaneously). However, sometimes, because of premature teaching of formulas, the order is switched: (1) numerical structuring, then (2) spatial structuring. This latter activation pattern is typical of procedural reasoning without proper grounding in conceptual understanding, as shown in the following example.

JK: The volume of a box is length times width times height.
Int: Do you know why that equation works?
JK: Because you are covering all three dimensions, I think. I'm not really sure. I just know the equation.

Example from Study 3: Spatial-Numerical Linked Structuring for Angles

For the computer-presented problem in Fig. 10.9, KS employed several spatial-numerical linked structurings to find the angle that rotates the green point onto the red point.

KS: I think it may be 40 [enters 40; green ray rotates to the 40° position in Fig. 10.10].
Int: So what are you thinking?
KS: So if this is 40 [angle in Fig. 10.10], I may have to go up maybe *20 more*.
Int: Okay, why 20 more?

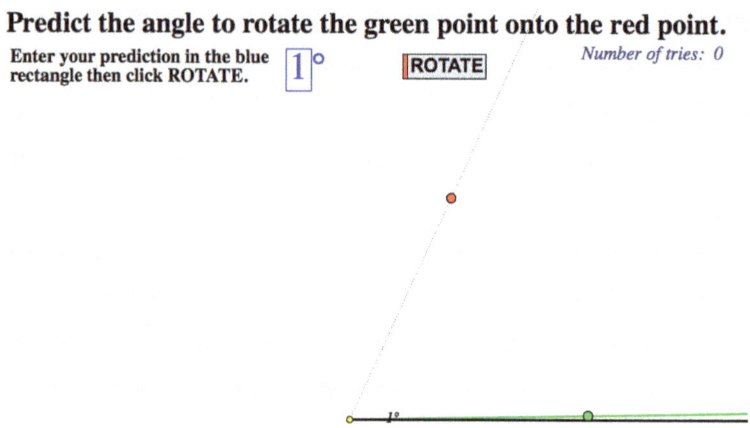

Fig. 10.9 Computer-presented problem for KS

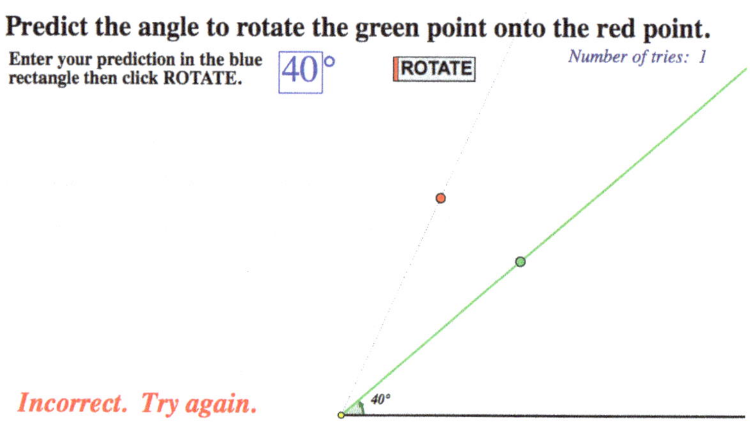

Fig. 10.10 KS's first angle prediction

KS: Cause, if this was 40 [pointing at the interior of the green 40° angle], then half of it is this [pointing to the interior of the angle between 40° and the target angle; enters 60°; Fig. 10.11].

Int: Very close, what are you thinking?

KS: Hum. So maybe with the other [computer page showing 5° iterations of a ray] it shows that they were really close together, so maybe it'd be 65 [enters 65°; Fig. 10.12].

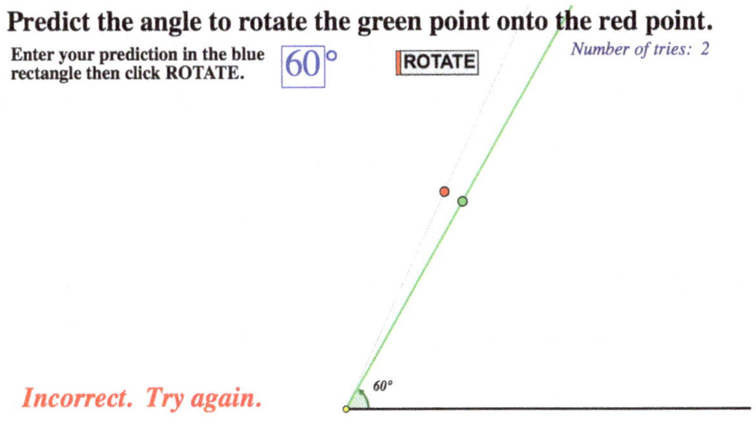

Fig. 10.11 KS's second angle prediction

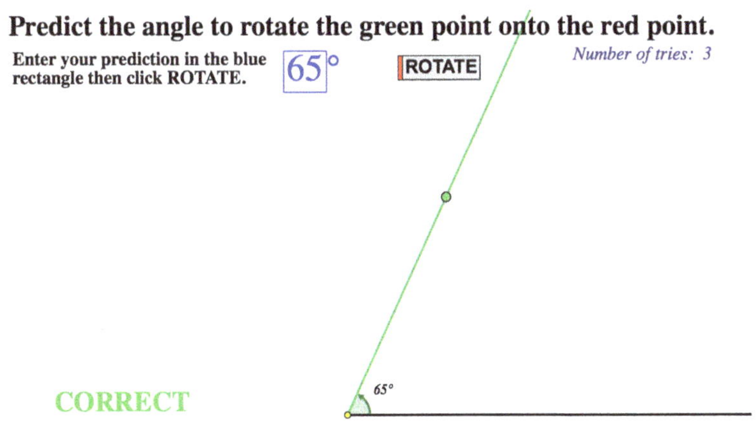

Fig. 10.12 KS's final angle prediction

The sequence of spatial-numerical linked structurings KS used in solving this problem are shown in Table 10.4.[5] After viewing the result of her first estimate, which is quite a bit off, KS reasoned that her original estimate was too small. This is an example of SNLS 1 in which KS implemented a smaller-angle-smaller-measure operation for angles. KS then, using SNLS 2, implemented a half-angle-half-measure operation for angles. Then, using SNLS 3, she added 20° to 40° to produce a second estimate of 60°. Finally, in her third estimate, KS used SNLS 4 followed by SNLS 3 to recall a previously viewed 5° angle and add it to her 60° estimate.

[5] The more formal expression for what students often describe as "Angle X equals Angle Y plus Angle Z" is "Angle X is congruent to Angle Y union with Angle Z."

Table 10.4 Definitions of types of angle spatial-numerical linked structuring

SNLS 1. [Bigger Angle ⇔ Greater Measure; Measurement Property 2]: If Angle X is bigger than Angle Y, then the measure of Angle X is greater than the measure of Angle Y. *Definition: Angle X is spatially "bigger" than Angle Y if the angles have the same vertex and Angle Y fits in the interior of Angle X.*
SNLS 2. [One-half angle ⇔ One-half measure]: If Angle X is one-half of Angle Y, then the measure of Angle X equals one-half of the measure of Angle Y. *Definition: Angle X is one-half of Angle Y if the initial side of Angle X coincides with the initial side of Angle Y and the terminal side of Angle X bisects Angle Y into two equal angles the same size as Angle X.*
SNLS 3. [Add angles ⇔ Add measures; Measurement Property 3]: If Angle X "equals" Angle Y "plus" Angle Z, then the measure of Angle X equals the measure of Angle Y plus the measure of Angle Z. *Definition: If point D is in the interior of Angle ABC, then Angle ABC equals Angle ABD "plus" [joined with] Angle DBC.*
SNLS 4. [Compare perceived angle to recalled angle; Measurement Property 1]: The student compares a perceived angle to the recalled visual image of a previously seen angle, and says that the two angles are congruent so their measures are equal.

Example from Study 4: Spatial-Numerical Linked Structuring for Length

To examine the way students use spatial-numerical linked structuring with length, we consider a student's work in a computer golf game (Fig. 10.13). Students "putt" a ball by entering a distance and angle. When students click the *PUTT* button, the ball travels to the right the entered distance, then arcs around counterclockwise as it sweeps out the entered angle, which is a multiple of 5° (Fig. 10.14). Students receive visual feedback on each of their estimates until they determine a correct putt angle and distance. SJ was working the problem in Fig. 10.13.

SJ: [Pointing along hash marks 0–140 on the number line with the cursor] These lines are the pixels right?

Int: Yep. So this [pointing with a finger] is 100 pixels. That's 200 [pointing]. So they might be counting by, what do you think, in those little ones [points to hash marks between 100 and 200 on the number line]?

SJ: 25s?

Int: So let's see. If this is 100 [pointing to 100]. That'd be 125 [pointing to 110], 150 [pointing to 120], 175 [pointing to 130], 200 [pointing to 140].

SJ: Aw, never mind.

Int: So what do you think?

SJ: 10, 15, [points along hash marks 110 to 190 on the number line] 45. No [goes back to 110 on the number line]. *Oh, tens!*

SJ understood that each hash mark indicated the same amount of linear extent (Measurement Property 1), but she could not immediately determine the correct numerical value of the distance between hash marks. When she estimated 25 as the distance, the interviewer used a correct SNLS that iterated by 25 starting at the

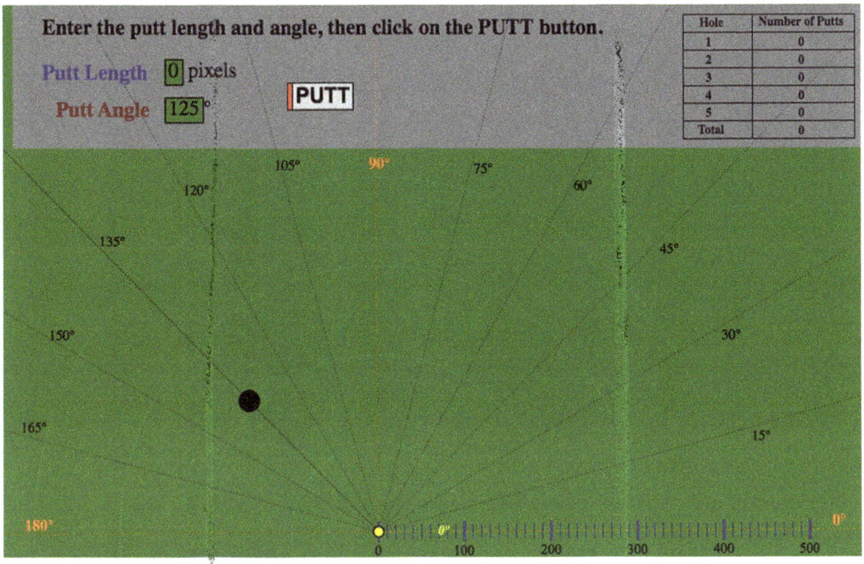

Fig. 10.13 Computer golf game

Fig. 10.14 Image of a putt in computer golf game

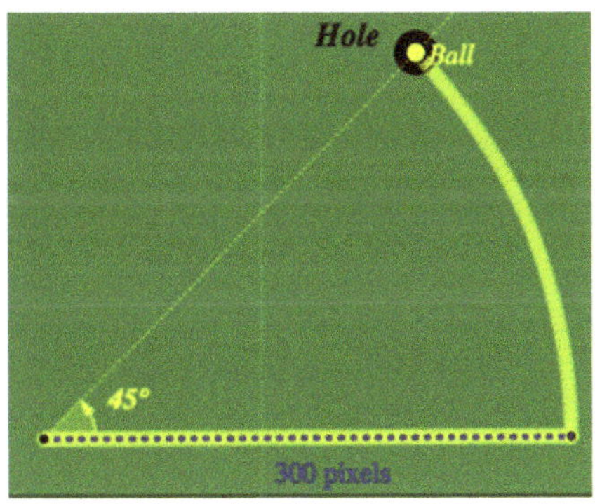

landmark for 100 so that SJ recognized that 25 was too large. After iterating with a value of 5 and realizing it was too small, she correctly concluded 10 as the distance between hash marks. This is an example of a student using the measurement properties and an iteration SNLS to develop an understanding of the coordinate system inscription embedded in the game.

Another example of SJ's SNLS while playing the golf game is presented below and shown in Fig. 10.15 and Fig. 10.16.

SJ: [For the problem in Fig. 10.15] Okay. This one is probably going to be 50 [points the cursor at 50 on the number line]. Because like 10, 20, 30, 40, 50 [counting on the 10–50 hash marks with the cursor]. Here's the 50 [moves from 50 toward the hole; Fig. 10.16]. Maybe even 60.

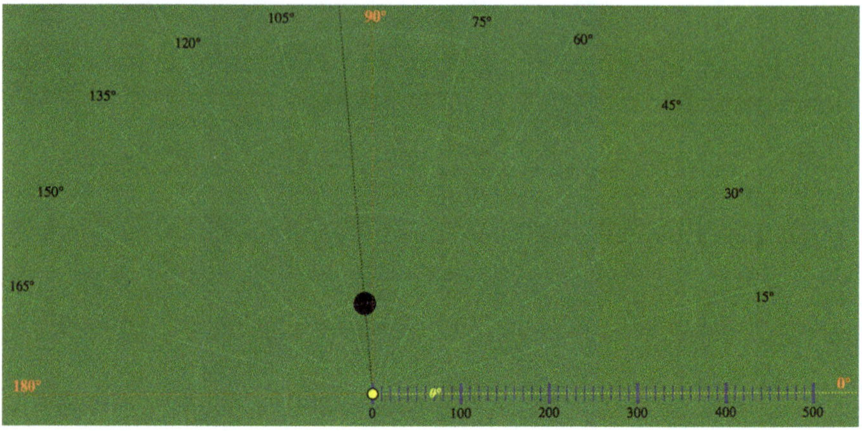

Fig. 10.15 Putt problem for SJ

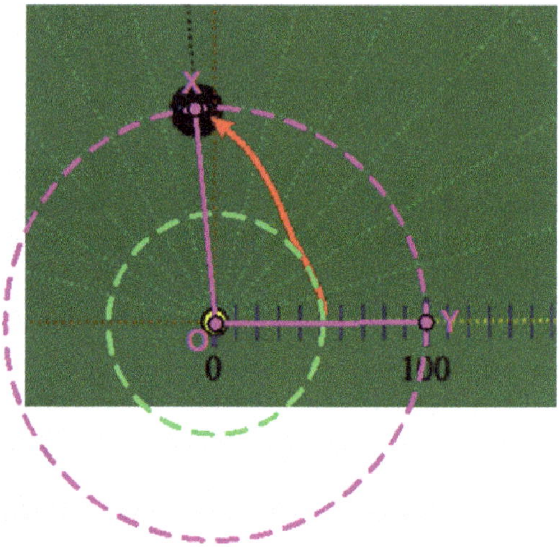

Fig. 10.16 SJ's spatial structuring of the rotation path of the ball

In this example, because SJ's spatial structuring of the rotation path of the ball is incorrect, her numerical choice for the length of the putt was incorrect. Furthermore, SJ did not seem to understand the meaning of the distance-arc inscriptions in this coordinate system. Because of her incorrect structuring of a point rotation, she did not recognize that every point on a 100-pixel circular distance arc is the same distance from the origin as the reference measurement on the number line. Similar to many elementary students using rectangular coordinates (Battista, 2007; Sarama, Clements, Swaminathan, McMillen, & González Gómez, 2003), SJ did not properly conceptualize the spatial-structural metric properties of the polar-coordinate-based game system.

Importantly, the spatial numeric linked structurings required to putt the ball are more complex than interpreting number lines because they require a coordinated understanding of angle and length in the context of a rotation. One aspect of spatial structuring for the golf game is to correctly determine the ball's rotation path. The rotation path from preimage (original) point to the image point is a portion of a circle centered at the origin. The circle is determined by the linear extent between the hole and the turn center (the circle's radius) and the amount of rotation, which is the angle between the hole and the x-axis. Thus, understanding the ball's rotation path requires a coordinated understanding of both angle and length measurement.

One crucial element of the spatial numeric linked structurings in this example is the recognition of the structure of the ball's path as the set of points equidistant from the origin and passing through 50 on the number line. Although SJ needed to spatially structure the set of points that are 50 units away from the origin as the green circle in Fig. 10.16, she instead structured this set of points as the path shown by the red arrow. Additionally, she needed to link that circular spatial structuring to view the length of the 0-to-50 segment on the reference number line as the radius of this circle. In this case, the correct spatial structuring is seeing the path of the ball as an arc on a circle with center the origin and radius the segment from the origin to the 100 mark on the x-axis. The correct corresponding numeric structuring is seeing this circle and arc as having a radius of 100 and the extent of the arc as determined by the 95° angle from the hole to the origin to point Y, as shown in Fig. 10.16. Because Segment YO can be rotated about the origin to segment XO, and because isometries preserve length, Segment XO is congruent to Segment YO (the spatial structuring). Therefore Length XO = Length YO =100 (the linked numeric structuring).

Example from Study 5: Spatial-Numerical Linked Structuring for Area

The ability to mentally construct an accurate spatial structure for rectangular arrays is a critical reasoning process for students finding areas of rectangles. But this array structure is surprisingly difficult for students to construct (Battista, Clements, Arnoff, Battista, & Borrow, 1998). For example, student CS was asked to determine the number of squares required to completely cover the inside of the rectangle in Fig. 10.17a (Battista et al., 1998). CS correctly counted the pre-drawn squares, but, because of inadequate structuring, her counting of interior squares was incorrect (Fig. 10.17b). Her spatial structuring of interior squares violated the measurement properties because the squares were either overlapping or were different sizes.

A somewhat more sophisticated level of spatial structuring and linked numerical structuring was exhibited by BI, who was enumerating the number of squares that cover the rectangle shown in Fig. 10.18a.

BI: First I count the bottom and there's 6. [Moving his hands inward as shown in Fig. 10.18b] So the top and bottom would equal 12. And these 2 [pointing to the middle squares on the right and left sides] would be 14. [Using fingers to estimate where individual squares were located] I'd say maybe 12 in the middle; 12 + 12 = 24. So I'd say 24.

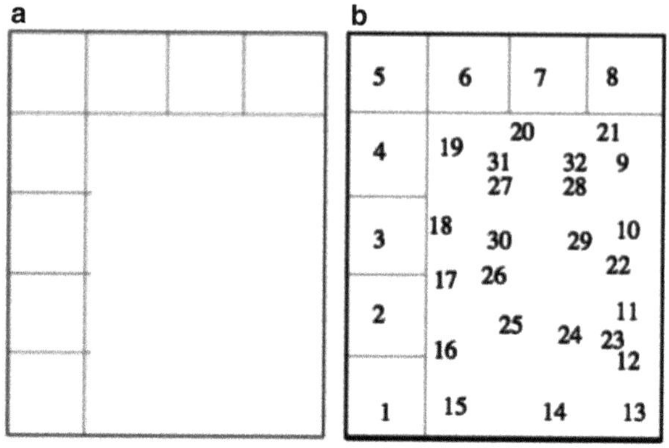

Fig. 10.17 CS's structuring and counting of imagined squares

Fig. 10.18 BI's reasoning

Unlike CS, BI structured squares in groups, which gave rise to the addends in his additive numerical structuring. However, BI was unable to spatially structure the interior squares, so he just guessed his final addend.

Spatial Numeric Linked Structuring (SNLS) used by BI:
Spatial structuring: [bottom row ∪ equivalent top row ∪ right/left side left-overs ∪ middle]
Linked numerical structuring: [6 + 6 + 2 + 12][6]

[6] Think of the "union" of two spatial objects as the spatial object gotten by joining the two objects.

PT enumerated the squares that cover the interior of the rectangle shown in Fig. 10.19a as follows.

PT: [Counting squares in the bottom row as in Fig. 10.19a] 1, 2, 3, 4, 5, because 1, 2, 3, 4, 5 [pointing to the pre-drawn squares as in Fig. 10.19b]. … Just bring them down to make one row [motioning to the bottom row with his fingers].
Int: Can you predict how many altogether?
PT: 5, 10, 15, 20 [pointing to successive rows, as in Fig. 10.19c].

PT inferred that there were five squares in the bottom row because he visualized moving the pre-drawn squares downward to establish their horizontal positioning in the bottom row. He then visualized that stacking four rows vertically composed the

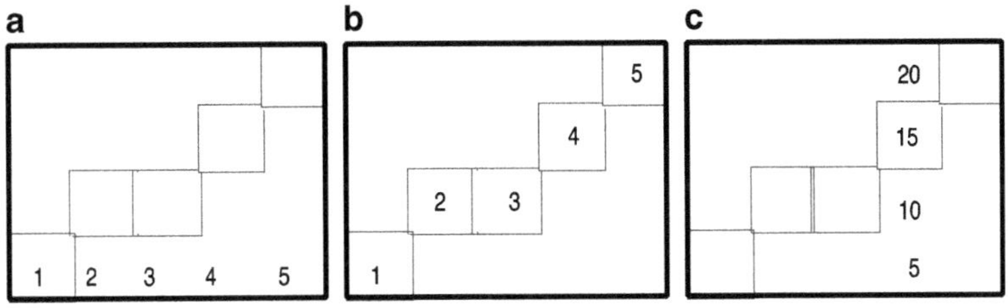

Fig. 10.19 PT's reasoning

whole rectangular array. Thus, PT spatially structured the squares that cover the rectangle into a row-by-column array. Then, because he decomposed the array into an iteration of row composites of five squares, he applied a multiplicative numerical structuring of skip counting by fives to enumerate the squares.

Spatial Numeric Linked Structuring (SNLS) used by PT:
Spatial structuring: [rectangular array: 4 rows, 5 in each row]
Linked numerical structuring: [5, 10, 15, 20]

As area problems become more difficult than counting unit-squares, the required spatial reasoning becomes more sophisticated as students have to start operating on composites of unit squares. Using iDGi area modules, KA, a sixth grader, developed an understanding of how to determine the area of a polygon drawn on a square grid by decomposing it into rectangles and right triangles (Battista, 2017b) (see Fig. 10.20). The SNLS for these problems is shown below.

Fig. 10.20 First type of area decomposition problem for KA

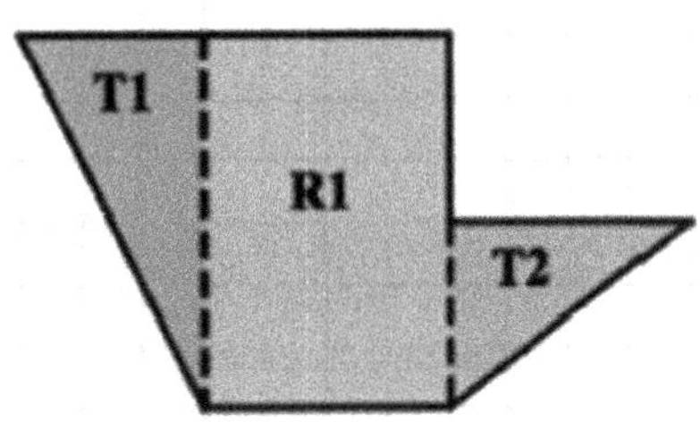

Spatial Numeric Linked Structuring (SNLS) used by KA:
Spatial structuring:
Whole Figure = Triangle T1 ∪ Rectangle R1 ∪ Triangle T2
Linked numerical structuring:

$$\begin{aligned}\text{Area Whole Figure} &= \text{Area Triangle T1} + \text{Area Rectangle R1} + \text{Area Triangle T2}\\ &= L\times W/2 \qquad + \qquad L\times W \qquad + \qquad L\times W/2\end{aligned}$$

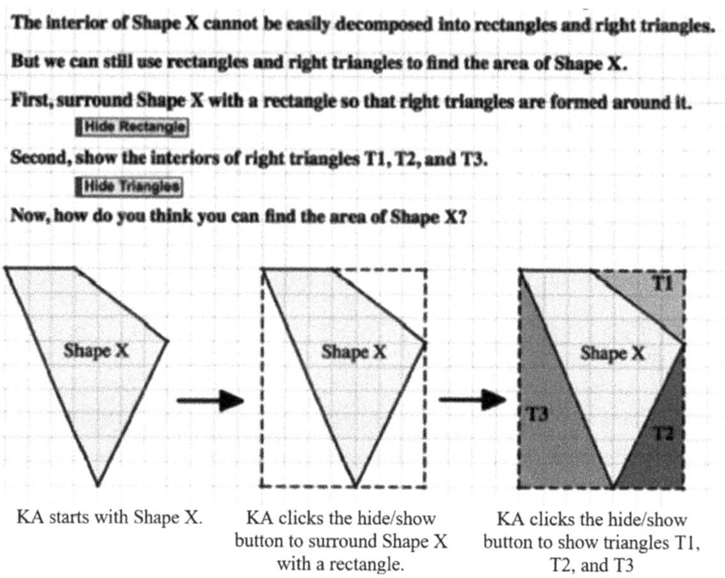

Fig. 10.21 Second type of area decomposition problem for KA

KA then proceeded to a more difficult decomposition strategy targeted by iDGi (see Fig. 10.21). First, she read the directions and clicked the hide/show buttons.

KA: So I have to figure out what is in here. [After counting squares along the T3's perpendicular sides] 4 times 9 is 36, divide that in 2, and that would be 18 here. Now for shape T1. … It would be 4 times 3 is 12. Divide that in half and get 6. Okay, so triangle 2 … 3 times 6 would be 18. Divide that in half, and it would be 9. [Entering numbers in a calculator] 33. Okay, I don't know if I have covered the whole thing of shape X, or I think I have covered the whole shape of shape X [long pause].

Although KA correctly found the areas of the three triangles external to shape X, she did not know how to proceed because she was trying to make sense of the situation using the SNLS reasoning she used for Fig. 10.20 in which she decomposed polygons into rectangles and triangles and *added* their areas (employing Measurement Property 3). KA did not activate the proper SNLS shown below in which the area of Shape X can be found by subtracting the areas of Triangles 1, 2, and 3 from the area of the surrounding rectangle. To some degree, KA reasoned like MR did when working rotation problems—she did not use spatial reasoning to properly guide her analytic strategy (see Fig. 10.21).

Spatial Numeric Linked Structuring (SNLS) to correctly determine area in Fig. 10.21*:*
spatial structuring: Outlined Rectangle = Triangle T1 ∪ Triangle T2 ∪ Triangle T3 ∪ Shape X
linked numerical structuring:

Step 1: Area Outlined Rectangle = Area Triangle T1 + Area Triangle T2 + Area Triangle T3 + Area Shape X
Step 2: Area Shape X = Area Outlined Rectangle – Area Triangle T1 – Area Triangle T2 – Area Triangle T3

Example from Study 6: Spatial-Numerical Linked Structuring for Volume

To find the volume of a rectangular box, students must create another SNLS that enables them to enumerate unit cubes in the box. This task is quite challenging for students, causing many of them to create a variety of incorrect SNLSs (Battista, 2004; Battista & Clements, 1996).

For the building shown in Fig. 10.22a, FR counted based on the spatial structuring shown in Fig. 10.22b. He said that there are 12 cubes on the front, then inferred 12 on the back; he counted 16 on the top, then inferred 16 on the bottom; finally, he counted 12 cubes on the right side, then inferred 12 on the left side. He then added these numbers. FR's numerical structuring of *12 + 12 + 16 + 16 + 12 + 12* corresponded to his spatial structuring of *(front ∪ congruent back) ∪ (top ∪ congruent bottom) ∪ (right side ∪ congruent left side).* So his spatial-numerical structuring violated Measurement Property 3—many of the cubes that he counted occupied the same space, that is, they intersected so their measures could not be added.

Below are three alternative SNLSs for the same cube building. In Fig. 10.23a, the spatial structuring *front ∪ what's left on right side ∪ (9 columns of 3)* corresponds to the numerical structuring of *12 + 9 + (repeat 9 times counting 3 cubes in a column).* In Fig. 10.23b, the column spatial structuring corresponds to the multiplicative numerical structuring of skip-counting *3, 6, 9 … 45, 48.* Another spatial structuring is horizontal layers (Fig. 10.23c) which students variously structure numerically as *16 + 16 + 16, 3 × 16*, or *skip counting 16, 32 48.*

Note that, unlike the first SNLS in Fig. 10.22b, the last three SNLSs produce correct answers for the total number of cubes in the building. Given that there are multiple correct SNLSs for this cube-building enumeration task, part of SNLS reasoning is consideration of enumeration efficiency and generality. The SNLS in Fig. 10.23a is correct but too cumbersome to be efficient and too unwieldy for large arrays—so it is not easily generalized. The SNLS in Fig. 10.23b could be conceptualized in terms of three cubes in each column times four columns in a horizontal row times four horizontal rows, leading to the standard volume formula, as could the layer structuring SNLS (Fig. 10.23c).

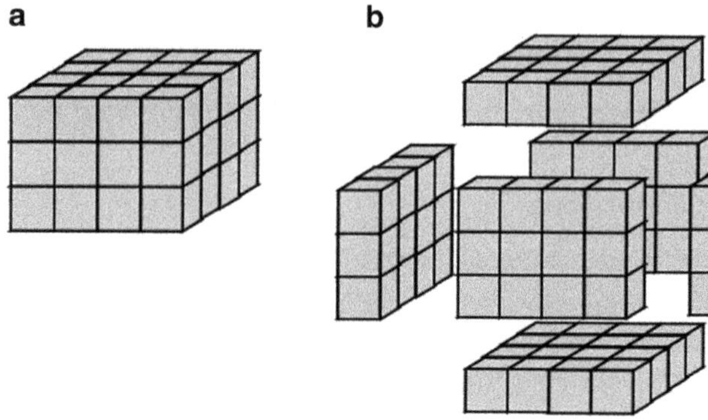

Fig. 10.22 FR's spatial structuring of unit cubes in box

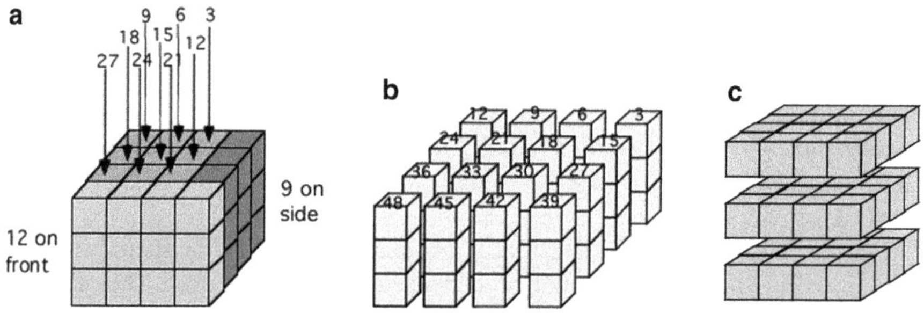

Fig. 10.23 Spatial structurings for SNLSs for cube building in Fig. 10.22

Additional SNLS Reasoning for Operating on Volume

The next example further illustrates how SNLS reasoning can be used to make sense of geometric measurement problems that deal with generalizations rather than enumeration. Consider the following problem (Battista, 2012a, 2017b): *The dimensions of a box are 4 cm by 3 cm by 2 cm. Give the dimensions of a box that has twice the volume.* The most common error that students make for this problem is to multiply all three dimensions by 2, a numerical structuring that is not connected to a proper spatial structuring. Visual SNLS reasoning can help students understand why the numerical structuring of *multiplying all the dimensions by 2* is incorrect and what correct numerical structurings are possible. For instance, Fig. 10.24a shows that doubling all three dimensions of a *4 cm by 3 cm by 2 cm* box gives eight times the original box volume, whereas Fig. 10.24b, c, d show that doubling any one of the dimensions of the box doubles its volume.

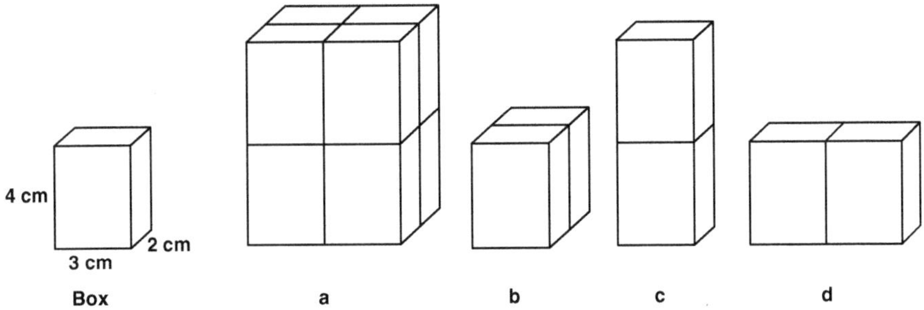

Fig. 10.24 Rectangular prisms made from copies of box

Spatial Numeric Linked Structuring (SNLS) for problem in Fig. 10.24*:*
Spatial structuring a: New Rectangular Prism = Box ∪ Box ∪ Box ∪ Box ∪ Box ∪ Box ∪ Box ∪ Box
Linked numerical structuring a: Volume of New Rectangular Prism = 8 × Volume of Box
Spatial structuring b, c, or d: New Rectangular Prism = Box ∪ Box
Linked numerical structuring b, c, or d: Volume of New Rectangular Prism = 2 × Volume of Box

Example from Study 7: Extending Volume Measurement Reasoning to Packing Problems

One way to extend and deepen students' volume-related spatial-numerical reasoning is to have them enumerate the number of cube packages that fit in a rectangular box without breaking packages apart (Battista & Berle-Carman, 1996, 2017b; Winer, 2010). For instance, two fifth graders, NA and PE, were predicting how many 2 by 2 by 3 cube packages fit in a 4 by 6 by 5 rectangular box (see Fig. 10.25a).

PE: I think 6 [counting 1–6 in Fig. 10.25a]. I knew that there was 6 because there isn't 6 rows going up [points to the height in the box picture].

PE saw that because the height of the package is 3 and the height of the box is 5 when oriented with the 3 dimension vertical, only one layer of the packages fit on the bottom of the box.

NA: I think that there are 9.
PE: How did you get 9?
NA: [Counting on the box picture in Fig. 10.25a] I counted the bottom and I got 6. And then there is 7 … 8 … 9.

PE: You can't break them up.

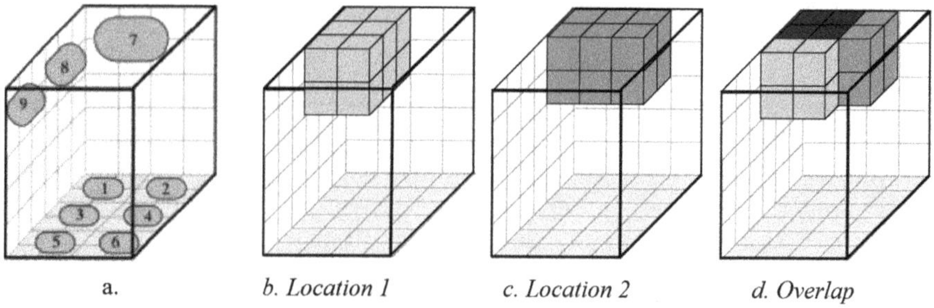

Fig. 10.25 NA and PE's enumeration structuring for cube packages that fit in a rectangular box

Fig. 10.26 PE's structuring explanation for enumerating packages

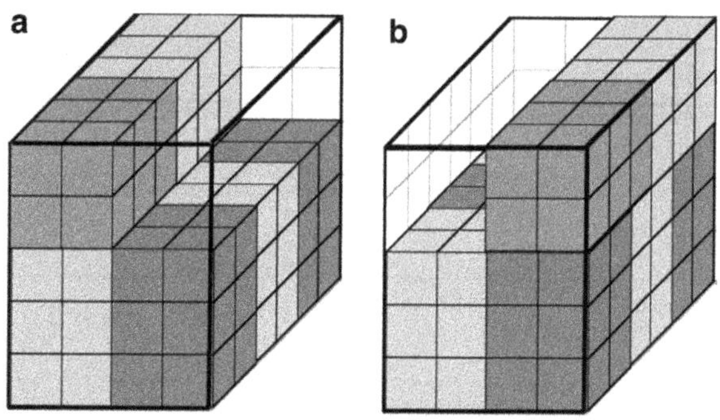

NA: No, you don't break them; you put them sideways like this [shows Location 1 (Fig. 10.25b)].

NA did not visually recognize that two of his count locations (7 and 8) actually occupy part of the same space (see Fig. 10.25b, c, d). As shown below, when the boys tested their predictions by filling the box with copies of the package, they found a different answer.

NA: Six [points to 6 packages he placed on the bottom of the box]. See 7, 8 [places 2 more packages in the box sideways (Fig. 10.26a)].
PE: [Staring at the 8 packages already in the box] It would be 10. See. Look there are 6 on the bottom and then there are 8 right here [pointing to the 2 packages that NA placed on top of the bottom 6]. Then 9, 10 [moving the 7 and 8 packages to the right (Fig. 10.26b)].
NA: Oh yeah, 10.

NA and PE could not correctly visualize and enumerate the packages in the box until they used concrete materials. The repeated checking of their ideas with concrete materials enabled the boys to develop more accurately structured mental models and visualizations.

In addition to the spatial processes needed for structuring unit-cube arrays elaborated earlier in the chapter, the cube-package problems required the following spatial processes to create a proper spatial structuring of the packages (Winer, 2010) (see Table 10.5):

- *Orienting* is the spatial mental process of choosing an orientation of a package. Orienting usually involves mentally rotating the object.
- *Orthogonal projecting* is the spatial mental process of projecting an array of cube faces from one orthogonal view of a package onto an array of squares that appears on a parallel interior side of the box.
- *Locating using orthogonal projection* is the spatial mental process of determining the exact location of a cube-package within a box, including its orientation, using orthogonal projection (Table 10.5).

Properties and Spatial Processes Used in Structuring Volume and Packing Problems

For students to correctly solve volume and packing problems, their spatial numerical linked structurings must satisfy the Measurement Properties and be supported by correct spatial processes. For a properly structured correct *unit-cube* enumeration (Fig. 10.23a), a student's mental model must locate the spatial positions of all the cubes so that they completely fill the rectangular prism space without gaps or overlaps, and it must organize the cubes in a way that makes correct enumeration possible. For a properly structured correct *cube-package* enumeration (Fig. 10.26),

Table 10.5 Spatial processes for structuring cube packages fitting in a rectangular box

Orienting: Student changes the orientation of the cube-package in space.	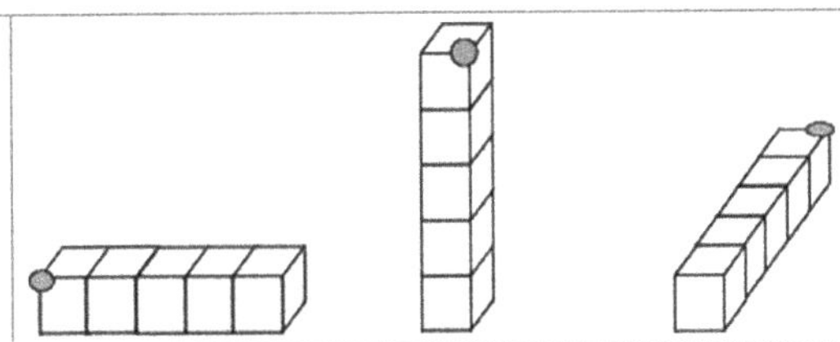
Orthogonal projecting: Student orthogonally projects 3 sides of a package onto corresponding arrays of squares on parallel sides of the box *Locating using orthogonal projection:* Student locates the package in the box by coordinating and integrating 2 or more orthogonal projections shown above	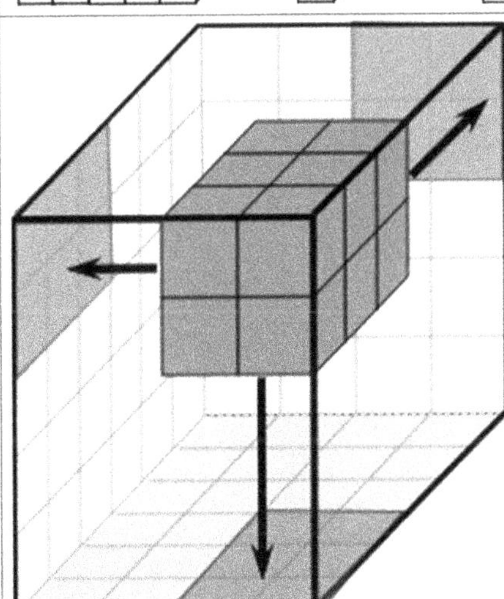

a student's mental model must structure the packages as filling the maximal amount of rectangular box space, which implies no gaps, other than where packages will not fit in the box, and no overlaps, using the spatial processes of orienting, projecting, and locating to ensure that the packages are properly located, and it must organize the packages in a way that makes correct enumeration possible.

Concluding Remarks

This chapter elaborates in great detail, for a variety of important geometric topics, the connection between spatial visualization and property-based spatial analytic reasoning. On the one hand, spatial visualization can help students organize and monitor steps in their spatial analytic reasoning. On the other hand, to be effective, spatial visualization and spatial analytic reasoning must be based on operable knowledge of relevant geometric properties of the spatial-geometric objects under consideration. We have seen numerous examples where spatial reasoning breaks down either because of poor visualization or because it is not properly connected to underlying geometric properties.

How exactly the specific reasoning processes we have described are tied to traditional spatial assessments is unknown. But our analysis suggests that, for the study of geometric reasoning, it might be worthwhile to go beyond traditional spatial tests to assessments that are more closely aligned with the actual spatial processes used in geometric reasoning.

References

Battista, M. T. (1990). Spatial visualization and gender differences in high school geometry. *Journal for Research in Mathematics Education, 21*(1), 47–60.

Battista, M. T. (1994). On Greeno's environmental/model view of conceptual domains: A spatial/geometric perspective. *Journal for Research in Mathematics Education, 25*(1), 86–94.

Battista, M. T. (1999). Fifth graders' enumeration of cubes in 3D arrays: Conceptual progress in an inquiry-based classroom. *Journal for Research in Mathematics Education, 30*(4), 417–448.

Battista, M. T. (2004). Applying cognition-based assessment to elementary school students' development of understanding of area and volume measurement. *Mathematical Thinking and Learning, 6*(2), 185–204.

Battista, M. T. (2007). The development of geometric and spatial thinking. In F. Lester (Ed.), *Second handbook of research on mathematics teaching and learning* (pp. 843–908). Reston, VA: National Council of Teachers of Mathematics.

Battista, M. T. (2008). Development of the shape maker's geometry microworld: Design, principles, and research. In G. W. Blume & M. K. Heid (Eds.), *Research on technology and the teaching and learning of mathematics: Vol 2. cases and perspectives* (pp. 131–156). Charlotte, NC: Information Age Publishing.

Battista, M. T. (2012a). *Cognition based assessment and teaching of geometric measurement (length, area, and volume): Building on students' reasoning*. Portsmouth, NH: Heinemann.

Battista, M. T. (2017a). *Reasoning and sense making in the elementary grades*: 3–5. National Council of Teachers of Mathematics.
Battista, M. T. (2017b). *Reasoning and sense making in the elementary grades: 6–8.* National Council of Teachers of Mathematics.
Battista, M. T., & Berle-Carman, M. (1996). *Containers and cubes*. Palo Alto, CA: Dale Seymour Publications.
Battista, M. T., & Clements, D. H. (1996). Students' understanding of three-dimensional rectangular arrays of cubes. *Journal for Research in Mathematics Education, 27*(3), 256–292.
Battista, M. T., Clements, D. H., Arnoff, J., Battista, K., & Borrow, C. V. A. (1998). Students' spatial structuring of 2D arrays of squares. *Journal for Research in Mathematics Education, 29*(5), 503–532.
Battista, M. T., & Frazee, L. M. (2018). Students' use of property knowledge and spatial visualization in reasoning about 2D rotations. In *ICME TSG-12 Secondary Geometry Monograph.* New York, NY: Springer Publishing.
Battista, M.T., Frazee, L.M., & Winer, M. L. (2017). Adolescents' understanding of isometries in a dynamic geometry curriculum. *Paper presented at National Council of Teachers of Mathematics (NCTM) Research Conference*, San Antonio, TX.
Battista, M. T. (2012b). *Cognition based assessment and teaching of geometric shapes: Building on students' reasoning*. Portsmouth, NH: Heinemann.
Calvin, W. H. (1996). *How brains think*. New York: Basic Books.
Clements, D. H., & Battista, M. T. (1992). Geometry and spatial reasoning. In D. A. Grouws (Ed.), *The handbook on research in mathematics teaching and learning*. New York, NY: Macmillan.
English, L. D., & Halford, G. S. (1995). *Mathematics education: Models and processes.* Mahwah, New Jersey: Lawrence Erlbaum Associates.
Gorgorió, N. (1998). Exploring the functionality of visual and non-visual strategies in solving rotation problems. *Educational Studies in Mathematics, 35*(3), 207–231.
Greeno, J. G. (1991). Number sense as situated knowing in a conceptual domain. *Journal for Research in Mathematics Education, 22*(3), 170–218.
Hegarty, M. (2004). Mechanical reasoning by mental simulation. *Trends in Cognitive Sciences, 8*(6), 280–285.
Hegarty, M. (2010). Components of spatial intelligence. *Psychology of Learning and Motivation, 52*, 265–297.
Johnson-Laird, P. N. (1983). *Mental models: Towards a cognitive science of language, inference, and consciousness (No. 6)*. Cambridge, MA: Harvard University Press.
Johnson-Laird, P. N. (1998). Imagery, visualization, and thinking. In J. Hockberg (Ed.), *Perception and cognition at century's end* (pp. 441–467). San Diego, CA: Academic Press.
Knauff, M., Fangmeier, T., Ruff, C. C., & Johnson-Laird, P. N. (2003). Reasoning, models, and images: behavioral measures and cortical activity. *Journal of Cognitive Neuroscience, 15*(4), 559–573.
Krantz, D. H., Luce, R. D., Suppes, P., & Tversky, A. (1971). *Foundations of measurement, Vol I: Additive and polynomial representations*. New York: Academic Press.
Markovits, H. (1993). The development of conditional reasoning: A Piagetian reformulation of mental models theory. *Merrill-Palmer Quarterly, 39(1)*, 131–158.
Mix, K. S., & Cheng, Y. L. (2012). The relation between space and math: Developmental and educational implications. *Advances in Child Development and Behavior, 42*, 197–243.
Mix, K. S., Levine, S. C., Cheng, Y. L., Young, C., & Hambrick, D. Z. (2016). Separate but correlated: The latent structure of space and mathematics across development. *Journal of Experimental Psychology: General, 145*(9), 1206–1227.
Moise, E. E. (1963). *Elementary geometry from an advanced standpoint.* Reading, MA: Addison-Wesley.
Newcombe, N., & Shipley, T. (2015). Thinking about spatial thinking: New typology, new assessments. In J. Gero (Ed.), *Studying visual and spatial reasoning for design creativity*. Dordrecht: Springer.

Sarama, J., Clements, D. H., Swaminathan, S., McMillen, S., & González Gómez, R. M. (2003). Development of mathematical concepts of two-dimensional space in grid environments: An exploratory study. *Cognition and Instruction, 21*(3), 285–324.

Steffe, L. P., & Thompson, P. W. (2000). Teaching experiment methodology: Underlying principles and essential elements. In R. Lesh & A. E. Kelly (Eds.), *Research design in mathematics and science education* (pp. 267–307). Hillsdale, NJ: Erlbaum.

Uttal, D. H., Meadow, N. G., Tipton, E., Hand, L. L., Alden, A. R., Warren, C., & Newcombe, N. (2013). The malleability of spatial skills: A meta-analysis of training studies. *Psychological Bulletin, 139*(2), 352–402.

van Hiele, P. M. (1986). *Structure and insight.* Orlando: Academic.

Winer, M. L. (2010). *Fifth graders' reasoning on the enumeration of cube-packages in rectangular boxes in an inquiry-based classroom.* (Master's thesis). Retrieved from http://www.ohiolink.edu/etd/.

Chapter 11
Learning Through and from Drawing in Early Years Geometry

Nathalie Sinclair, Joan Moss, Zachary Hawes, and Carol Stephenson

Abstract This chapter focuses on the relations between spatial reasoning, drawing and mathematics learning. Based on the strong link that has been found in educational psychology between children's finished drawings and their mathematical achievement, and the central importance of diagramming in mathematics thinking and learning, we wanted to study children's actual drawing process in order to gain insight into how the movements of their hands and eyes can play a role in perceiving, creating, and interpreting geometric shapes and patterns. We pay particular attention to the interplay between children's drawings and their gestures, to the role of language in modulating children's perceptions, and to the back and forth that drawing seems to invite between two-dimensional and three-dimensional perceptions of geometric figures. We seek to forge new ways of including drawing as part of the teaching and learning of geometry and offer new ways of thinking about and analyzing the types of spatial/geometric reasoning young children are capable of.

Keywords Diagram · Gesture · Geometry · Verbal · Visual · Triangle · Square · Symmetry · Congruence · Dimensional deconstruction · Pointing · Tracing · Structure · Transformation · Array · Quick draw · Segment · Diagonal

From its earliest roots in Ancient Greece, drawing has been a significant practice of geometers. The drawings of geometers, whether in the sand, on papyrus, or on paper, have involved the use of tools that produce one-dimensional marks on two-dimensional surfaces. With the widespread use of textbooks and worksheets, drawings that used to be made by hand are now offered to students ready-made, thereby reducing the amount of drawing they do themselves. However, during the past decade, we have seen a renewed interest in promoting drawing and diagramming

N. Sinclair (✉)
Faculty of Education, Simon Fraser University, Burnaby, BC, Canada
e-mail: nathsinc@sfu.ca

J. Moss · Z. Hawes · C. Stephenson
Ontario Institute for Studies in Education/University of Toronto, Toronto, ON, Canada
e-mail: joan.moss@utoronto.ca; carol.stephenson@utoronto.ca

K. S. S. Mix, M. T. Battista (eds.), *Visualizing Mathematics*, Research in Mathematics Education, https://doi.org/10.1007/978-3-319-98767-5_11

in school mathematics, both amongst educational psychologists and mathematics educators. This "return to drawing" is likely due in part to an increased awareness of and research on the importance of spatial reasoning and, more broadly, the role of embodied cognition in mathematics thinking and learning (Mix & Cheng, 2012). Indeed, a growing body of research points to sensorimotor activities, such as moving, gesturing and drawing, as being fundamental spatial processes in the learning and communication of mathematics (e.g., see Lakoff & Núñez, 2000).

In this chapter, we turn to children's drawings—both the act and artefact—as a means of eliciting and working towards a better understanding of children's geometric and spatial reasoning. More specifically, we examine how children can learn *through* and *from* drawing, with a particular focus on spatial processes and key concepts of primary school geometry. In examining the complexity of the drawing process, we aim to show how drawing is not a static or innate skill, but one that can be worked on and improved through the intermediaries of language, gestures and spatial visualisation. Broadly speaking, we aim to shed light on the mathematical learning and instructional opportunities afforded through the act of geometric drawing.

We begin our study by first providing a brief review of research on children's drawings. Working within the theme of this book, we draw sharp parallels in the ways in which the disciplines of psychology and mathematics education (and mathematics more generally) have traditionally made children's drawings a focus of their respective work. Our own work attempts to build on the longer tradition of research on drawing in educational psychology but in a way that is more sensitive to the disciplinary and pedagogical values of geometry education. In this way, we see our study as being mutually informed by research from psychology and mathematics education, and also as informing both disciplines going forward.

What we Know About Drawing: A Review of the Literature from Psychology and Mathematics Education

Early psychological studies of children's drawings focused on stages of development and viewed drawings as markers of cognitive maturation. For example, one of the first papers on the subject was published by Ebenezer Cooke in 1885. He offered a qualitative analysis of children's drawings in terms of age-related stages of representation. Beginning in the early twentieth century, researchers became increasingly interested in how children's drawings were related to developmental change, and also, more specifically, how differences in children's drawings were related to associated differences in general intelligence (Ivanoff, 1909; Kamphaus & Pleiss, 1992). Indeed, early research efforts indicated correlations between objective measures of children's drawings and intelligence (Goodenough, 1926a; Ivanoff, 1909). Interestingly, it was during this time that the longstanding tradition began—and continues to this day—of assigning quantitative scores to children's drawings and using these scores as metrics and correlates of intellectual functioning (Claparede, 1907; Ivanoff, 1909).

A major impetus in this movement was Goodenough's publication of *Measurement of Intelligence by Drawings* in 1926, where he introduced the Goodenough Draw-A-Man test, an assessment of children's intelligence based on how well they were able to represent and draw the human figure. Spanning from its development in 1926 and continuing to the late 1950s, the Draw-A-Man Test was consistently one of the most popular tests used by clinical psychologists in the United States (Sundberg, 1961). Although the popularity of this test waned thereafter, the central ideas about what children's drawings can tell us about their general cognitive development remained strong. For example, Piaget drew on this earlier work and incorporated the developmental stages of drawing into his own developmental framework. Children's drawings were taken as evidence of developmental shifts in core cognitive competencies and aligned with the belief that children's drawings follow a consistent, universal and sequential progression (Kellogg, 1970). Accordingly, children were only able to progress to a more complex representational stage once earlier, more basic representations were mastered.

According to this view, children's drawing behaviour unfolds naturally and the role of adults and education in this process is minimal to non-existent (e.g., see Brooks, 2009). Moreover, this view conceives of children's drawings as stable indicators of general cognitive development, or, said differently, artefacts or outcomes of general cognition. In the words of Goodenough (1926b), "the nature of the drawings made by children in their early years is conditioned by their intellectual development" (pp. 185). The psychological literature is bereft of studies looking at how the very act of drawing may provide an essential vehicle for cognitive development and learning. Given how spatial reasoning improves through practice, the act of drawing could also improve one's performance, both in drawing and spatial reasoning, especially if it was adequately supported, which would challenge the idea of using drawing as an indicator of intelligence.

The scientific study of children's drawings has changed over the past few decades and there is increasing acknowledgement of drawing as a driver of cognition rather than a mere indicator or outcome of cognition. More recently, researchers in the psychological sciences have begun to reveal more nuanced connections between children's drawings and domain-specific knowledge and performance (Brooks, 2009; Malanchini et al., 2016). Researchers have revealed especially strong connections between children's drawing skills and mathematics performance (e.g., see Carlson, Rowe, & Curby, 2013). For example, children's abilities to accurately draw human figures at 4½ years of age have been found to significantly correlate with teacher ratings of these same children's mathematics skills (i.e., numbers, geometry, measurement, data, and applied problems) at the age of 12.

The ability to draw a human figure has been theorised to implicate a number of underlying mathematical concepts (Case & Okamoto, 1996; Malanchini et al., 2016). For example, awareness of number of body parts, proportional reasoning, appropriate use of space (e.g., depth cues), and symmetry are all mathematically relevant features inherent in the drawing of human figures. In fact, fundamental to drawing most anything is the need to consider spatial relations within and between objects. This is true of self-generated images, but is also, as we will see below, involved in the act of copying a static image or design. The close relations between drawing and mathematics might thus be explained by the geometrical and spatial reasoning involved in the drawing process.

Copying an image requires the drawer to attend to the geometric and spatial relations present. Researchers have consistently found correlations between children's abilities to copy simple geometric designs and mathematics performance (e.g., see Carlson, Rowe, & Curby, 2013; Grissmer, Grimm, Aiyer, Murrah, & Steele, 2010; Kulp, 1999). For example, Fig. 11.1 shows an example of the types of items from the Visual Motor Integration test (VMI). This assessment involves having children copy simple geometric designs and children are assigned a score based on accuracy. Children's performance on this assessment has been shown to be a reliable predictor of both concurrent and future mathematics achievement (Kulp, 1999; Kurdek & Sinclair, 2001; Pieters, Desoete, Roeyers, Vanderswalmen, & Van Waelvelde, 2012; Sortor & Kulp, 2003). For example, research by Pieters et al. (2012) found that 7- to 9-year-olds' performance on the VMI explained a substantial proportion of variance in both number fact retrieval and procedural calculation. Furthermore, children with mathematics learning disabilities performed significantly worse on the measure compared to their peers.

The relation between children's drawings and mathematics is even more telling when we consider how children use drawings and other marks to both represent and understand mathematics problems. We see examples of this when children make discrete marks on a piece of paper in order to keep track of and represent the two addends of an addition problem. Children use drawings as a means to represent and bring meaning to fractions problems (e.g., shading in 1/3 of a rectangular array). Drawings also assist in the comprehension and solutions to mathematical word problems. In fact, this is an area of study where psychologists have made significant headway in recent years. Researchers have found that children's representational drawings of mathematical word problems provide important insight into individual

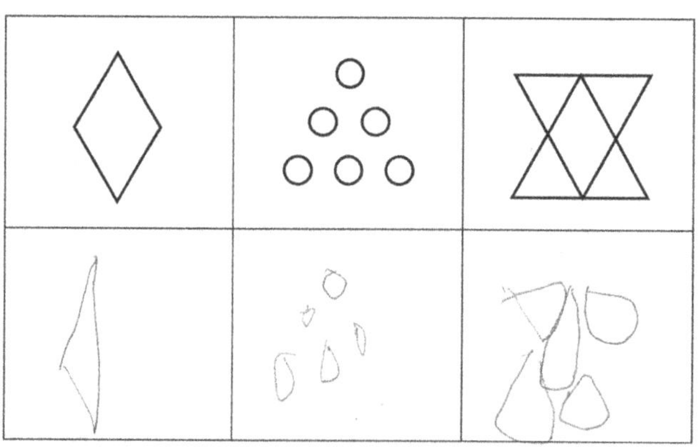

Fig. 11.1 Example of types of items and responses found on tests of Visual-Motor Integration (VMI; e.g., see Beery & Beery, 2010)

differences in solution accuracy (Boonen, van der Schoot, van Wesel, de Vries, & Jolles, 2013; Boonen, van Wesel, Jolles, & van der Schoot, 2014; Hegarty & Kozhevnikov, 1999). Children who focus their drawing efforts on detailing the visual-spatial relations of the word problem tend to outperform children who focus their drawings on the more pictorial aspects of the problem (e.g., including extraneous details rather than just the essential mathematical relations). This suggests that drawing is linked to mathematical abstraction.

From a mathematics education point of view, there has been a growing focus on drawing, and especially on the use and creation of diagrams. Perhaps because of the seminal nature of Polyà's (1957) work on problem solving (one of the heuristics in the first stage of "Understanding the Problem" is "Draw a diagram"), much of the literature has focussed on the diagrams that students create while solving problems (Bremigan, 2005; Diezmann & English, 2001; Nunokawa, 2006; Yancey, Thompson, & Yancey, 1989).

Increasingly, researchers have also realised that diagrams are not always transparent for learners, who might perceive and interpret them in different ways, especially diagrams that carry a significant amount of cultural encoding, such as the Cartesian coordinate system. For example, Steenpaß and Steinbring (2014) focus on students' subjective interpretations of mathematical diagrams, offering the distinction between object-oriented (where the focus is on the visible elements of the diagram) and system-oriented (where the focus is on the relation between the elements of the diagram) as two different ways that students may interpret diagrams.

Some mathematics education research has begun to attend to the interplay between diagramming and gestures in mathematical activity—an interest fuelled in part by recent theories of embodied cognition. Although most often studied separately, there is a natural relation between the two as they both involve actions with the hand. Indeed, using the work of the philosopher of mathematics Gilles Châtelet, who studied the pivotal role of diagramming in mathematical inventions, de Freitas and Sinclair (2012) examined the interplay of gesturing and diagramming in undergraduate students' drawings, highlighting the way in which these drawings can be seen as gestures in "mid-flight" and thus capturing on the page the mobile actions of the hand. Also with an attention to the interplay between gestures, diagrams, and speech, Chen and Herbst (2013) compared the interactions of two groups of high school students: one working with a diagram that contained relevant labels (for vertices and angles) and another working with a diagram that contained no labels. While the students in the first group only used pointing gestures, those in the other group made gestures that extended the existing diagram (by extending a segment, for example) and thus created new geometric elements. These students talked about hypothetical objects and properties, which led the authors to conclude that the "gestures played a crucial role in engaging students in reasoned conjecturing" (p. 303) because the hypothetical objects enabled the students to make and justify conjectures.

Theoretical Considerations

Given that we are working in the context of geometry, it is important to underscore the fact that geometry is a complex activity that does not just involve shapes or the use and creation of visual images, but instead centrally involves the interplay between seeing and saying; that is, between visualisation and the language for stating and deducing properties. Indeed, Duval (2005) argues that much of the school geometry curriculum encourages prototypical thinking, in which shapes are recognised based on their visual properties. The shape becomes an iconic representation. This approach fails to develop coordination between the visual and discursive registers of geometry. Duval proposes that children should engage in more construction tasks, which would better support the crucial process of *dimensional decomposition.* This process involves the passage from one dimension to the other, for example, from the two-dimensional square to the one-dimensional line segments that constitute the square. When going from seeing a shape to describing a shape, there is usually a reduction of dimension. Dimensional deconstruction involves both seeing a basic shape (such as a square) as being constructed from a network of lines and points and seeing that different shapes can emerge from that network of lines (such as two triangles).

While drawing can be a powerful way of engaging in dimensional deconstruction, since it involves the creation of lines and points that form two–dimensional shapes, we also follow Duval in stressing the importance of the coordination between the visual and the discursive, which we pursue in our tasks by asking children to describe what they see and how they would draw. To this verbal/visual interplay, we follow also recent work in embodied cognition in focussing on the gestural as an important means through which people think, communicate and invent. Gestures can act as effective replacements for speech, and can therefore participate in the interplay between the visual and the discursive. Gestures are also effective at communicating the spatial and temporal aspects of mathematical concepts (e.g., Núñez, 2003; Sinclair & Gol Tabaghi, 2010). Finally, drawing itself can be seen as a kind of manual gesture (Streeck, 2009), which can easily transform into a gesture "in the air" (sometimes called a drawing gesture, when people gesture as if they are holding a pen and drawing on paper—therefore not leaving a visible trace). Indeed, the temporal nature of drawing connects it strongly to the temporal nature of gesturing.

Methods

In this chapter, we report on a study in which we examined the interplay of gestures, diagrams and speech as young children were engaged in various drawing tasks. The tasks we chose were based on Wheatley's Quick Draw Program. Accounts of this program in classrooms, and past research conducted on the effectiveness of the program (Hanlon, 2010; Tzuriel & Egozi, 2010; Weckbacher & Okamoto, 2015;

Wheatley, 1997; Yackel & Wheatley, 1990), have consistently shown that it encourages spatial thinking, improves students' recognition of and ability to name shapes and improves their general geometric vocabulary. Given their value in prior research, we wanted to build on these tasks and adapt them to more complex situations that would involve both gesturing and talking as well as drawing.

Participants and Setting

Eleven kindergarten students (ages 5.9–6.2 years) from a classroom of 22 children participated in the study. The children were attending a fee-paying laboratory school in Toronto that is open to all learners serving students from nursery to sixth grade. Seven percent of the population receive financial assistance and up to 15% of students receive some form of special academic assistance. A central mission of the school is to serve as a model for inquiry-based teaching and learning. The 11 students selected for the study were chosen by their classroom teacher (fourth author) because, as a group, they offered a range of mathematics ability.

The students in this kindergarten classroom had all been exposed to an enriched geometry curriculum involving composition and decomposition, shape, spatial transformations and symmetry along vertical, horizontal and diagonal lines. In addition, and directly relevant to this study, early in January all of the children in the class, as part of their regular mathematics program, had participated in two half-hour lessons of quick image drawings, which involved them in drawing, from memory, geometric shapes embedded in squares. Specifically, in each of these two lessons the teacher held up for 3 s an image of a geometric shape embedded in a square (e.g., a diamond (rhombus) with a horizontal line through the midpoint), and asked the children to "take a picture in their minds" and only once the image was removed from view, they were instructed to draw what they remembered of the image. To facilitate the process for the young children, each child had been provided with a response sheet with four squares, thereby allowing them to draw the image(s) directly into the square(s). The letters L and R were written on the outer edges of each square to indicate left and right sides respectively to help the children describe the images they created. In the first of the two lessons, the students did not receive any feedback on their drawings but were given a second quick look at the target image to make any changes they thought necessary.[1] In the second lesson, after the students had drawn their image they were encouraged to describe what they drew—e.g., how they got started, where they placed their pencils, which lines they drew first. Over the two lessons, the children were shown a total of 12 geometric drawings in squares, which were either part of, or adapted from Wheatley's (2007) most recent Quick Draw program. Full descriptions of the quick image drawing lessons can be found as part of a new spatial geometry curriculum (see Moss et al., 2016, pp. 126–129).

[1] In the original Quick Images, students were then shown the image and asked to comment on it and their drawings. As stated below, this was done in the interviews as well.

Materials: The Interviews

Two different exploratory open-ended clinical interviews were designed for the present study: Every child in the study participated in one or the other of the two interviews. The design for the interviews was based both on the way the kindergarten children responded to the two short lessons taught to them earlier in the school year, and also on our observations over the years of the many teachers and some hundreds of students participating in *Can You Draw This* lessons. These lessons have been part of our Math for Young Children (M4YC) project, an ongoing spatial and geometry professional development program that began in 2011 (e.g. Hawes, Tepylo, & Moss, 2015)

Thus, for Interview 1 (see Fig. 11.2 for images) we designed an exploratory, open-ended protocol to probe the potential of *Can You Draw This* activities. While we continued to use images of geometric shapes embedded in squares, rather than presenting them as quick image, as outlined in the first of the two pilot lessons we describe above, we followed Duval's work in varying the activities to promote more visual/discursive synergies. For example, we described images for children to draw instead of presenting the image visually. We also did the reverse: asking the children to describe an image for the interviewer to draw; in addition we also had the children study an image and consider its symmetries, congruence, composition and structure prior to their drawing.

One of the things we noticed in the two pilot lessons conducted in January with all of the kindergarten children was the difficulty a number of students had in reproducing an isosceles triangle in the square. As part of the earlier lessons in the kindergarten class, we included two isosceles triangle challenges. In the first, the triangle was oriented upwards with the "point" touching the top edge of the square. In the second, the triangle was oriented towards the left with the point to the left edge of the square. In trying to reproduce the first of the two triangles, many children's drawings did not go into the corners of the square or touch the top lines. The drawings of a number of the children produced for the isosceles triangle oriented to the left often did not resemble any kind of triangle. We were intrigued by the challenges faced by many children in their attempt to reproduce the triangles. Thus, Interview 2 (see Fig. 11.3 for images) was designed with a specific focus on triangles. For example, the first task asked the students to comment on, and compare

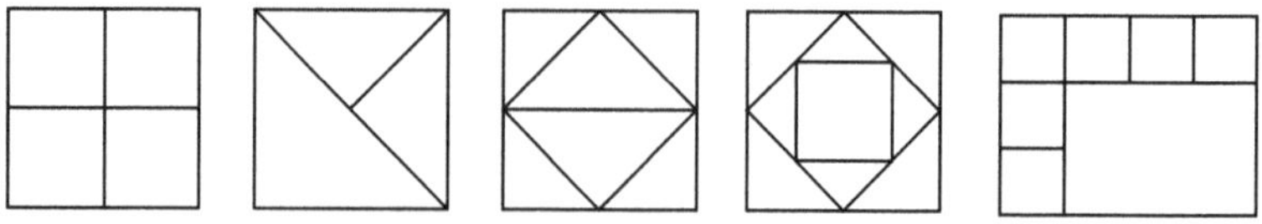

A B C D E

Fig. 11.2 Items used in Interview Protocol 1. For each item, children were asked the following (abbreviated) questions: (**A**) Can you draw this?; (**B**) Can you draw this from what I tell you?; (**C**) Can you describe how to draw it?; (**D**) Can you draw this?; (**E**) Can you complete the grid?

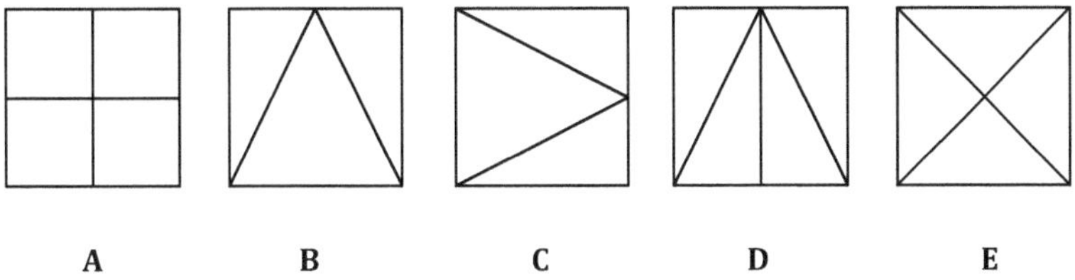

Fig. 11.3 Items used in Interview Protocol 2. For each item, children were asked the following (abbreviated) questions: (**A**) Can you draw this?; (**B**) Can you spot the difference between the two pictures?; (**C**) Can you draw this?; (**D**) Can you draw this?; (**E**) Can you draw this?

an incorrectly drawn triangle (incorporating typical errors by students) with the correct image (see Fig. 11.3B). Other questions required students to draw triangles in a different orientation or with a vertical line through the midpoint. Interview 2 was administered to two of the 11 students, both of whom had had significant trouble with the pilot lesson drawings.

Procedure

The interviews were conducted one-on-one in March 2017. The second, third, and fourth authors each conducted three or four interviews, each for about 30 min, in small rooms adjacent to children's classrooms. Children were provided with pencils and a small booklet, each page containing a drawing of a single square. All interviews were videotaped using one camera to capture the drawing sequences and the gestures, which helped facilitate the classification of children's responses. Administration followed a semi-structured individual interview protocol, with the task order and procedures remaining constant.

Analysis

This was an exploratory qualitative study and the analysis had a number of steps. To analyse the videotaped interviews each author looked at all of the interview videos and firstly attended closely not only to the actual drawings that the children were making (both the process of drawing and the final product), but also the language they were using (and being offered by the interviewer) and the gestures they made. We each highlighted particular segments of the videos in which we noticed shifts in the way that the children were seeing the drawings. Second, we each identified places in each video where different mathematical concepts arose. Some of these were planned in developing the interview protocol (symmetry, congruence) but others emerged from our attending to the unplanned aspects of the videos such as the back and forth between global and local features of the images. Once we had generated a list of excerpts that were related to these two main categories, we met again to discuss the excerpt and select which ones to use for this chapter. In the next two sections, we therefore focus on the two aspects of drawings alluded to in our title, which are the ways that children learn *through* the process of drawing (and talking and gesturing) and the concepts they can learn *from* the drawings they have produced.

Figure 11.2 provides the images on which Interview Protocol 1 was based: 9 of the 11 children received this interview. In the first task, shown in Fig. 11.2A, the children were shown the image for a few seconds, then the image was taken away and the children were asked to draw it. For the second task, the interviewer described the image shown in Fig. 11.2B in a step-by-step manner and the children had to draw it accordingly. For example, the interviewer said "Start at the top left corner and draw a line segment to the bottom right corner. Then draw a line segment from the top right segment to the middle of the line segment you just drew." For the third task, the children were shown the image in Fig. 11.2C and asked to describe how to draw it, as if to another person who could not see it. The image in Fig. 11.2D was shown to the children, who were asked questions about what they saw, about symmetry and about congruence. Then the image was taken away and the children were asked to draw it. Finally, in Fig. 11.2E, the children were given the unfinished grid and asked to complete it. Although this instruction could be interpreted in many ways, all of the children assumed they had to produce an array.

All the children succeeded in tasks 1, 2 and 3. Eight of the nine children who were asked to draw Fig. 11.2D did so correctly, though one child drew the diamond and then was prompted to continue with the interviewer who asked "do you remember how many squares you said you saw?" For the final task, five of the children completed the grid using lines and the four others completed it using squares or parts of squares.

Two children were given interview protocol 2, with the tasks based on the images shown in Fig. 11.3.

Learning Through Drawing: Focus on Verbal, Gestural and Visual Interplay

Across all the interviews (for both interview protocols 1 and 2), the interviewers offered and probed for spatial language, introducing words such as middle, top, side, bottom and diagonal. The children all made extensive use of gestures, though this was not explicitly requested by the interviewers. In this section, we are interested in how students used gestures and speech, and how this might help us better understand what they are perceiving and how they are visualising when they draw.

In this short episode (3:29–4:15), Neva was asked to describe how to draw Fig. 11.2C. She went around with her pencil. Carol asked her to, "Tell me where you would start?" She began by placing her pencil around the middle of the bottom side. Following the interview protocol, the interviewer intervened before Neva continued and asked for precision, thus eliciting the words "bottom" and "middle" from Neva.

Carol: So where is that?
Neva: On the bottom.
Carol: Anywhere on the bottom?
Neva: In the middle.
Carol: Bottom middle. And then where are you going to go? Tell me with your words before you do it.

Neva then traced out a segment with her right thumb before responding to Carol:

Neva	Neva	Neva
Puts thumb on bottom middle	*Traces segment with thumb*	"here"
		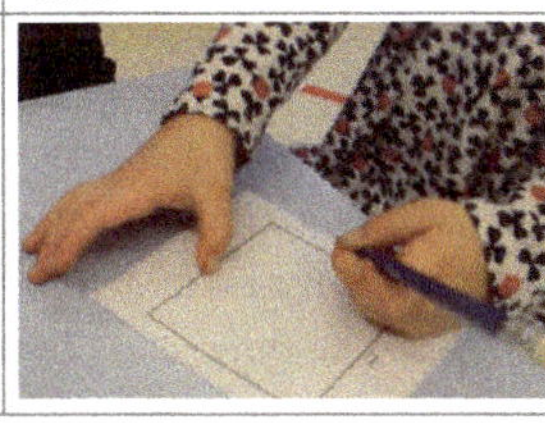

Again, Carol elicited more words, this time "diagonal" and tried to get Neva to describe how she would draw her segment.

Carol: What kind of line will that be?
Neva: A diagonal (*starts to draw, from bottom middle*).
Carol: Where are you going to stop?
Neva: Right here (*places RH thumb on midpoint of side and lifts pencil*).
Carol: How would you describe that point?
Neva: Right at the R.[2]
Carol: R? Okay! Great! (*Neva finishes drawing the segment*). And now where would you go? What would you tell people to do next?

Carol seemed to have expected the use of words such as "middle" and "side" but accepted Neva's description of "how would you describe that point?" as sufficiently precise. When Neva was asked to describe where she would go next, she responded with the deictic "right here" and put the eraser end of her pencil on the top line, but decidedly not in the middle.

[2] Neva says "R" because the letters R and L had been placed near the middle of the right and left sides of the square, respectively, to help provide orientation for the children.

Neva	Carol	Neva	Neva	Carol	Neva
Right here	Where's that?		At the top	Anywhere at the top?	No

Carol	**Neva**	**Carol**	**Carol**		**Neva**
Where at the top	Right here.	Okay. Let's see that.	And what kind of line did you just make?		Another diagonal.
	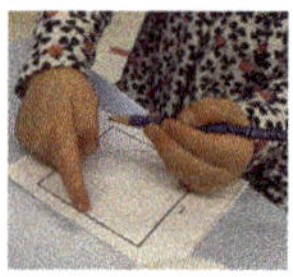				

We highlight this excerpt because it shows two places where the use of language, of gestures and of drawing was imbricated. In the first place, the gesture Neva used to trace out the desired segment is followed by the oral description, which suggests that the gesture enables Neva to describe the drawing, perhaps by actualising it on the paper. She only said "here" once she had completed tracing the segment with her thumb. In the second place, Carol's question about "anywhere in the middle" prompted Neva to make a pointing gesture with her RH to the middle of the top side, whereas she had previously been pointing (both with the eraser end of her pencil and then with the pencil tip) towards the right side of the square. We hypothesise that although Neva didn't say "middle", Carol's question oriented her to showing the middle with her finger, after which she drew the correct segment. The middle of the side had become an anchor point around which gesturing, talking and drawing occurred, even though there was no midpoint actually visible on the sheet of paper.

Without the gesturing (tracing, pointing) and the talking (using words introduced by Carol), Neva would have had a more difficult time describing what she was seeing in the original drawing and would likely have not drawn the diagonal segment correctly. We find this significant because it displaces drawing tasks from being uniquely about seeing, remembering and reproducing, and shows how gestures and language can change how children see and draw. In line with the findings of Hu, Ginns, and Bobis (2015), it would seem that the acts of tracing and pointing, particularly when it is close to the paper, can enhance learning in the context of geometry. Of interest in our study, unlike that of Hu et al., is that Neva pointed and traced spontaneously, without being asked to do so by the interviewers.

The second example also relates to the third task (Fig. 11.2C), but we focus on a slightly different phenomenon, which involves the interplay between seeing, drawing/gesturing and saying. John had already described the image as "a square that's tilted with a line in the square" when Zack asked him to explain how to draw the image "step by step".

Zack: Okay, where would I put my pencil first (*Zack places his pencil on the page and indicates he is ready to follow the instructions)?*

John: Start at the (*places pencil at the top middle*) mmmiddle (*turns to Zack*) of the top, middle of the page.

Zack: Okay.

John: At the top of the square, start there, and then draw a diagonal (*traces pencil along diagonal of* Fig. 11.2C *and places pencil on the middle of the left side*) line to the top (*looks down at his booklet, where the R and L legend can be seen*) left corner.

As with Neva, John traced the line first, before he announced where it would end. Though it was evident through gestures, John struggled to describe the location of the endpoint of the first diagonal, which should have been the left middle (and not the left corner), referring back to the booklet for help. His choice of the word "corner" is difficult to interpret, but it could be that in looking at the booklet, he saw that a kind of corner was formed at the left middle where two diagonals meet. Zack then asked for clarification.

Zack: Top left corner? Can you tell me a little more what you mean by the top left? Cause I started at the top middle and then where does my pencil go?

John: To the top, to the bottom (*looks down*) to the middle (*eraser end down, traces from top middle to the left middle, then looks down*) left.

Zack: Okay, gotcha. Now what?

John: Then you draw (*eraser end on the left middle*) a line (*moves eraser end to top middle and retraces segment from top middle to left middle*) to the bottom of the (*looks at booklet and puts the eraser end on the bottom middle*) square.

Zack: Okay, where does the…

John: Middle

In this subsection of the transcript, John was struggling to describe the drawing of the second diagonal. Before he did so, he began by re-tracing (with the eraser end of the pencil) the first diagonal, as if the second diagonal was a continuation of the first one. This suggests that John was seeing the square in terms of four repeated actions so that the drawing of the first diagonal was a rehearsal for the drawing of the second one, a rehearsal that also included the verbal descriptions of top, left, bottom and middle. He might also have been seeing a continuous path consisting of a sequence of segments that needed to be drawn from the beginning. In either case, it would seem that John was thinking of the square as a whole, and not seeing four independent segments. Certainly, the opportunity to perform the drawing gesture seemed important to John's way of seeing and describing the image. Across all the interviews, we saw this phenomenon repeatedly, where the children would gesture-draw a segment before describing it and/or before saying where the segment would end. We see in this phenomenon the dual nature of the perception of a segment, which can be seen as a single object to be apprehended all at once or as the process of moving from one point to another (in a straight line). The images can privilege the former, while the act of drawing privileges the latter.

A final, third example of children gesturing/drawing in response to a request to describe an image involved Mathias, on the fourth task of interview protocol 2 (Fig. 11.3D). The interviewer asked Mathias how he had drawn the image and he began by putting his pencil down on the table and placing his right index finger on the bottom left corner of the square, then tracing his finger along the diagonal line up to the top middle. Then the interviewer asked him, "How would you say this? How would you describe it [to your teacher]?"

Interviewer	**Mathias**	**Interviewer**	**Mathias**
The bottom left?		And then where did you go?	
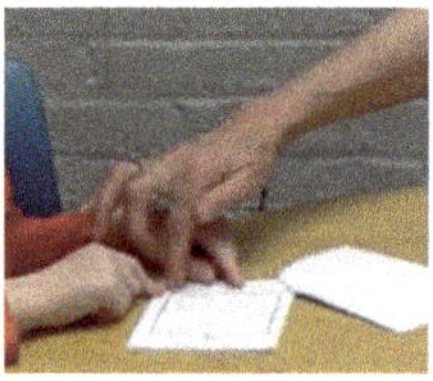	*Nods*		 *Takes pencil and traces it along diagonal*
Interviewer	**Mathias**	**Interviewer**	**Mathias**
You went up to the?	Top	Where at the top was it? Was it on the side?	In the middle top

Here again we see the act of tracing the drawing occurring first, with the child perhaps taking the question literally (what did you do?) but also perhaps using the gesture-drawing with his index finger to bring to focal attention the segment that he wants to talk about, the destination that the segment will arrive at and to movement between the two endpoints. Mathias thus makes the diagonal line three times, once by drawing it, once by tracing it with his index finger and a third time by gesture-tracing it with his pencil. It is only after the third time, with prodding of the interviewer, that Mathias describes the segment in terms of the location in the square, and where we see the beginning of the visual and language registers working together.

Learning Mathematics from Drawing

Instead of seeing the act of drawing as an end in itself, we want to highlight the various ways in which the kinds of drawing tasks that we used in the interview can give rise to significant mathematical ideas, especially in geometry, in which the actual act of drawing (instead of using given images) plays a pivotal role. We have already shown how the invitations to describe the images enabled the children to develop more geometric language, not so much in terms of the names of shapes, but in terms of position (on, in, middle), property (straight) and parts (side, corner, etc.).

In this section, we focus on particular mathematical concepts that were pursued through the different tasks. These are: congruence, symmetry and structure.

Congruence

The interviewers asked about congruence on several occasions (the language used was "are these the same?" or, when considering similar shapes, "which of these are the same size?"), never as a way to guide drawing, and always as a way to describe a given drawing. All but three of the children correctly identified shapes that were congruent as being "the same" (in all cases reflected shapes). We found this interesting especially in relation to some of the early fraction work that some teachers and researchers have done in which it is taken for granted that the different parts of a whole are of the same size, but children are clearly not always seeing it that way.

Sometimes, the interviewers asked the children how they might "prove" that two given shapes were the same, which the children did in a variety of ways involving both transformations and measurement. For example, Leo said, in reference of Fig. 11.2D, that you could cut the four outer triangles up and then "pile them up" one on top of the other (making a gesture as if holding a deck of cards that needs to be lined up). Christine also suggested cutting the four triangles in Fig. 11.2D up and making a pile. Neva suggested cutting the triangles up and putting them beside each other. Diana and Sara both suggested that you could fold the piece of paper to show that the triangles were the same. In terms of measurement, Elka referred to the fact that the two triangles in Fig. 11.3B were the same by showing that each of the corresponding sides were the same size. Maya used her fingers—using her finger width as a unit of measure—to count the lengths of the different sides for Fig. 11.2D. In relation to Fig. 11.2D, Diana moved her pencil along the sides of the diamond asserting that "all of the lines are the same" and then moved her pencil around the inner square asserting that "all of the lines are the same". When asked how that helped her see that the smaller, inner triangles were the same, she again traced the sides of the inner square (which are the hypotenuses of the inner triangles) saying they were the same, and then traced the two other sides of one inner triangle, asserting that they were the same and then repeating for the other three triangles. When asked whether the two triangles making up the rectangle in Fig. 11.3D were the same Leo asserted they are because there is a line dividing them in half (gesturing a cutting action).

The use of transformations (cutting and flipping or rotating or piling up) focusses on the shapes as a whole, while the measuring strategies engage in dimensional deconstruction in that the children are attending to the lengths of the segments that make up the shapes. We hypothesise that the shifting of attention to the segments and their lengths arose out of the drawings that the children made, where they had to attend to the one-dimensional properties of the image more than the two-dimensional properties.

Symmetry

The children were frequently asked, after having drawn or described an image, to say whether the image was symmetrical. Symmetry had already been part of the children's classroom activities in which the children were challenged to both identify and create symmetrical images in a variety of ways over the course of 3 weeks, so it is perhaps not surprising that they were all able to identify at least one line of symmetry in the images they were shown. Each child used a gesture to indicate the lines of symmetry (for Fig. 11.2D, for example, first starting with a vertical one, then horizontal and, for some, also diagonal). While some children used their whole hand to indicate a line of symmetry, most used a drawing gesture to do so. For example, Maya used a drawing gesture (with her RH index on the page) (Fig. 11.4A) and said "if you put it down this way" and did so for each other line of symmetry. In another case using Fig. 11.2C, Sandro raised his hand in the air, which was holding a pencil, oriented the pencil so that it was pointing towards the top middle of the picture and moved it down by several centimetres (Fig. 11.4B), saying "I think so" in a very tentative manner. When the interviewer provided encouraging feedback, he shifted the pencil to the top middle of the square and moved it downwards—but not all the way, a similar short amount as he had done previously "in the air". Christine was asked to "use your pencil to show me where a line of symmetry might be" and placed the eraser end of her pencil on the top middle of the paper, moving it down to the bottom middle (Fig. 11.4C)—saying nothing as she did this. Also saying nothing, Sarah used her whole pencil (Fig. 11.4D) to indicate the horizontal line of symmetry. One student, when asked whether the square had symmetry or was symmetrical, responded affirmatively, explaining that "if you fold it that way it would work" and making a whole hand horizontal gesture starting at the top and then moving to the bottom of the square (Fig. 11.4E).

As in the discussion of segments in the previous section, the line of symmetry was seen both as an object (with a whole hand gesture or a whole pencil one) and a process of moving from one point to another (with finger or pencil). It may be that the presence of the pencil in their hands encouraged more process conceptions of lines of symmetry.

There were also several instances in which the students made use of symmetry in their drawings, without being prompted. For example, when describing to the

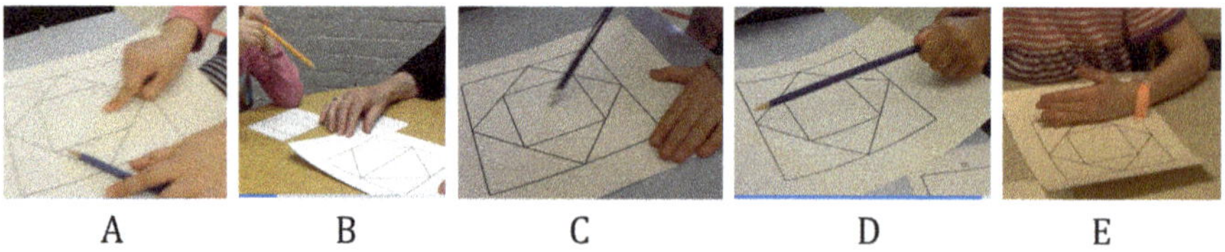

Fig. 11.4 (**A**) Maya tracing line of symmetry with index finger; (**B**) Sandro pointing to line of symmetry with pencil tip; (**C**) Christine gesture-drawing the line of symmetry; (**D**) Sarah using the pencil as a line of symmetry; (**E**) Diana using the whole hand to indicate folding over a line of symmetry

interviewer how to draw Fig. 11.2C, Leo told the interviewer to draw the first diagonal line going from the top middle to the left middle and then to "jump back" and "do the same thing on the other side", which suggests that he was seeing symmetry in the image. One other child, Diana, did a similar thing, telling the interviewer that he could "do the same" for the bottom part of the diamond, after having drawn the two segments making up the top part, adding that "it doesn't matter which side you do" and "you can start from either side". This leads nicely to the next type of geometric thinking that we observed, which was the children's movement back and forth from the local to the global, that is, from focusing on particular one-dimensional objects (lines, points, corners) to two-dimensional shapes (squares, diamonds, triangles).

Intrafigural Structure

In the first task for interview protocol 1, when the children were asked what they saw, they all referred to global shapes such as a T or a cross, and a window or four boxes. The latter two descriptions include the outer square and are more two-dimensional in nature whereas the former two ignore the outer square and thus focus more on the relation of the lines inside the outer square. In general, upon being prompted to describe what they saw, the children used a global approach (e.g., "I see a cross"), but when asked to deconstruct the image or draw the image based on a description, their attention shifted to one-dimensional parts. This is perhaps unsurprising, given that the pencil is a tool for creating one-dimensional objects and even when drawing a triangle or a square, these 2D shapes have to be enclosed in segments. Following Duval (1998), as well as Whiteley (2002), what is important in geometry is the ability to move back and forth between the local and the global, depending on the context of the problem. The geometer must be able to see the image in Fig. 11.2B both in terms of three triangles, but also in terms of midpoints, diagonals and angles. When Duval encourages educators to place more emphasis on one-dimensional objects in early geometry education, it's because most curricula focus children's attention on identifying prototypical two-dimensional shapes. While the move from global to local was common in the interviews, we also saw more complex and dynamic shifts in the children's ways of seeing.

For example, Leo (interview protocol 1), who was first asked to draw the image shown in Fig. 11.2A, did so by drawing two line segments (first horizontal and then vertical). Then, when he was asked, "what do you see?" he responded, "four boxes". In this case, even the act of drawing the lines did not shift his attention away from the two-dimensional shapes. However, 2 min later, after he had followed the instructions of the interviewer to draw the image shown in Fig. 11.2B (which he did correctly) and was asked "what do you see, in shapes?", he used his pencil to point to and then gesture-trace the two line segments he had drawn in his own booklet, saying "diagonal, diagonal". He then gesture-traced the right side of the square, the short diagonal and half of the long diagonal, and when he got to the top right corner said, "triangle". Then he moved his pencil to the bottom side, gesture-tracing out the congruent triangle on the bottom and said "triangle" again. (Given that neither triangle is in its prototypical orientation or shape, this identification is not insignificant.) From the first task to the second, Leo thus drew one-dimensional objects, described two-dimensional objects, drew one-dimensional objects, described one-dimensional objects then described two-dimensional shapes, thereby going back and forth between seeing the image in terms of one-dimensional objects and seeing it in terms of two-dimensional shapes.

A different shift from one to two dimensions, and one that is highly relevant to the prototypical tendencies that children have in two-dimensional shape identification, arose in the discussion of the images shown in Fig. 11.2C and D. Most of the children began by referring to the inscribed squares as diamonds. For example, when shown the image in Fig. 11.2C, Maya describes it as "a diamond and a straight line" (thus naming both a two-dimensional shape and a one-dimensional object). However, several of the children also referred to the same shape as a square, either after having rotated the booklet around or after having been asked what different shapes they noticed (this was especially true for the image shown in Fig. 11.2D, which the children described as having three squares). For example, when Diana was asked, "What do you see?" after being shown the image in Fig. 11.2C, she leaned forward, put her pencil on the page and said "there's a diamond" then traced her pencil along the horizontal line and said, "and split in two, which means two triangles, so there's one, two, three, four, five, six triangles." When asked what other shape the diamond could make, Diana responded, "square", then turned the paper around.

Given the tenacious way in which children identify shapes through prototypical means (what does it look like?), we hypothesise that the flexibility that these children showed may have arisen out of their drawing activities and, in particular, their attending to line segments through drawing, gesturing and describing, here and in previous related activities in class. This would be consistent with Duval's (2005) theory.

Thus far, we have considered structure in terms of the shifts between local and global—and particularly between one- and two-dimensional geometric objects—but the theme of structure arose quite intentionally with the fifth task of interview protocol 1, when students were asked to complete a grid (see Fig. 11.2E). This question has previously been used by Mulligan and Mitchelmore (2009) as a way of assessing young children's understanding of mathematical array structure. They found a strong correlation with general mathematics performance. Children who performed poorly on this task produced laborious and mathematically inefficient drawings. For example, a child who completed the grid by creating individual squares, as seen in Fig. 11.5a, was seen as having only partial structural awareness (Mulligan and Mitchelmore identified two classes of drawings that showed even less structural awareness). On the other hand, children who complete the grid by drawing three lines to extend the columns and rows, as seen in Fig. 11.5b, were seen as demonstrating fully developed structural awareness. In their study, 27% of the 103 grade 1 students tested produced drawings like the one in Fig. 11.5a while 24% produced a drawing like the one in Fig. 11.5b. The remaining students produced

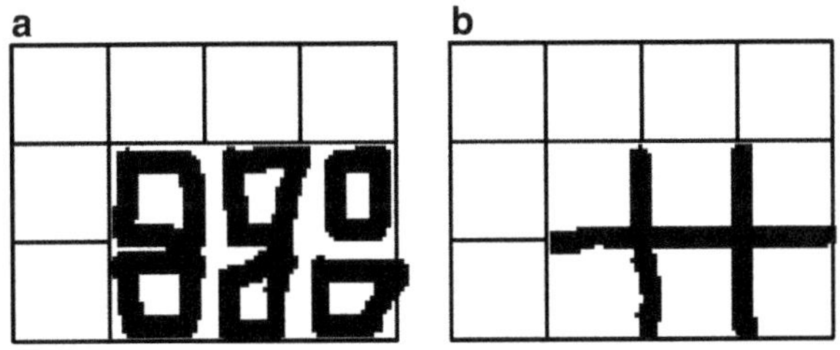

Fig. 11.5 (**a**) Demonstration of partial structural awareness; (**b**) demonstration of full structural awareness

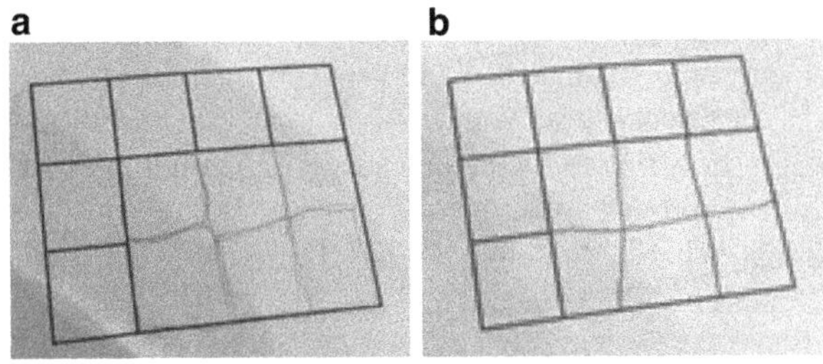

Fig. 11.6 (**a**) Example of partial/full structural awareness; (**b**) example of full structural awareness

drawings with comparatively less structure and were classified as demonstrating either pre-structural (11%) or emerging structural (38%) awareness in their grid completion. Other studies have found that it is not until about fourth grade that most children learn to construct the row-by-column structure of rectangular arrays (Battista, Clements, Arnoff, Battista, & Borrow, 1998; Outhred & Mitchelmore, 2000).

Against these findings, we were interested in seeing how children in the current study would perform on the task and what might be revealed about their structural awareness. Given Carol's extensive and continued focus on geometrical structures in her teaching, we predicted that her students might demonstrate a relatively high level of structural awareness. Of the nine children who were given this task in our study, none produced drawings like the one in Fig. 11.5a; four produced drawings that we consider partial/full (see Fig. 11.6a); five produced drawings like the one in Fig. 11.5b (see Fig. 11.6b). Thus, all students appeared to demonstrate at least some level of structural awareness, indicating higher levels of performance than previously reported despite being much younger in age (e.g., see Mulligan & Mitchelmore, 2009). Given the small sample size and selected population, it is difficult to interpret these findings and more research is needed to understand the effects of learning conditions on children's development of structural awareness. However, this finding does raise the intriguing possibility that Carol's extensive focus on geometrical structure (through building, graphing, drawing and grid activities—as well as Can You Draw This activities)—may have positively influenced her students' performance.

Some evidence that performance on the task is flexible and prone to immediate improvements can be seen in the case of Elka who was asked to complete the task both with and without prompts. When first asked to finish completing the grid, Elka did so by drawing individual squares (see Fig. 11.7a). Although her drawing was accurate in that it produced a 4 × 3 grid of adjoining cells, her approach to the task was inefficient and imprecise (i.e., her cells were of varying proportions and drawn in piecemeal). Here we see that when unprompted, Elka did not demonstrate full structural awareness. However, as revealed next, Elka's second attempt at the task provides insight into her performance and potentially demonstrates the immediate impact of instruction.

Zack: What if you did it like this? So, watch my finger. What if you just took your pencil here and went, zoom (*uses finger to gesture the drawing of a horizontal line across the page*). And then you took your pencil and you went, zoom (*uses finger to gesture drawing vertical line that create new row*). And then you took your pencil and you went, zoom (*uses finger to gesture drawing of another vertical line to create new rows*). Do you think that would work too?

Elka: (*Nods.*)

Zack: What if you try it? What if you try to do it with long lines, as long as you possibly can?

Even before the interviewer has finished the question, Elka had completed a horizontal line that created the second and third rows. She then quickly and efficiently completes the grid by drawing two vertical lines that complete the grid (see Fig. 11.7b). In comparing Elka's first and second attempts at this task, it is clear that the second attempt was much more accurate and according to Mulligan and Mitchelmore (2009), was representative of a more sophisticated understanding of mathematical structure. In this example we see how a few short prompts may have been enough to facilitate drawing performance. However, it is also possible that Elka's improvements are not indicative of increases in structural awareness per say, but a result of simply copying the interviewers gestures or even as a result of repeated practice on the same task. This finding, along with the general finding of high levels of structural awareness amongst Carol's students, is deserving of further research as it addresses the important question of the extent to which children's structural awareness, and more broadly geometric drawing performance, is influenced by developmental constraints but also malleable and subject to improvements given proper instruction. Research of this sort will help reconcile the

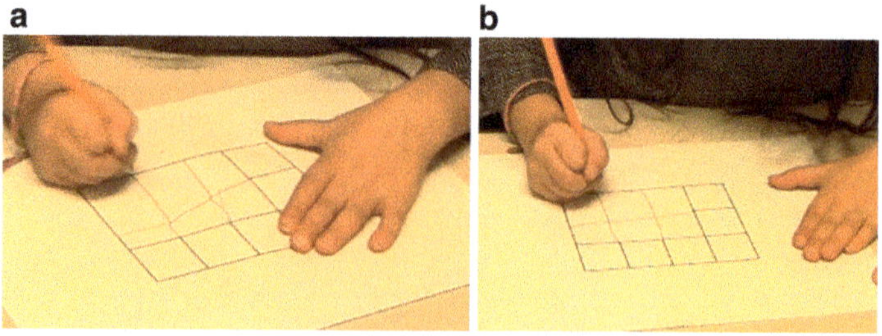

Fig. 11.7 (**a**) Elka's first attempt. (**b**) Elka's second attempts at the grid task

presumably false dichotomy evident in the larger research literature; that is, the contrast in views of children's drawings as relatively stable outcomes of cognitive development versus the perspective of children's drawings as a dynamic process that not only presents an artefact *of* learning but also an act *for* learning. More broadly, a better understanding of the impact that instruction has on children's geometric drawing behaviour and related mathematical insights will go a long way in helping us better understand the potential role of drawing in early mathematics education.

Discussion

In this chapter, we have drawn on educational psychology research to motivate investigation into children's spatial reasoning more broadly, and their drawing in particular. Instead of studying children's finished drawings for how they indicate cognitive development or intelligence or even correlation to mathematical ability, we have focussed on their drawing processes in order to gain insight into how children might learn through and from drawing, and how this might relate to the development of their spatial reasoning. We developed tasks in which the children's drawing processes involved instructional prompts and included both language and gesture. This was done in part to investigate how language and gesture might interplay with drawing and in part in order to pursue drawing as a geometrical activity, following the work of Duval. This latter point signals a shift away from the more commonly used van Hiele (1986), in which there is less emphasis both on the coordination of visual and language registers and on the dimensional deconstruction involved in drawing and seeing. It also differs from the focus that is found in Clements and Sarama's (2011) learning trajectories of composing and decomposing, particularly in relation to the centrality of drawing and of using/naming one-dimensional geometric objects.

In analysing the drawing processes of the children participating in our study, we were able to identify a significant number of situations in which the use of language or gesture changed the way that the students saw, drew and described (that is, the way they spatially reasoned) the images shown in Figs. 11.2 and 11.3. Gestures were used extensively to mark out future line segments, including their endpoints either at the corner or in the middle of the sides of existing line segments. The use of words such as "middle" and "top" or "side" also oriented the children's drawings and enabled them to successfully describe images to the interviewers. Based on our analyses, we propose that providing children with opportunities to draw while also talking and gesturing can improve their performance on Quick draw-type tasks and increase their spatial reasoning.

Instead of focussing solely on finished drawings, our interview protocols also included questions that enabled us to probe mathematical concepts that could be relevant to the drawings the children had produced. As we described above, these concepts include congruence, symmetry and structure. While these concepts can be taught and assessed without a drawing component, we suggest that, based on our analyses, the very act of drawing, and of gesturing and tracing, changes the way that students see (spatially reason about) geometric images. The actual drawing of the segments that make up a triangle or square seems to prompt children to attend to two-dimensional properties of shapes, which they can use in order to reason about whether or not the shapes are the same. The act of drawing may also help children become aware of symmetry (of drawing the same thing on both sides), as well as structures such as grids.

Despite increased interest in, and current research findings showing the importance of spatial reasoning for mathematics education, and despite the call for increased focus and time spent on geometry in the mathematics curricula (e.g. NCTM) very little progress has been made either within mathematics education research or in instructional material design (apart from Moss et al., 2016). In particular, in current standards and curricula for early school geometry, drawing has had extremely limited attention. Our chapter seeks to forge new ways of including drawing as part of the teaching and learning of geometry and offers new ways of thinking about and analysing the types of spatial/geometric reasoning young children are capable of. In this regard, our study takes a different view of the potential of drawing and identifies productive ways in which drawing could support spatial reasoning in the context of geometry. Our study specifically looks at both gesturing and drawing together, a focus arising from the study of mathematical activity (Châtelet, 2000), and one that we see as very productive in future research in spatial reasoning—and drawing in particular—in both educational psychology and mathematics education.

References

Battista, M. T., Clements, D. H., Arnoff, J., Battista, K., & Borrow, C. V. A. (1998). Students' spatial structuring of 2D arrays of squares. *Journal for Research in Mathematics Education, 29*(5), 503–532.

Beery, K., & Beery, N. (2010). *The Beery–Buktenica developmental test of visual motor integration* (6th ed.). Bloomington, MN: Pearson.

Boonen, A. J., van der Schoot, M., van Wesel, F., de Vries, M. H., & Jolles, J. (2013). What underlies successful word problem solving? A path analysis in sixth grade students. *Contemporary Educational Psychology, 38*(3), 271–279.

Boonen, A. J., van Wesel, F., Jolles, J., & van der Schoot, M. (2014). The role of visual representation type, spatial ability, and reading comprehension in word problem solving: An item-level analysis in elementary school children. *International Journal of Educational Research, 68*, 15–26.

Bremigan, E. G. (2005). An analysis of diagram modification and construction in students' solutions to applied calculus problems. *Journal of Research in Mathematics Education, 36*(3), 248–277.

Brooks, M. (2009). Drawing, visualisation and young children's exploration of "big ideas". *International Journal of Science Education, 31*(3), 319–341.

Carlson, A. G., Rowe, E., & Curby, T. W. (2013). Disentangling fine motor skills' relations to academic achievement: The relative contributions of visual-spatial integration and visual-motor coordination. *The Journal of Genetic Psychology, 174*(5), 514–533.

Case, R., & Okamoto, Y. (1996). The role of central conceptual structures in the development of children's thought. *Monographs of the Society for Research in Child Development, 61* (Nos. 1–2).

Châtelet, G. (2000). *Les enjeux du mobile*. Paris: Seuil English translation by R. Shore & M. Zagha: *Figuring space: Philosophy, mathematics, and physics*. Dordrecht: Kluwer.

Chen, C.-L., & Herbst, P. (2013). The interplay among gestures, discourse, and diagrams in students' geometrical reasoning. *Educational Studies in Mathematics, 83*(2), 285–307.

Claparede, E. (1907). Plan d'experiences collectives sur le dessin des enfants. *Archives de Psychologic, 6*, 276–278.

Clements, D. H., & Sarama, J. (2011). Early childhood teacher education: The case of geometry. *Journal of Mathematics Teacher Education, 14*, 133–148.

Cooke, E. (1885). Art teaching and child nature. *London Journal of Education.*

de Freitas, E., & Sinclair, N. (2012). Diagram, gesture, agency: Theorizing embodiment in the mathematics classroom. *Educational Studies in Mathematics, 80*(1), 133–152.

Diezmann, C, M. & English, L. D. (2001). Promoting the use of diagrams as tools for thinking. In A. A. Cuoco & F. R. Curcio (Eds.), *The roles of representation in school mathematics* (pp. 77–89). Reston, VA: National Council of Teachers of Mathematics.

Duval, R. (1998). Geometry from a cognitive point of view. In C. Mammana & V. Villani (Eds.), *Perspectives on the teaching of geometry for the 21st Century: an ICMI study* (pp. 37–52). Dordrecht: Kluwer.

Duval, R. (2005). Les conditions cognitives de l'apparentissage de la géométrie: Développement de la visualisation, differenciation des raisonnement et coordination de leurs fonctionnements. *Annales de didactique et sciences cognitives, 10*, 5–53.

Goodenough, F. L. (1926a). *Measurement of intelligence by drawings*. New York: World Book.

Goodenough, F. L. (1926b). A new approach to the measurement of the intelligence of young children. *The Pedagogical Seminary and Journal of Genetic Psychology, 33*(2), 185–211.

Grissmer, D., Grimm, K. J., Aiyer, S. M., Murrah, W. M., & Steele, J. S. (2010). Fine motor skills and early comprehension of the world: Two new school readiness indicators. *Developmental Psychology, 46*(5), 1008–1017.

Hanlon, A. E. C. (2010). *Investigating the influence of Quick Draw on pre-service elementary teachers beliefs, in concordance with spatial and geometric thinking: A mixed methods study.* (Doctoral dissertation, Oklahoma State University).

Hawes, Z., Tepylo, D., & Moss, J. (2015). Developing spatial thinking: Implications for early mathematics education. In B. Davis & Spatial Reasoning Study Group (Eds.), *Spatial reasoning in the early years: Principles, assertions and speculations* (pp. 29–44). New York, NY: Routledge.

Hegarty, M., & Kozhevnikov, M. (1999). Types of visual–spatial representations andmathematical problem solving. *Journal of Educational Psychology, 91*(4), 684–689.

Hu, F. T., Ginns, P., & Bobis, J. (2015). Getting the point: Tracing worked examples enhances learning. *Learning and Instruction, 35*, 85–93.

Ivanoff, E. (1909). Recherches experimentales sur le dessin des ecoliers de la Suisse Romande: Correlation entre l'aptitude an dessin et les autres aptitudes. *Archives de Psychologic, 8*, 97–156.

Kamphaus, R. W., & Pleiss, K. L. (1992). Draw-a-person techniques: Tests in search of a construct. *Journal of School Psychology, 29*(4), 395–401.

Kellogg, R. (1970). *Analyzing children's art*. Palo Alto, CA: Mayfield.

Kulp, M. T. (1999). Relationship between visual motor integration skill and academic performance in kindergarten through third grade. *Optometry & Vision Science, 76*(3), 159–163.

Kurdek, L. A., & Sinclair, R. J. (2001). Predicting reading and mathematics achievement in fourth-grade children from kindergarten readiness scores. *Journal of Educational Psychology, 93*, 451–455.

Lakoff, G., & Núñez, R. (2000). *Where mathematics come from: How the embodied mind brings mathematics into being*. New York, NY: Basic books.

Malanchini, M., Tosto, M. G., Garfield, V., Dirik, A., Czerwik, A., Arden, R., … Kovas, Y. (2016). Preschool drawing and school mathematics: The nature of the association. *Child Development, 87*(3), 929–943.

Mix, K. S., & Cheng, Y. L. (2012). The relation between space and math: Developmental and educational implications. *Advances in Child Development and Behavior, 42*, 197–243.

Moss, J., Bruce, C., Caswell, B., Flynn, T., & Hawes, Z. (2016). *Taking Shape: Activities to develop geometric and spatial thinking*. Don Mills, ON: Pearson Education Canada.

Mulligan, J., & Mitchelmore, M. (2009). Awareness of pattern and structure in early mathematical development. *Mathematics Education Research Journal, 21*(2), 33–49.

Núñez, R. (2003). Do real numbers really move? Language, thought, and gesture: The embodied cognitive foundations of mathematics. In R. Hersh (Ed.), *18 unconventional essays on the nature of mathematics* (pp. 160–181). New York: Springer.

Nunokawa, K. (2006). Using drawings and generating information in mathematical problem solving processes. *Eurasia Journal of Mathematics, Science and Technology Education, 2*(3), 34–54. https://doi.org/10.12973/ejmste/75463.

Outhred, L. N., & Mitchelmore, M. C. (2000). Young children's intuitive understanding of rectangular area measurement. *Journal for Research in Mathematics Education, 31*, 144–167.

Pieters, S., Desoete, A., Roeyers, H., Vanderswalmen, R., & Van Waelvelde, H. (2012). Behind mathematical learning disabilities: What about visual perception and motor skills? *Learning and Individual Differences, 22*(4), 498–504.

Polya, G. (1957). *How to solve it* (2nd ed.). Princeton, NJ: Princeton University Press.

Sinclair, N., & Gol Tabaghi, S. (2010). Drawing space: Mathematicians' kinetic conceptions of eigenvectors. *Education Studies in Mathematics, 74*(3), 223–240.

Sortor, J. M., & Kulp, M. T. (2003). Are the results of the Beery-Buktenica Developmental Test of Visual–Motor Integration and its subtests related to achievement test scores? *Optometry and Vision Science, 80*, 758–763.

Streeck, J. (2009). *Gesturecraft: The manu-facturing of meaning*. Amsterdam: John Benjamins.

Steenpaß, A., & Steinbring, H. (2014). Young students' subjective interpretations of mathematical diagrams— elements of the theoretical construct "frame-based interpreting competence". *ZDM—The International Journal on Mathematics Education, 46*(1), 3–14.

Sundberg, N. D. (1961). The practice of psychological testing in clinical services in the United States. *American Psychologist, 16*, 79–83.

Tzuriel, D., & Egozi, G. (2010). Gender differences in spatial ability of young children: The effects of training and processing strategies. *Child Development, 81*(5), 1417–1430.

van Hiele, P. M. (1986). *Structure and insight: A theory of mathematics education*. New York: Academic Press.

Weckbacher, L. M., & Okamoto, Y. (2015). Discovering space in the elementary classroom. *Journal of Education and Learning, 4*(1), 35.

Wheatley, G. H. (1997). Reasoning with images in mathematical activity. In L. D. English (Ed.), *Mathematical reasoning: Analogies, metaphors, and images* (pp. 281–297). Mahwah, NJ: Erlbaum.

Wheatley, G. H. (2007). *Quick draw: Developing spatial sense in mathematics* (2nd ed.). Bethany Beach, DE: Mathematics Learning.

Whiteley, W. (2002). Teaching to see like a mathematician. Retrieved from http://www.math.yorku.ca/~whiteley/Teaching_to_see.pdf.

Yackel, E., & Wheatley, G. H. (1990). Spatial sense: Promoting visual imagery in young pupils. *Arithmetic Teacher, 37*(6), 52–58.

Yancey, A. V., Thompson, C. S., & Yancey, J. S. (1989). Children must learn to draw diagrams. *Arithmetic Teacher, 36*(7), 15–19.

Chapter 12
The Interaction Between Spatial Reasoning Constructs and Mathematics Understandings in Elementary Classrooms

Tom Lowrie and Tracy Logan

Abstract Numerous studies from cognitive and educational psychology research have highlighted the strong association between spatial reasoning and mathematics performance. This chapter examines this relationship from a mathematics education perspective, with a focus on elementary classrooms. Three spatial constructs critical to mathematics instruction and learning are identified: namely, spatial visualization; mental rotation; and spatial orientation. These constructs are described in relation to student's encoding and decoding of mathematics information and the increasing influence these constructs have on mathematics assessment. The extent to which spatial training can enhance student's math performance is also considered in relation to these three constructs. Implications highlight the potential of explicitly focusing on spatial reasoning in math classrooms, given the malleability of instruction and ongoing affordances of technology.

Keywords Spatial reasoning · Mathematics · Space · Spatial visualization · Mental rotation · Spatial orientation · Encoding · Decoding · Curriculum · Classroom · Elementary · Geometry · Graphics · Graphical languages · STEM · Assessment · Digital · Technology · Spatial training · Australia

Introduction

This chapter considers the role and nature of spatial reasoning in students' mathematics understanding, specifically in the elementary classroom. Although spatial reasoning has been widely and deeply investigated in the cognitive and educational psychology literature, its impact on classroom practices has been more subtle. The strongest association between school mathematics and spatial reasoning has been described within the geometry and measurement strands of the mathematics curriculum (Battista, 2007) and linked to a general notion of geometric reasoning—that

T. Lowrie (✉) · T. Logan
Faculty of Education, University of Canberra, Bruce, Australia
e-mail: thomas.lowrie@canberra.edu.au; tracy.logan@canberra.edu.au

K. S. S. Mix, M. T. Battista (eds.), *Visualizing Mathematics*, Research in Mathematics Education, https://doi.org/10.1007/978-3-319-98767-5_12

is, how one "reasons *about* objects; one reasons *with* representations" (Battista, 2007, p. 844). There is also a view that spatial reasoning has important domain significance in mathematics, especially in one's capacity to decode graphs and diagrams (Postigo & Pozo, 2004). Current school-based practices, from both curriculum and assessment perspectives, have moved toward more visual and graphic forms of representation. This is unsurprising given the increased use of graphics in society and the ongoing challenge of representing burgeoning amounts of information in visual and graphic forms (Lowrie & Diezmann, 2009). From a young age, students are exposed to visual forms of communication with more intensity and engagement, whether playing computer games or navigating web pages. This ease of engaging with, and accessibility to, visual information heightens the need to understand the role of spatial reasoning in how children encode and decode information in their increasingly visual world.

Defining Spatial Reasoning

Spatial concepts tend to develop through engagement in our inherently spatial world or through activities that promote spatial skills or understandings. The interactions with our environment are associated with thinking *about* space, thinking *in* space, and thinking *with* space (National Research Council, 2006). Hegarty and Stull (2012) argued that visuospatial reasoning involves either thinking about space at the smaller scale of objects (including object recognition) or the larger scale of an environment (including orientation in unfamiliar contexts). We tend to interact with, and manipulate, small-scale objects from the perspective of oneself from a stationary vantage point. Interaction with large-scale objects and environments tends to be undertaken with oneself moving around or within the environment or context (Battista, 2007), with mental models developed over time (Hegarty & Stull, 2012). As Battista (2007) indicated, we engage in large-scale spaces, while on small-scale spaces. In fact, there is evidence to suggest that the encoding and decoding of information from small- and large-scale objects activates different parts of the brain (Kosslyn & Miller, 2013). Although the process of reasoning visually and mentally can take the form of both symbolic and picture-like representations (Slotnick, Thompson, & Kosslyn, 2005), spatial reasoning includes, but is not limited to, the generation and manipulation of images and the navigation of space.

Spatial reasoning is not a single unitary construct; rather it consists of several dimensions. Even though there is some agreement regarding the multidimensionality of spatial ability, the composition of the sub-components is not well defined. Elsewhere we have identified a set of spatial constructs that align well to both mathematics curricula and the assessment of school mathematics (Ramful, Lowrie, & Logan, 2016). Three spatial constructs are defined, namely, spatial visualization, mental rotation, and spatial orientation. The definitions for the respective constructs could not be taken from one source (e.g., Carroll, 1993), since such definitions do not encompass school-based mathematics content. Consequently, we defined spatial visualization as multistep "manipulations of spatially presented information" (Linn & Petersen, 1985, p. 1484); mental rotation as the ability to mentally rotate two- or three-dimensional figures and to imagine their positions after they are rotated around an axis (Linn & Petersen, 1985); and spatial orientation as "the ability to imagine how a stimulus array will appear from another perspective" (Kozhevnikov, Hegarty, & Mayer, 1999, p. 4). In our spatial orientation definition, we broaden the definition to include movement in space. Kozhevnikov and Hegarty (2001) classify the distinction between mental rotation and spatial orientation as often task-dependent, determined by the degree of rotation required to complete a task. That is, where a task requires a certain degree of rotation, the main spatial process will shift from a mental rotation, where the task taker moves the object in their mind, to that of orientation, where the task taker changes their spatial perspective. In other instances, the small-scale manipulation of objects is classified within the spatial construct of mental rotation, while large-scale engagement is classified as spatial orientation.

Although the presence of spatial reasoning is well established in childhood (Newcombe & Frick, 2010), spatial skills and understandings are not generally taught in explicit ways in classrooms, and it has certainly not been the responsibility of the mathematics teacher to provide such instruction (Bishop, 2008). As such, spatial skills develop through experiences outside of school contexts, or through reasoning experiences inside of and outside of school that inadvertently promote such thinking. These examples include engagement with multiplicative arrays (Clements, Battista, Sarama, & Swaminathan, 1997), the interpretation of maps (Lowrie & Logan, 2007), the affordances of technologies (Sinclair & Bruce, 2015), as well as geometric reasoning.

The recognition and categorization of objects feature prominently in the elementary curriculum, established initially through the recognition of objects by color, shape, or size. The categorization of objects increases complexity. For example, the classification of different types of quadrilaterals requires considering some features of an object while ignoring others (e.g., focusing on internal angles of the objects rather than their size or color). This type of spatial thinking is classified as spatial visualization. The presence of spatial reasoning in a wide range of curriculum content is evident with some investigation. However, to date there is no specialized guidance for spatial reasoning instruction.

Overview of the Chapter

From curriculum, assessment, and instruction perspectives, we describe the role and nature of spatial reasoning in elementary school mathematics. Throughout the chapter, we consider the role spatial reasoning has in students encoding and decoding of mathematics information. We identify spatial reasoning in relation to three spatial constructs, namely, spatial visualization, mental rotation, and spatial orientation. These constructs are used to categorize tasks within large-scale national and international assessment exercises as a way of highlighting the indirect influence spatial reasoning has in the mathematics curricula. Although no current elementary school curriculum has a content strand assigned to spatial thinking, spatial skills are increasingly required to understand mathematics, in part, due to technology. The subsequent section examines the role of spatial training on students' mathematics performance. It analyzes data from a program that explicitly introduced the three spatial constructs to students through a pre- and post-test design with control and intervention classes. The changing relationships between the students' spatial reasoning and performance on the mathematics tasks are a focus of the analysis. Finally, conclusions and implications are drawn.

The (Potentially Changing) Nature of Geometry in the Elementary Classroom

Geometry as the Focus of Spatial Reasoning

Across the world, students' engagement with geometric understandings in elementary grades can be limited to the recall of two- and three-dimensional objects and the definitions of these objects. As Sinclair and Bruce (2015) noted, school geometry continues to concentrate on the naming and sorting of shapes by property. Even with such attention to the naming of shapes, students encounter difficulties when trying to identify figures and objects represented in an "atypical" form, most likely because students rarely encounter objects that are not displayed in the typical orientation or elevation (Ho & Lowrie, 2014; Sarfaty & Patkin, 2013). To some degree this is understandable, since elementary teachers' pedagogical content knowledge of geometry understandings is generally poor (Clements & Sarama, 2011) they tend to rely on standard textbook orientations to expose students to content. Advances in technology have afforded new opportunities for the exploration of 2D shapes and 3D objects, with mobile tablets being the device most likely to influence classroom practices if teachers are to become less reliant on typical representations (Ng & Sinclair, 2015). In fact, recent curriculum reconceptualization is based on opportunities afforded by digital technologies, and the necessity to promote transformational geometry in ways that promote mentally manipulating 2D figures and 3D objects (Sinclair & Bruce, 2015).

It could be argued that although geometry has "survived" in the past, it has the potential to "thrive" in the elementary curriculum due to its association with spatial ability. In their seminal work, Clements and Battista (1992) maintained that the geometry curriculum afforded opportunities for teachers to encourage students to reason spatially and develop mathematics logic in ways that could not be accessed elsewhere in the curriculum. They also highlighted ways in which tools (such as Logo) could foster constructivist approaches that encouraged students to utilize visualization skills as they translated, rotated, and reflected 2D objects (Battista &

Clements, 1988). In a post-Piagetian era, Battista, Clements and colleagues demonstrated that students in the early years could engage with geometric content in ways that promoted spatial structuring (Battista, Clements, Arnoff, Battista, & Borrow, 1998) and dynamic imagery (Clements et al., 1997). Spatial visualization skills were developed through engagement with concepts associated with 2D space, area, and coordinate systems, with demonstrable benefits for students' capacity to understand numerical units and arrays (necessary for multiplicative thinking).

More recently, there has been increasing momentum by educators to rethink how geometry is presented in elementary schools, with a stronger emphasis on the nexus between geometry and spatial reasoning (e.g., Ministry of Education, Curriculum Planning and Development Division, 2006; Ontario Ministry of Education, 2008). Such curriculum and support documents (i.e., from Singapore and Canada) have raised practitioners' awareness of spatial reasoning; however, the association between geometric reasoning and spatial reasoning is both vague and inexplicit. In part, this is because learning opportunities remain content focused, framed within the content of properties and shapes, geometric relationships, and location and arrangement. More recent documents have successfully established clear links that afford classroom teachers opportunities to enhance students' spatial reasoning skills. For example, the Ontario Ministry of Education's (2014) *Paying Attention to Spatial Reasoning* support document identifies where spatial visualization is critical for understanding mathematics ideas and concepts. The document outlines learning opportunities that enhance spatial visualization through the exploration of geometry-based tasks, including the composition and decomposition of shapes, and net and paper-folding activities. Activities that highlight the role of spatial reasoning in number sense and the decoding of graphs are also provided. Disappointingly, vague and misleading information is provided when describing other spatial skills including mental rotation and perspective taking. The OME (2014) support document is the most comprehensive available to classroom teachers to date, and yet it lacks a compelling framework that situates the cognitive and educational psychology foundations of spatial reasoning into practice. Consequently, the underlying philosophy of the document inadvertently suggests that the more one emphasizes geometry in the mathematics curriculum, the more likely students' spatial reasoning will improve. We propose that a framework based on spatial reasoning processes and classroom practices is needed to move away from the traditional definitions of the construct that are grounded in their measurement.

Spatial Reasoning Supports Encoding and Decoding of Mathematics Information

Spatial reasoning is influential in mathematics problem solving, across a range of contexts that are not geometry based. In such situations, spatial reasoning involves the use of "internal [encoding] or external [decoding] of visual or spatial representations" (Shah & Miyake, 2005, p. xi) which typically involve imagery or diagrammatic and abstract reasoning. Students with high spatial visualization ability perform at a much higher level in mathematics than those students who possess medium or low spatial visualization ability due to the cognitive demands of simultaneously encoding and decoding information (Shea, Lubinski, & Benbow, 2001). To encode graphic information, an individual typically draws pictures or diagrams, or represents and manipulates images mentally. In elementary school mathematics, encoding is commonly accessed when students draw pictures to represent information usually contained in word problems. Decoding is utilized in situations when students are required to interpret graphic information.

Interpreting graphics information requires spatial reasoning since the understanding of the representation involves engagement with a visual symbol system and perceptual processes. The symbol system is composed of (a) *visual elements*, such as shapes, that represent objects or ideas and (b) the *spatial relationships* among the elements within the graphic (e.g., one shape inside another). Perceptual processes are used to distinguish between embedded graphics information such as position, length, angle, area, volume, density, color saturation, color hue, texture, connection, containment, and shape (Cleveland & McGill, 1984). Mackinlay's (1999) work on graphical languages provides a framework for understanding the various techniques required to solve specific types of graphics representations (see Table 12.1).

Elsewhere we (and colleagues) have incorporated Mackinlay's framework into mathematics contexts that are commonly found in elementary school curricula and assessment (Diezmann & Lowrie, 2009). Although there are numerous types of graphics, the surface detail and item structure align well to spatial features associated with axes, position, and arrangement (see Table 12.2).

In a study that considered the relation between 347 Grade 5 (10- and 11-year-olds) students' performance across graphics-rich mathematics tasks and nonverbal reasoning, Lowrie, Diezmann, and Logan (2012) found moderately strong

Table 12.1 Mackinlay's (1999) graphical languages by encoding technique

Graphical languages	Examples	Spatial encoding process?
Axis languages	Number line, scale	Encodes information by the placement of a mark on an axis
Opposed-position languages	Line chart, bar chart, plot chart	Information is encoded by a marked set that is positioned between two axes
Retinal-list languages	Graphics featuring color, shape, size, saturation, texture, orientation	Retinal properties are used to encode information. These marks are not only dependent on position
Map languages	Road map, topographic map	Information is encoded through the spatial location of the marks
Connection languages	Tree, acyclic graph, network	Information is encoded by a set of node objects with a set of link objects.
Miscellaneous languages	Pie chart, Venn diagram	Information is encoded with additional graphical techniques (e.g., angle, containment)

Table 12.2 Example mathematics items classified under Mackinlay's (1999) graphical languages

Axis item	Estimate where you think 1.3 should go on this number line 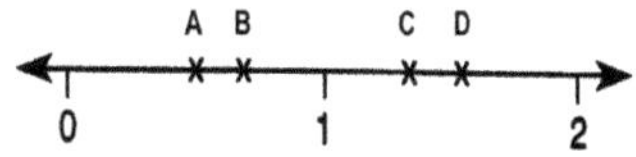
Apposed-position item	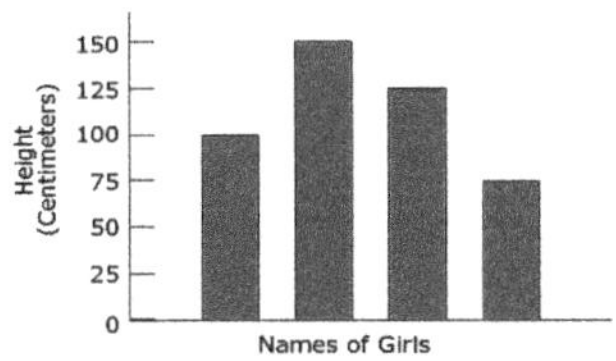 The graph shows the heights of four girls The names are missing from the graph. Debbie is the tallest. Amy is the shortest. Dawn is taller than Sarah How tall is Sarah?
Retinal-list item	When the last piece is put into the puzzle it shows three triangles Which piece is missing from this puzzle? 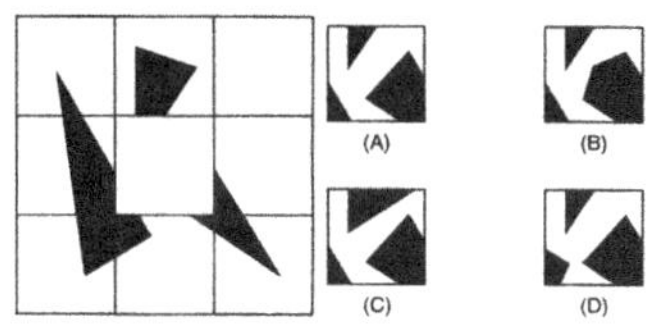
Map item	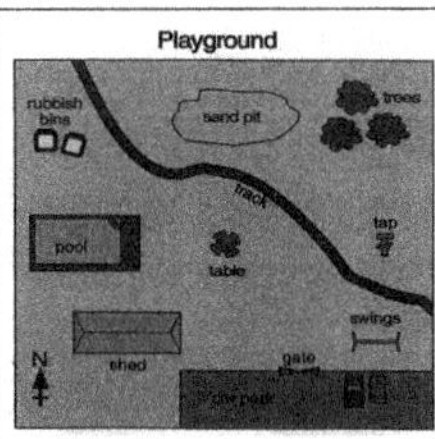 Ben went from the gate to the tap, then to the shed, then to the rubbish bins. How many times did he cross the track?
Connection item	The diagram shows what some animals eat. What are some of the animals that these birds eat? 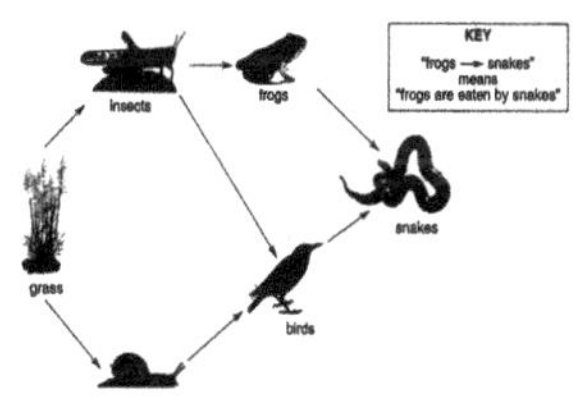
Miscellaneous item	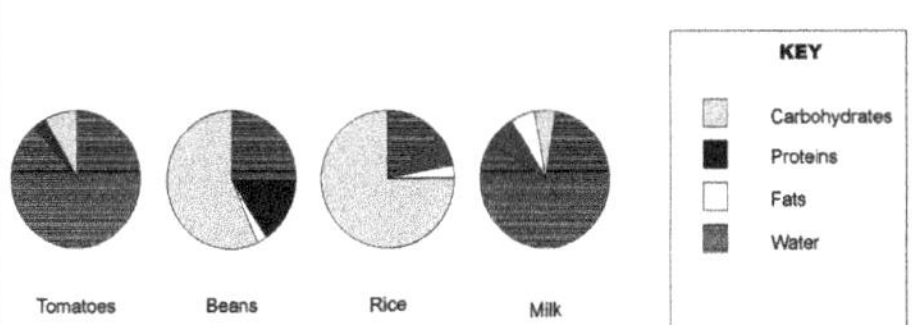 These graphs show the proportion of carbohydrates, proteins, fats and water in some foods. Which food has the highest proportion of carbohydrates?

relationships among performance on the respective graphical language tasks. Nonverbal reasoning (as measured by the Ravens Test) was significantly correlated with each of the six graphic language categories (see Table 12.3).

The relationships between nonverbal reasoning measured by the Ravens test (of which spatial reasoning is one type) and the six respective graphics languages were more highly (and positively) correlated than the associations among the six graphics

Table 12.3 Correlations between graphics language tasks and nonverbal reasoning

Axis	Opposed-position	Retinal-list	Map	Connection	Miscellaneous
0.435**	0.532**	0.401**	0.402**	0.458**	0.490**

Note: ** denotes $p < 0.001$

languages. We concluded that the spatial demands required to decode information from the tasks were critical, since elementary-aged students were not taught explicit graphic conventions in school. Despite the teaching of mathematical concepts such as charts, number lines, and maps, nonverbal reasoning was still the most highly correlated.

Encoding and Decoding Graphics

For a sustained period, mathematics educators and cognitive psychologists have been studying representations—imagery, picture, graphic, or diagram related. These representations are generally defined within two systems, namely *internal* and *external* representations (Goldin & Shteingold, 2001). Internal representations can be classified as pictures "in the mind's eye" (Kosslyn, 1983) and include various forms of concrete and dynamic imagery. According to Bishop (2008), internal representations are associated with personalized, idiosyncratic ideas and images; although it is also likely that these internal representations are influenced by metonymies and prototypes. External representations can be encountered by students or generated by them. Such representations can be characterized as physical objects, potentially text- or graphic-based, in a digital or non-digital form, that require a degree of interpretation. These may include conventional symbolic systems of mathematics (such as an algorithmic computation) or graphical representations (such as maps and pie charts). These two systems are intertwined and are "a two-sided process, an interaction of internalization of external representations and externalization of mental images" (Pape & Tchoshanov, 2001, p. 119). Elsewhere, it has been suggested that these two forms of representations provide different affordances (Lowrie & Diezmann, 2009). Internal representations are often used to encode information, while external mathematics and graphic representations presented to students require decoding (what someone else has encoded).

Encoding generally occurs when students construct their own representations to solve a task, whether a heuristic or a self-generated scaffold. Encoding techniques include drawing diagrams, using visualization or gesture, and combinations of different representations to support understanding. Encoding techniques support the manipulation, storage, and retrieval of information in ways that separate information into more easily understood forms, for example, drawing an array to solve a multiplication task. In some countries, such as Singapore, the explicit teaching of heuristics is still practiced, with students encouraged to represent information in an encoded form (Ho & Lowrie, 2014). In contrast, decoding techniques are typically utilized to make sense of, discern and interpret information from visual or graphic tasks or features, such as interpreting a bar graph or reading a map to determine location of objects based on given directions.

The presented representation of a task influences the degree to which students evoke encoding and decoding techniques. When a task has an embedded graphic, decoding is more likely to be utilized. By providing a graphical representation to scaffold thinking, a whole new set of skills and practices is brought to the foreground. By contrast, when a task does not have a graphic, more opportunities are afforded to utilize encoding techniques.

To highlight this case, two probability tasks are presented, one (re)presented as a traditional word problem (Fig. 12.1) and the other designed with text and a graphical display (Fig. 12.2), where information essential to generating a solution is presented in the graphic. The two tasks are drawn from Australia's national assessment program (ACARA, 2009a, 2010a), with both items classified as probability tasks suitable for elementary students in Grade 3.

Figure 12.1 presented students with a probability task designed to elicit students' understanding of the word *impossible*. Although elementary students might use encoding techniques such as visualizing the box or drawing the box with the marbles in it, for this question most students logically reasoned that there were no white marbles in the box; hence, it is impossible to take one out (Lowrie, 2012).

The second task required the same conceptual understanding of chance (see Fig. 12.2), that is, to assess students' understanding of the concept *impossible*. Representationally, this task contained an embedded graphic along with written text. The embedded graphic provided both contextual and mathematical information associated with the type and number of pegs in the bag.

Grade 3 students find the non-graphic task (Fig. 12.1) easier to solve than they do the graphic task (Fig. 12.2)—64 and 55%, respectively (Queensland Studies Authority, 2009, 2010). In 27 interviews with students of this age, the increased difficulty between the two tasks is due predominantly to the challenges of decoding the graphics—that is, the increased spatial challenges presented in interpreting the diagram. Noteworthy, 68% of Grade 5 students are able to solve the graphics-embedded task. Thus, a 2-year age difference is necessary to gain similar performance to the task without the graphic. To this point, the visual representation of the context (e.g., the bag) adds an element to the task, potentially increasing the cognitive demands of the task.

A box contains 6 red marbles, 10 blue marbles and 4 yellow marbles.

Which colour marble is **impossible** to take from the box?

red	blue	white	yellow
○	○	○	○

Fig. 12.1 A word-based probability task (ACARA, 2009a)

Jess takes 2 pegs out of this bag at the same time.

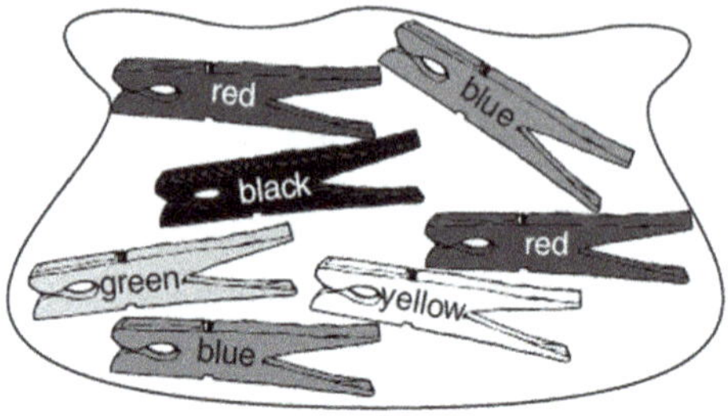

Which of these is **impossible**?

- a blue peg and a black peg
- a red peg and a red peg
- a green peg and a green peg
- a yellow peg and a black peg

Fig. 12.2 A probability task with a graphic that provides essential information (ACARA, 2010a)

In the following sections, we consider the influence of spatial reasoning across both encoding and decoding techniques.

Spatial Reasoning and Encoding

Encoding techniques, such as drawing a diagram or re-representing a graphic, are considered an effective way of solving mathematics tasks, especially when encountering novel or complex tasks. In fact, the use of specific problem-solving strategies or heuristics such as "draw a diagram" has long been advocated in the mathematics education literature (Polya, 1965) and it is still embedded within some national curriculum documents (e.g., Ministry of Education, Curriculum Planning and Development Division, 2006, Singapore).

In a study with Grade 6 students in Singapore, we presented participants with mathematics tasks drawn from Singapore's national assessment, the Primary School Leaving Examination. One of the tasks is presented below.

> *The chairs in a hall were arranged in rows. Each row had the same number of chairs. Weiming sat on one of the chairs. There were 5 chairs to his right and 5 chairs to his left. There were 7 rows of chairs in front of him and 7 rows of chairs behind him. How many chairs were there in the hall?* (SEAB, 2009)

Figure 12.3 illustrates six different representations students used to solve the Weiming Chair Task. In the first two representations, Fig. 12.3a, b, the encoded external pictorial representations are clearly spatial in nature, with the respective students representing chairs in a spatial array. In Fig. 12.3a, the student represents

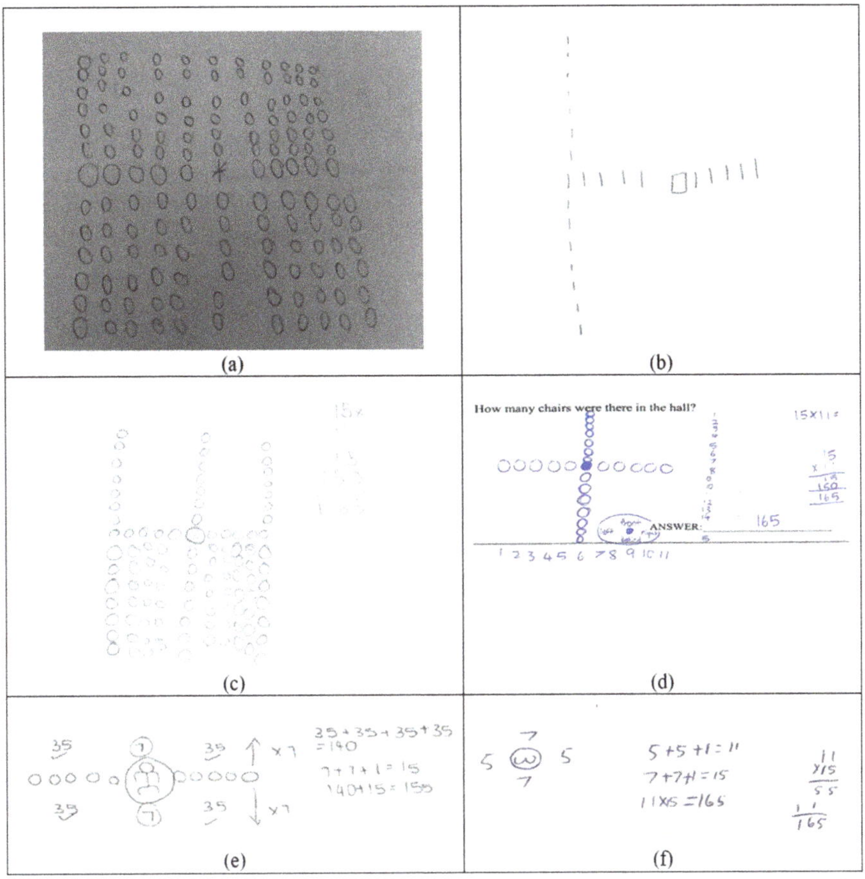

Fig. 12.3 Six representations by students of the Weiming Chair Task (SEAB, 2009)

Weiming as an "X" in the center of the rows of chairs. In Fig. 12.3b, the student represents Weiming as a "box" but does not attempt to draw all of the chairs within the rows. It could be argued that this encoding technique requires a more sophisticated level of spatial reasoning and is less concrete in nature. The next two representations, Fig. 12.3c, d, illustrate instances when students used pictorial reasoning to initially represent the problem scene before using more analytic strategies to complete the task successfully. In Fig. 12.3c, the student has almost completely represented the spatial arrangement of the array before generating an equation to solve the task. In Fig. 12.3d, the student moved toward analytic reasoning much faster, presumably when the pattern had been understood. Even students who relied on analytic reasoning to solve the task used some form of spatial reasoning to represent the tasks. One of these students generated an incorrect solution (see Fig. 12.3e). The student encoded information by drawing the center line of the array, with five circles on each side of Weiming (represented as a stick figure) to represent chairs.

Unfortunately, this student failed to consider Weiming's row in the generation of the solution, an example of an error in spatial perspective taking. The sixth representation (Fig. 12.3f) highlights pre-algebraic reasoning. Even with this relatively sophisticated thinking for a Grade 6 student, the representation included evidence of spatial encoding. This student explained they did not need to physically draw the chairs, but still evidenced an array-like internal spatial representation.

In contrast to the previous problem, in an example from an Australian assessment item (ACARA, 2009b) students were asked to solve the following question, which does not have explicit spatial content:

> *Lin is packing 34 cakes into boxes. Each full box holds 5 cakes. What is the smallest number of boxes Lin needs to pack all the cakes?*

Nevertheless, spatial encoding techniques can support understanding when analytic reasoning is not accessed. For example, in Fig. 12.4, the student drew boxes to represent the cakes rather than generating a solution through calculations. In this instance, the spatial encoding process scaffolded problem solving. As such, spatial reasoning is critical—especially if students need to encode the task to support their reasoning. Despite the lack of graphical information, in the absence of appropriate mathematical proficiency, visualization (and consequently spatial thinking) is consistently a fall-back strategy.

Students' spatially encoded representations often highlight their conceptual understanding, and to some degree provide insights into a student's readiness to solve more complex tasks. When students can consistently represent solutions in an analytic (but non-rote) form, it is likely they have developed a sound understanding of the concept within a domain. Those students who do not yet possess sufficient analytic reasoning skill must draw on spatial visualization skills to support their representations of the problem space.

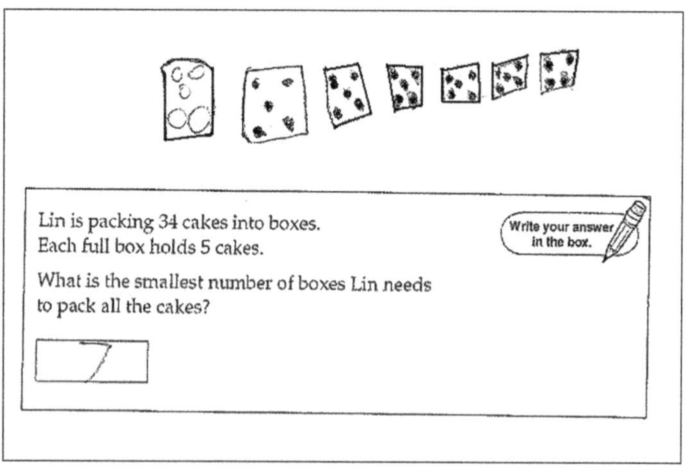

Fig. 12.4 Example of a student using an encoding technique (ACARA, 2009b)

Spatial Reasoning and Decoding

Decoding techniques are most commonly utilized in situations where (in which) students need to interpret information graphics (i.e., graphical elements of assessment items that contain information relevant to the task) generated by others, rather than their own encoded representations. To decode a graphic, an individual must contend with multiple sources of information that may include text, keys or legends, axes and labels (Kosslyn, 2006), as well as perceptual elements of retinal variables (e.g., depth of shading and pattern; Bertin, 1967). For elementary-aged students, the decoding of the actual graphic within a task can be as demanding as the interpretation of the mathematics content associated with the task. As Lowrie and Diezmann (2009) indicate "the actual mathematics of a given task is not likely to be the critical aspect of reasoning and problem solving if the student is not able to access and interpret the information effectively" (p. 146).

Elsewhere, we have argued that there are a variety of spatial demands placed on students when decoding mathematical task information, especially in graphics-rich tasks (Lowrie et al., 2012). In this 2012 study, we identified specific aspects of a task that either promoted or hindered understanding. These task elements included the graphic, the mathematics content (including the task context), and the associated literacy demands of the task. As discussed previously, graphical languages were used to categorize the tasks. The category of task that was most challenging for students were the Retinal-list items—that is, items that required the students to use spatial visualization (see Fig. 12.5a) or mental rotation (see Fig. 12.5b). Such items proved to be the most difficult for students to solve. In general, these findings are consistent with national test results across Australia, where items that require high spatial demands, are commonly among the most difficult to solve. In Australian national assessment, spatial tasks are included under the heading of numeracy and not distinct from mathematics.

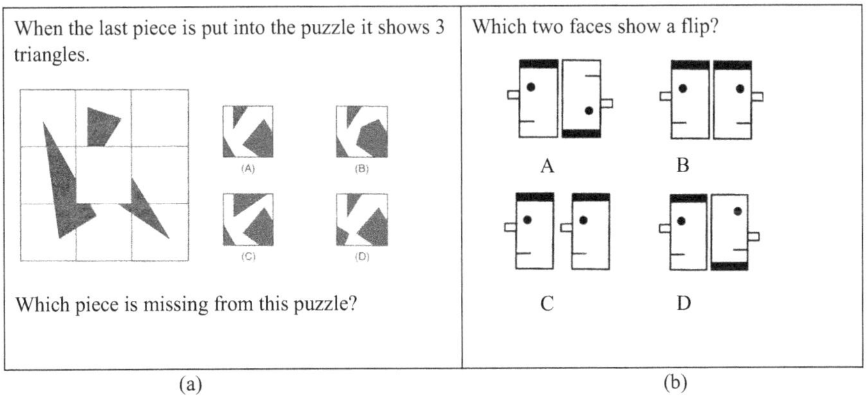

Fig. 12.5 (**a**) Puzzle task requiring spatial visualization (Educational Testing Centre, 2002). (**b**) Faces task requiring mental rotation (Queensland Studies Authority, 2001)

Assessment, Learning, and Spatial Reasoning

Increased levels of teacher accountability and the seemingly relentless desire to compare and measure students' performance have dramatically shaped curriculum design and classroom practices this century. An unintended consequence of this heightened testing regime is a re-emphasis on *number and operations* within the elementary curriculum. As Jones (2000) argued, there is insufficient space to include interesting geometry in the overcrowded curriculum. In a similar vein, Porter (1989) argued that the lack of balance in the elementary curriculum resulted in geometric understandings being treated in isolation and without depth. To some degree, the narrow scope afforded to geometry in the curriculum remains.

Given the influence that assessment has on teaching practices, it is vital to understand the changes occurring in assessment procedures and how they are related to spatial reasoning. In addition, the current flexibility and availability of technology is changing the nature of assessment with little consideration for the cognitive demands of such change. In fact, national and international assessment agencies have already begun to represent 3D objects in technically enhanced and dynamic ways without regard for the different processing requirements placed on students (e.g., Australian Curriculum, Assessment and Reporting Authority; Programme for International Student Assessment). Interpreting and making sense of these objects requires different types of reasoning—that is, thinking with perceptual and relational elements concurrently. Such reasoning is critical for success when students encounter graphics-rich mathematics tasks as highlighted by recent research (Lowrie & Diezmann, 2007; Lowrie & Logan, 2007). In fact, spatial reasoning is implicitly evident in many facets of educational assessment—consequently, not understanding the ways that spatial reasoning may be involved in assessment tasks lessens the likelihood that the tasks will assess what is intended.

Spatial Reasoning in National and International Summative Testing

In this section, we present a content analysis of spatial items in international-, national-, and state-based mathematics assessments. The data are sourced from released items from the following instruments: the Trends in International Mathematics and Science Study (TIMSS); National Assessment Program Literacy and Numeracy (NAPLAN); the Smarter Balanced Assessment Consortium; and the Partnership for Assessment of Readiness for College and Careers (PARCC).

The Australian instrument (NAPLAN) has a much higher proportion of spatial items than all other instruments analyzed, with between 14 and 28% of all items requiring the decoding of specific spatial information. The sample from the international TIMSS instrument included 16% with spatial items. By contrast, the United States state-based instruments contained very few spatial items, with almost no

Table 12.4 Percentage of test items with a spatial construct

	Assessment	Grade 3 Grade 4 (TIMSS)	Grade 5
International	TIMSS (2011)	16% ($n = 73$)	N.A
USA	California (2016)	6% ($n = 31$)	10% ($n = 31$)
	New York (2015, 2016)	0% ($n = 56$)	1012% ($n = 50$)
Australia	NAPLAN (2009–2014)	14–26% ($n = 35$)	12–23% ($n = 40$)

Note: California data based on Smarter Balanced Assessment Consortium mathematics practice tests. New York data based on the Partnership for Assessment of Readiness for College and Careers (PARCC) released items. Data for New York (2015) assessments include both the Performance-Based Assessment and End-of-Year assessments

items in the Grade 3 tests (see Table 12.4). Examples of the spatial items within the respective instruments, categorized by the three spatial constructs (namely, spatial visualization, mental rotation, and spatial orientation) are presented in Figs. 12.6, 12.7, and 12.8 respectively.

The Influence of Spatial Reasoning across Assessment Modality

The influence of spatial reasoning on problems with graphics content in assessment is not limited to the category of the mathematical content. The mode of assessment (i.e., digital versus traditional paper and pencil) has been shown to differentially impact performance on identical tasks (Lowrie & Logan, 2015). In the current section, we consider the use of spatial reasoning on students' interpretation of mathematics items that are presented across different modes.

We have previously examined the question of spatial demands when solving digital versus paper and pencil graphic tasks[1] (Lowrie, Ramful, Logan, & Ho, 2014). There were significant differences between iPad and pencil-and-paper test modes for items that required spatial visualization and spatial orientation. For the spatial visualization task (see Fig. 12.9a), there was a small effect size of Cohen's $d = 0.16$ with higher mean scores on the digital version. By contrast, the spatial orientation task (see Fig. 12.9b) mean scores were higher for the pencil and paper version (Cohen's $d = 0.26$). We then categorized students' spatial visualization ability to determine whether this measure had an impact on performance. Scores on the paper folding test were used to divide the students into low, medium, and high spatial

[1] The digital and non-digital versions of the tasks were structurally similar, with care taken to ensure the fidelity of the items so they did not look different or need to be answered differently in the digital format.

Spatial Visualization

TIMSS (Grade 4)	California (Grade 5)	ACARA (Grade 5)
Ariana found the following patterns to make containers. Which pattern actually makes the container shown beside it? 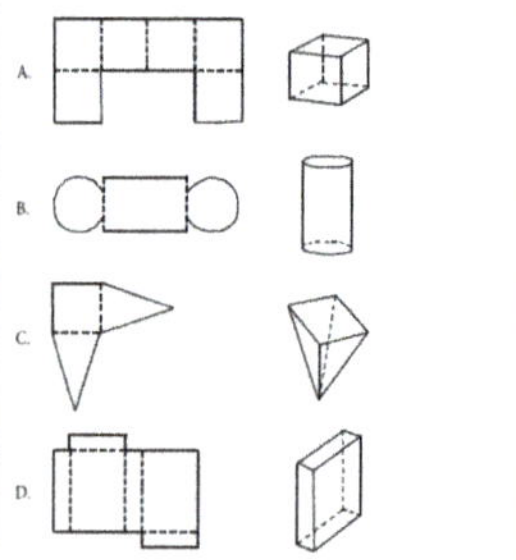 (Item #M041265, 2011)	Megan arranges Box A and Box B on her study table. • The dimensions of Box A are 10 by 5 by 4 inches. • The dimensions of Box B are 5 by 3 by 4 inches. 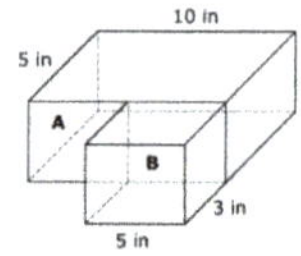 Enter the combined volume, in cubic inches, of both boxes. (Item #13, 2016)	Kim made this large cube using 27 small cubes.  The large cube has 6 faces. Kim removed some of the small cubes. The remaining object had 12 faces. What was the **smallest** number of small cubes Kim could have removed?

Fig. 12.6 Assessment items on spatial visualization

Mental Rotation/Imagery

TIMSS (Grade 4)	California (Grade 5)	ACARA (Grade 5)
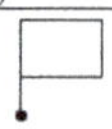 Which of the following shows the position of the shape above after a half turn or 180° rotation? 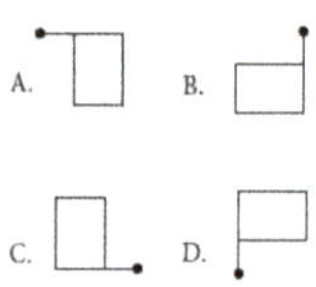 (Item #M03107, 2011)	N/A	The arrow is pointing South-East. 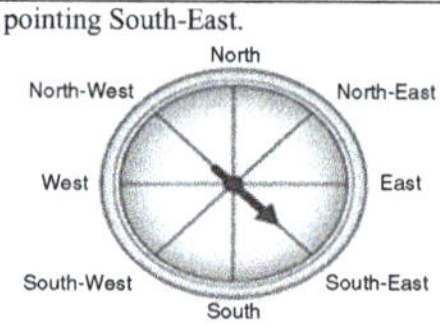 Nikita turns the arrow a quarter-turn clockwise. Which direction is the arrow pointing after the turn? South-West South-East South North-East

Fig. 12.7 Assessment items on mental rotation

Spatial Orientation

TIMSS (Grade 4)	California (Grade 5)	ACARA (Grade 5)
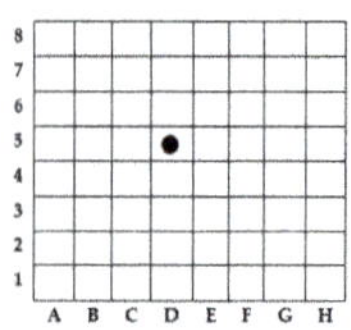 James is playing a board game. His counter is on square D5. Which of these moves would put his counter on square G7? A. 2 squares to the right and 3 squares up B. 2 squares to the left and 3 squares up C. 3 squares to the right and 2 squares up D. 3 squares to the left and 2 squares up (Item #M031088, 2011)	Calvin and Shana are in the same school. 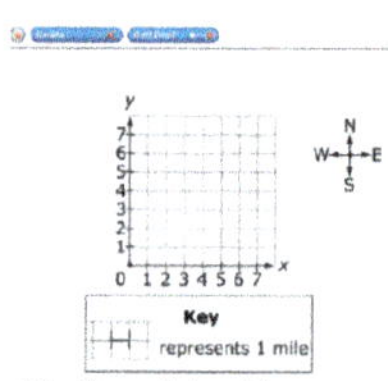 • The school is located at (0,0). • Calvin travels 2 miles east and 1 mile north to get home from school. • Shane travels twice the number of miles east and the same number of miles north as Calvin to get home. Use the Add Point tool to plot Shana's house. (Item #27, 2016)	A school's website shows these two plans of the school. 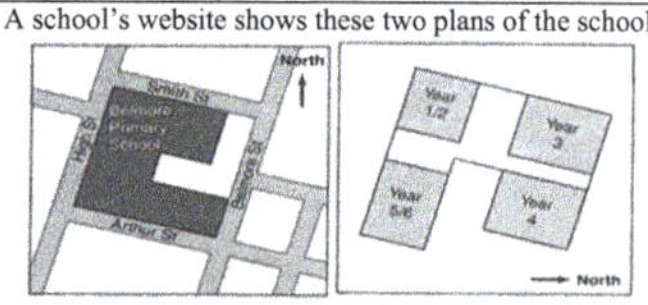 Which classroom is by the corner of High Street and Smith Street? Year 1/2 Year 3 Year 4 Year 5/6

Fig. 12.8 Assessment items on spatial orientation

Symmetry Task	Street Map Task
Ron paints these letters on a piece of paper. 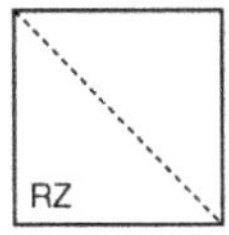 While the paint is still wet, he folds the paper along the dotted line. When Ron unfolds the paper, what will it look like? 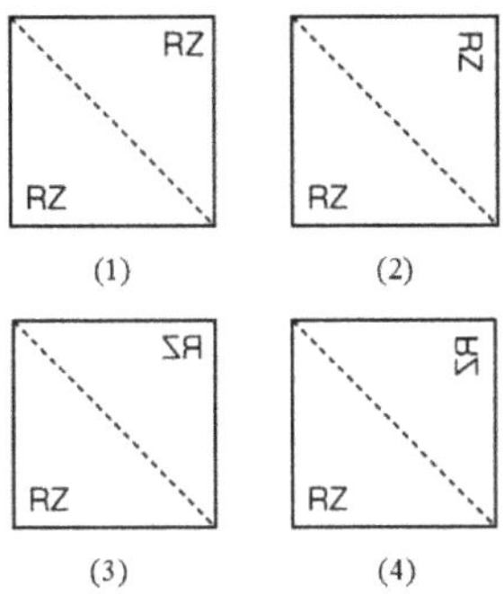 **S1**: I folded the paper along the dotted line. **S2**: I visualized/imagined folding the paper along the dotted line. **S3**: I did not use any of the above methods – I attempted the task in this way.	A car is travelling north-east along Don Road. The car is about to turn right into Plum Road. In which direction will the car be travelling after it turns right? 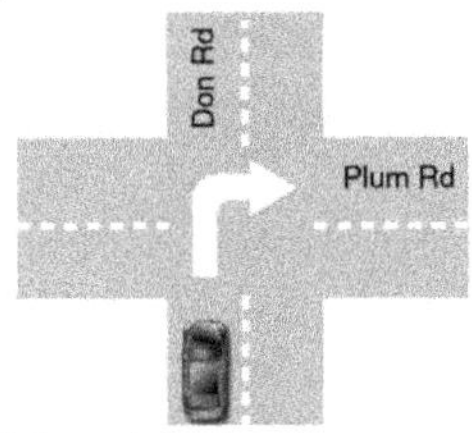 **S1**: I solved the task by drawing a compass indicating the North direction on the graphic. 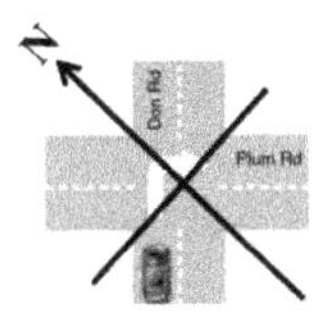 **S2**: I solved the task by turning the page so that the North direction is vertical to the edge of my table. **S3**: I visualized/imagined where the compass indicating the North direction will be on the graphic.) **S4**: I knew that a right hand turn is 90°, therefore 90° from north-east is South-East. **S5**: I did not use any of the above methods – I attempted the task in this way.
(a)	(b)

Fig. 12.9 Symmetry and Street Map tasks with frequently used strategies (ACARA, 2010b, 2010c)

visualization categories (see Lowrie et al., 2014). There were no significant differences in performance between the iPad and pencil-and-paper modalities for the spatial visualization task for each of the three levels of spatial visualization ability. By contrast, for the spatial orientation task, differences were significant for students with low and medium spatial visualization ability.

For the Symmetry task, the most common solution strategy was S2, irrespective of whether the students solved the task in the digital or pencil-and-paper mode. Although the pencil-and-paper format afforded students opportunities to solve the problem in a variety of ways (especially S1), the mental encoding strategy (S2) was most common. We were unable to explain why students who utilized this strategy on the iPad were more likely to solve the task successfully. It may be the case that the limited options for strategy selection reduced the fidelity. The Street Map task required students to superimpose and rotate a visual compass from its prototypical North position on the given graphic. For this task, the pencil-and-paper mode tended to encourage students to draw a compass on the diagram, with S1 the most common strategy. By contrast, the most common strategy on the digital device was S3, prompting students to evoke an internal encoded representation of a compass indicating the North direction. Given the higher success rate on the pencil-and-paper version, it would seem the ability to concretely encode through drawing was highly beneficial. These examples highlight the influence of modality in students' representations and interpretations of mathematics tasks. In the coming years, studies that consider the influence of spatial reasoning on mathematics will need to contend with the instructional and assessment mode in which these tasks are presented.

Malleability of Spatial Reasoning in the Elementary Classroom

There has been a shift toward spatial training in the educational and psychological research, with several large-scale projects being undertaken across multiple countries (e.g., Ireland, Australia, United States; Bowe, Nevin, Carthy, Seery, & Sorby, 2016; Lowrie, Logan, & Ramful, 2017; Taylor & Hutton, 2013; Uttal, Miller, & Newcombe, 2013). It is known that elementary school students' spatial reasoning tends to improve through curricula instruction (Newcombe, 2013) and that specific training in spatial reasoning can improve students' quantitative skills (Cheng & Mix, 2014).

Recent work by Lowrie et al. (2017) has demonstrated the effectiveness of training on spatial reasoning for improving mathematics performance. This program was delivered by classroom teachers in place of standard geometry and measurement lessons for 10 weeks. While the training lessons aligned to the curriculum, there was no explicit geometry teaching for the duration of the training program. A control group received standard instruction. The mathematics assessment used to evaluate the transfer from spatial reasoning to mathematics included items that covered a range of mathematics content and representations. Subsequent analysis of the items resulted in the categorization of three item types: (1) number items containing no graphics, (2) graphic items that required visualization, and (3) other graphic items that required decoding. Examples of each are presented in Fig. 12.10.

The changes in scores for these three types of items were compared across intervention and control groups. For the number items there was no significant improvement for the intervention group compared with controls, $F(1, 149)$[2] $= 0.23$, $p = 0.63$. For the visualization items there was a significant difference in the improvement of the two groups, $F(1, 162) = 10.99$, $p = 0.001$, $d = 0.58$, with the intervention group showing greater improvement than the controls. For the additional graphics items there was no significant difference, $F(1, 162) = 2.57$, $p = 0.11$. However, there was a slightly larger improvement in the control group relative to the treatment group. Given the other graphics items specifically addressed the curriculum content, it was expected that the control group should improve. It is noteworthy that the removal of this curriculum content from the treatment group did not disadvantage intervention

[2] The reduced degrees of freedom are due to 13 students being unable to complete these items.

Number items (3 items)	Visualization Items (4 items)	Other graphics (5 items)
Ben has 2 identical pizzas. He cuts one pizza equally into 4 large slices. He then cuts the other pizza equally into 8 small slices. A large slice weighs 32 grams more than a small slice. What is the mass of one whole pizza?	These isometric drawings of some rectangular prisms are labelled A, B, C and D. 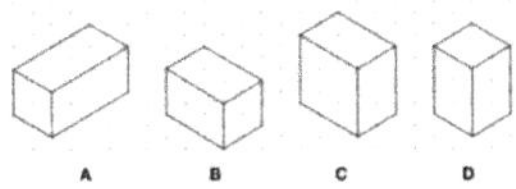 Which two drawings are of the same rectangular prism?	Lucy made 4 tree designs using sticks. There is a pattern in the way the trees grow. 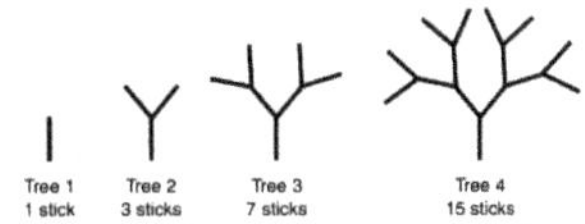 Lucy continues the pattern in the same way. How many sticks will Tree 5 have?
(a)	(b)	(c)

Fig. 12.10 Examples of mathematics tasks in the spatial training instrument

students in any significant way, while the findings for the visualization items demonstrate that without the intervention program the control students had no means of developing these skills within the current curriculum.

There is potential for ongoing work in spatial training and its influence on students' mathematics understanding given both the necessary function of spatial reasoning in the curriculum and the promising findings generated from the recent research.

Conclusion and Implications

There are numerous examples of spatial reasoning being promoted as an important general cognitive competence for the study of STEM topics in schools. For example, in the United States the National Research Council (2006) described the importance of spatial reasoning in terms of awareness, representation, and reasoning; while in Canada a Province-based support document for K-12 education was devoted to spatial reasoning (ServiceOntario Publications, 2014). In Australia, spatial reasoning is described as a general numeracy competence that should be taught across the curriculum (ACARA, 2011) or in the case of Singapore, a specific skill that needs to be addressed in mathematics (Curriculum Planning and Development Division, 2006). However, even though spatial reasoning has been consistently linked to success in school mathematics and other STEM-related subjects (Kell, Lubinski, Benbow, & Steiger, 2013; Nath & Szücs, 2014), few (if any) national school curricula describe spatial reasoning in ways that promote spatial awareness in terms of content or pedagogical practices. Almost all mention of the term is framed around the notion of being able to visualize, recognize shapes (especially in geometry), or interpret diagrams. Apart from the Ontario publication, no document considers constructs that form the cognitive dimensions that make up spatial reasoning. Consequently, it is unlikely that classroom teachers will gain insights into how spatial reasoning can be developed from a cognitive perspective, beyond the commonly held view that mathematics requires spatial thinking.

Our work, and the work we have undertaken with colleagues, has aligned spatial constructs to both mathematics content and pedagogical practices. The respective constructs—namely, spatial visualization; mental rotation; and spatial orientation—can be assigned to specific mathematics assessments tasks (Ramful et al., 2016). Although current curriculum documents typically address spatial reasoning in general terms only, there is some evidence to suggest that assessment bodies—across national and international jurisdictions—are requiring elementary students to solve spatially-specific mathematics tasks. For example, as much as 28% of mathematics items in Australia's nationally assessment program require students to solve tasks that are directly associated with spatial visualization, mental rotation, or spatial orientation. Internationally, 15–20% of TIMMS items are associated with these spatial constructs. In the United States, state-based instruments from California and New York have much lower representation—approximately 10% in Grade 5 and much less in Grade 3 instruments. Across these instruments, most items were associated with the spatial visualization construct, with almost none from the spatial orientation construct. It will be interesting to monitor these (and other) prominent and increasingly public national and international assessment trends over time. In Australia, for example, the number of mathematics items with embedded information graphics increased from 15 to 75% of the Grade 3 elementary national assessment in a 10-year period (Lowrie & Diezmann, 2009). These graphic items require the decoding of spatial information, even though not directly attributed to one of the three spatial constructs.

Aside from the proportion of graphic and spatial items in tests, the nature of these items differs considerably across country. For example, most graphics-based tasks in the Singaporean PSLE require students to decode and compute data from tables, number lines, and diagrams. By contrast, Australian tasks require graphics to be manipulated; including rotations, translations, and graphic comparisons. Thus, the processing requirements of the Australian graphic tasks were more spatially demanding (Lowrie, Logan, & Ramful, 2016). Although Singaporean students outperform Australian students on most number-based TIMMS items by a substantial margin, performance differences are generally minor or in favor of Australian students in the application component of the geometry content domain.

Over time, it will be interesting to monitor the extent to which national and international mathematics instruments rely less on traditional word problems and represent mathematics information in different ways. We have found moderately-high correlations between students' spatial reasoning and all types of graphic-language representations. In fact, high levels of spatial visualization ability are increasingly important when students encounter mathematics tasks in nontraditional digital modes (Lowrie et al., 2014). Spatial reasoning can also be important when encoding mathematics tasks, especially when the tasks are novel or complex. With the movement to tailored and online assessment, students' internal encoded representations (visual mental models) will become more critical—as traditional pencil-and-paper modes are replaced.

Given the positive effects of our spatial intervention program on students' spatial reasoning and mathematics performance (Lowrie et al., 2017), there is potential for classroom teachers to explicitly teach the three constructs as part of a program outside the business of the usual mathematics curriculum. The extent to which these constructs align to assessment and mathematics curricula vary across country, nevertheless, spatial reasoning is likely to become more influential and necessary in school performance and learning outside of school learning.

Several implications can be drawn from the research described in this chapter.

- Spatial reasoning will become increasingly important to school mathematics and beyond. Associations between spatial reasoning and geometric thinking will be at the forefront of this increased emphasis on spatial thinking.
- Technological innovations have afforded opportunities for mathematics ideas to be represented in different ways. These technologies (including hand-held devices) allow us to engage with mathematics in multiple modalities. In addition, technology advances have dramatically shaped how we assess and test mathematics knowledge.
- With respect to large-scale assessment, some international and national tests have given less attention to traditional word-based problems. As such, opportunities to utilize encoding techniques are restricted. We need to be mindful of the importance of encoding techniques, especially when students encounter novel or complex tasks.
- By contrast, technological advances have afforded the opportunity for mathematics tasks to be graphics rich. With this change it is imperative that teachers pay attention to the structure and conventions associated with specific types of graphics. The decoding of graphics is highly spatial and will impact student understanding in mathematics and most other curriculum areas (e.g., mapping in geography or analyzing chemical structures in science).
- In our work, defining spatial reasoning in relation to cognitive constructs has been beneficial in identifying the alignment of mathematics ideas to spatial understandings. Our colleagues in educational and cognitive science have provided great insights into how these constructs can be developed and better understood within the mathematics curriculum.
- Since spatial reasoning is malleable, specific programs that focus on these spatial constructs can improve students' mathematics understandings.

References

Australian Curriculum, Assessment and Reporting Authority [ACARA]. (2009a). *National assessment program literacy and numeracy: Numeracy year 3 2009*. Sydney: Author.

Australian Curriculum, Assessment and Reporting Authority [ACARA]. (2009b). *National assessment program literacy and numeracy: Numeracy year 5 2009*. Sydney: Author.

Australian Curriculum, Assessment and Reporting Authority [ACARA]. (2010a). *National assessment program literacy and numeracy: Numeracy year 3 2010*. Sydney: Author.

Australian Curriculum, Assessment and Reporting Authority [ACARA]. (2010b). *National assessment program literacy and numeracy: Numeracy year 5 2010*. Sydney: Author.

Australian Curriculum, Assessment and Reporting Authority [ACARA]. (2010c). *National assessment program literacy and numeracy: Numeracy year 7 2010*. Sydney: Author.

Australian Curriculum, Assessment and Reporting Authority [ACARA]. (2011). *Australian Curriculum v7.5: General capabilities: Overview*. Australian Curriculum, Assessment and Reporting Authority, Commonwealth Government. Retrieved on February 29, 2017, from: http://v7-5.australiancurriculum.edu.au/generalcapabilities/numeracy/organising-elements/using-spatial-reasoning.

Battista, M. T. (2007). The development of geometric and spatial thinking. In F. K. Lester Jr. (Ed.), *Second handbook of research on mathematics teaching and learning* (pp. 843–908). Charlotte, NC: Information Age Publishing.

Battista, M. T., & Clements, D. H. (1988). A case for a logo-based elementary school geometry curriculum. *Arithmetic Teacher, 36*, 11–17.

Battista, M. T., Clements, D. H., Arnoff, J., Battista, K., & Borrow, C. V. A. (1998). Students' spatial structuring of 2D arrays of squares. *Journal for Research in Mathematics Education, 29*(5), 503–532.

Bertin, J. (1967). *Semiology of graphics: Diagrams, networks, maps*. Wisconsin: University of Wisconsin Press. (first published in French in 1967 translated by William J. Berg in 1983).

Bishop, A. J. (2008). Spatial abilities and mathematics education–a review. In P. Clarkson & N. Presmeg (Eds.), *Critical issues in mathematics education* (pp. 71–81). New York: Springer.

Bowe, B., Nevin, E., Carthy, D., Seery N., & Sorby, S. (2016). *National spatial skills report - phase 1 preliminary findings*. CREATE Dublin Institute of Technology (Contributions to Research in Engineering and Applied Technology Education) 2016.

Carroll, J. B. (1993). *Human cognitive abilities: A survey of factor-analytic studies*. New York, NY: Cambridge University Press.

Cheng, Y. L., & Mix, K. S. (2014). Spatial training improves children's mathematics ability. *Journal of Cognition and Development, 15*(1), 2–11.

Clements, D. H., Battista, M., Sarama, J., & Swaminathan, S. (1997). Development of students' spatial thinking in a unit on geometric motions and area. *The Elementary School Journal, 98*(2), 171–186.

Clements, D. H., & Battista, M. T. (1992). Geometry and spatial reasoning. In D. A. Grouws (Ed.), *Handbook of research on mathematics teaching and learning* (pp. 420–464). New York: Macmillan.

Clements, D. H., & Sarama, J. (2011). Early childhood teacher education: The case of geometry. *Journal of Mathematics Teacher Education, 14*(2), 133–148.

Cleveland, W. S., & McGill, R. (1984). Graphical perception: Theory, experimentation, and application to the development of graphical methods. *Journal of the American Statistical Association, 79*(387), 531–554.

Curriculum Planning and Development Division. (2006). *Mathematics syllabus primary*. Retrieved from the Singapore Ministry of Education: http://www.moe.edu.sg/education/syllabuses/sciences/files/maths-primary-2007.pdf.

Diezmann, C. M., & Lowrie, T. (2009). An instrument for assessing primary students' knowledge of information graphics in mathematics. *Assessment in Education: Principles, Policy and Practice, 16*(2), 131–147.

Educational Testing Centre NSW. (2002). *Primary school mathematics competition, year 4*. Sydney, Australia: University of New South Wales.

Goldin, G., & Shteingold, N. (2001). Systems of representations and the development of mathematical concepts. In A. A. Cuoco & F. R. Curcio (Eds.), *The roles of representation in school mathematics* (pp. 1–23). Reston, VA: National Council of Teachers of Mathematics.

Hegarty, M., & Stull, A. (2012). Visuospatial thinking. In K. J. Holyoak & R. G. Morrison (Eds.), *The Oxford handbook of thinking and reasoning* (pp. 606–630). London: Oxford University Press.

Ho, S. Y., & Lowrie, T. (2014). The model method: Students' performance and its effectiveness. *Journal of Mathematical Behavior, 35*, 87–100.
Jones, K. (2000). Critical issues in the design of the geometry curriculum. In B. Barton (Ed.), *Readings in mathematics education* (pp. 75–90). Auckland, New Zealand: University of Auckland.
Kell, H. J., Lubinski, D., Benbow, C. P., & Steiger, J. H. (2013). Creativity and technical innovation spatial ability's unique role. *Psychological Science, 24*(9), 1831–1836.
Kosslyn, S., & Miller, G. W. (2013). *Top brain, bottom brain: Surprising insights into how you think*. New York, NY: Simon & Schuster.
Kosslyn, S. M. (1983). *Ghosts in the mind's machine*. New York: Norton.
Kosslyn, S. M. (2006). *Graph design for the eye and mind*. New York: Oxford University Press.
Kozhevnikov, M., & Hegarty, M. (2001). A dissociation between object manipulation spatial ability and spatial orientation ability. *Memory & Cognition, 29*(5), 745–756.
Kozhevnikov, M., Hegarty, M., & Mayer, R. (1999). *Students' use of imagery in solving qualitative problems in kinematics*. Washington DC: US Department of Education (ERIC Document Reproduction, 433 239).
Linn, M., & Petersen, A. C. (1985). Emergence and characterization of sex differences in spatial ability: A meta-analysis. *Child Development, 56*(6), 1479–1498.
Lowrie, T. (2012). Visual and spatial reasoning: The changing form of mathematics representation and communication. In B. Kaur & T. T. Lam (Eds.), *Reasoning, communication and connections in mathematics: Yearbook 2012, Association of Mathematics Educators* (pp. 149–168). Singapore: World Scientific.
Lowrie, T., & Diezmann, C. M. (2007). Solving graphics problems: Student performance in the junior grades. *Journal of Educational Research, 100*(6), 369–377.
Lowrie, T., & Diezmann, C. M. (2009). National numeracy tests: A graphic tells a thousand words. *Australian Journal of Education, 53*(2), 141–158.
Lowrie, T., Diezmann, C. M., & Logan, T. (2012). A framework for mathematics graphical tasks: The influence of the graphic element on student sense making. *Mathematics Education Research Journal, 24*(2), 169–187.
Lowrie, T., & Logan, T. (2007). Using spatial skills to interpret maps: Problem solving in realistic contexts. *Australian Primary Mathematics Classroom, 12*(4), 14–19.
Lowrie, T., & Logan, T. (2015). The role of test-mode effect: Implications for assessment practices and item design. In C. Vistro-Yu (Ed.), *In pursuit of quality mathematics for all, Proceedings of the 7th ICMI-East Asia Regional Conference on Mathematics Education* (pp. 649–655). Cebu, Philippines: Philippine Council of Mathematics Teacher Educators (MATHTED), Inc..
Lowrie, T., Logan, T., & Ramful, A. (2016). Cross cultural comparison of grade 6 students' performance and strategy use on graphic and non-graphic tasks. *Learning and Individual Differences, 52*, 97–108.
Lowrie, T., Logan, T., & Ramful, A. (2017). Visuospatial training improves elementary students' mathematics performance. *British Journal of Educational Psychology, 87*(2), 170–186.
Lowrie, T., Ramful, A., Logan, T., & Ho, S. Y. (2014). Do students solve graphic tasks with spatial demands differently in digital form? In J. Anderson, M. Cavanagh, & A. Prescott (Eds.), *Curriculum in focus: Research guided practice, Proceedings of the 37th annual conference of the Mathematics Education Research Group of Australasia* (Vol. 2, pp. 429–436). Sydney: MERGA.
Mackinlay, J. (1999). Automating the design of graphical presentations of relational information. In S. K. Card, J. D. Mackinlay, & B. Schneiderman (Eds.), *Readings in information visualization: Using vision to think* (pp. 66–81). San Francisco, CA: Morgan Kaufmann.
Ministry of Education, Curriculum Planning and Development Division. (2006). *Mathematics syllabus primary*. Retrieved from the Singapore Ministry of Education website: http://www.moe.edu.sg/education/syllabuses/sciences/files/maths-primary-2007.pdf.
Nath, S., & Szücs, D. (2014). Construction play and cognitive skills associated with the development of mathematical abilities in 7-year-old children. *Learning and Instruction, 32*, 73–80.
National Research Council. (2006). *Learning to think spatially: GIS as a support system in the K—12 curriculum*. Washington, DC: National Academy Press.
Newcombe, N. S. (2013). Seeing relationships: Using spatial thinking to teach science, mathematics, and social studies. *American Educator, 37*(1), 26–40.

Newcombe, N. S., & Frick, A. (2010). Early education for spatial intelligence: Why, what, and how. *Mind, Brain, and Education, 4*(3), 102–111.

Ng, O., & Sinclair, N. (2015). Young children reasoning about symmetry in a dynamic geometry environment. *ZDM Mathematics Education, 47*(3), 421–434.

Ontario Ministry of Education [OME]. (2008). *Geometry and spatial sense, grades 4 to 6: A guide to effective instruction in mathematics, kindergarten to grade 6*. Ontario: Queen's Printer for Ontario.

Ontario Ministry of Education [OME]. (2014). *Paying attention to spatial reasoning: Support document for paying attention to mathematics education*. Ontario: Queen's Printer for Ontario.

Pape, S. J., & Tchoshanov, M. A. (2001). The role of representation(s) in developing mathematical understanding. *Theory Into Practice, 4*(2), 118–127.

Polya, G. (1965). *Mathematical discovery; on understanding and teaching problem solving*. Hoboken: Wiley.

Porter, A. (1989). A curriculum out of balance: The case of elementary school mathematics. *Educational Researcher, 18*(5), 9–15.

Postigo, Y., & Pozo, J. I. (2004). On the road to graphicacy: The learning of graphical representational systems. *Educational Psychology, 24*(5), 623–644.

Queensland Studies Authority. (2001). *Aspects of numeracy test, year 3*. Camberwell, Australia: Australian Council for Educational Research.

Queensland Studies Authority. (2009). *3579 Test reporting handbook: 2009 national assessment program—Literacy and numeracy*. Brisbane, Australia: Author.

Queensland Studies Authority. (2010). *3579 Test reporting handbook: 2010 national assessment program—Literacy and numeracy*. Brisbane, Australia: Author.

Ramful, A., Lowrie, T., & Logan, T. (2016). Measurement of spatial ability: Construction and validation of the spatial reasoning instrument for middle school students. *Journal of Psychoeducational Assessment., 35*, 709. https://doi.org/10.1177/0734282916659207

Sarfaty, Y., & Patkin, D. (2013). The ability of second graders to identify solids in different positions and to justify their answer: Original research. *Pythagoras, 34*(1), 1–10.

ServiceOntario Publications. (2014). *Paying attention to spatial reasoning, K-12: Support document for paying attention to mathematics education*. Retrieved from http://www.edu.gov.on.ca/eng/literacynumeracy/LNSPayingAttention.pdf.

Shah, P., & Miyake, A. (2005). *The Cambridge handbook of visuospatial thinking*. New York: Cambridge University Press.

Shea, D. L., Lubinski, D., & Benbow, C. P. (2001). Importance of assessing spatial ability in intellectually talented young adolescents: A 20-year longitudinal study. *Journal of Educational Psychology, 93*, 604.

Sinclair, N., & Bruce, C. (2015). New opportunities in geometry education at the primary school. *ZDM – The International Journal on Mathematics Education, 47*(3), 319.

Singapore Examinations and Assessment Board. (2009). *PSLE Examination Questions Mathematics 2005–2009* (Part A, Item 21, p.6). Singapore, SEAB.

Slotnick, S. D., Thompson, W. L., & Kosslyn, S. M. (2005). Visual mental imagery induces retinotopically organized activation of early visual areas. *Cerebral Cortex, 15*(10), 1570–1583.

Taylor, H. A., & Hutton, A. (2013). Think3d!: Training spatial thinking fundamental to STEM education. *Cognition and Instruction, 31*(4), 434–455.

Uttal, D. H., Miller, D. I., & Newcombe, N. S. (2013). Exploring and enhancing spatial thinking links to achievement in science, technology, engineering, and mathematics? *Current Directions in Psychological Science, 22*(5), 367–373.

Chapter 13
Geometric Modeling of Mesospace Objects: A Task, its Didactical Variables, and the Mathematics at Stake

Patricio Herbst and Nicolas Boileau

Abstract For decades, mathematics educators have been interested in engaging K-12 students in the practice of creating and using mathematical models. What might this look like in the context of geometry? Inspired by claims that students come to secondary school with knowledge of three-dimensional space that can be leveraged to engage them in modeling, we wondered what it would take to have them do so. We designed a communication task aimed at engaging teenagers in the geometric modeling of mesospace objects—three-dimensional objects of scale comparable to that of the human body. Specifically, we asked a group of teenagers to plan and enact the movement of furniture through a narrow staircase in a residential home. In this paper, we present our original design considerations, an analysis of the teens' work, and a set of didactical variables that this analysis led us to believe need to be considered to ensure that such an activity engage teenagers in the geometric modeling of mesospace objects. The paper concludes with a discussion of the implications for research on a modeling approach to the teaching and learning of geometry.

Keywords Mesospace · 3D geometry · Modeling · Communication · Game · Diagram · Task · Didactical variable · Design · Calculation · Sketch · Angle · Rotation · Staircase · Instructions · Experiential world · Multimodal modeling · Scale · Conception of figure · Milieu · Devolution · Macrospace · Microspace · Furniture · Moving · Movers · Boxspring · Couch · Tabletop

P. Herbst (✉) · N. Boileau
Educational Studies Program, University of Michigan, Ann Arbor, MI, USA
e-mail: pgherbst@umich.edu; nboilea@umich.edu

K. S. S. Mix, M. T. Battista (eds.), *Visualizing Mathematics*, Research in Mathematics Education, https://doi.org/10.1007/978-3-319-98767-5_13

Introduction

Children accrue a great deal of experience managing three-dimensional space and shape throughout their lives. High school geometry instruction, mainly occupied with plane geometry, usually provides names for solids and formulas for calculating their surface areas and volumes (Chávez, Papick, Ross, & Grouws, 2011; González & Herbst, 2006), but it could do more. Specifically, high school geometry could provide occasions to model three-dimensional space. The use of geometric knowledge to model spatial management problems may not only motivate students to learn geometry, but may also provide them with intellectual resources to manage their spatial orientation and support or correct their development of solutions to practical problems. Yet, U.S. students do very little, if any, modeling of real-world phenomena in school (Meyer, 2015).

In this chapter, we illustrate what it might mean to have students model the experiential world with geometry, using the example of having teens write instructions for how to move various pieces of furniture. Our ultimate goal in using this example, however, is not to recommend the specific task as one that should be used in schools, but rather to explore our design of it and the geometry that it promoted the teens[1] to do. We do that in order to document the role played by the didactical variables that we believe are instrumental to creating a task space in which students would have to confront the decision to model three-dimensional space.

Mathematical Modeling and Spaces of Different Scale

A Modeling Perspective on Geometry

Mathematical modeling plays a crucial role in how humanity understands and interacts with the world (e.g., Skovsmose, 2000, p. 4., has said that mathematics has a "formatting power"), even if, at the level of individual action, such understanding is, for the most part, tacit or embodied in how we interact with or harness the power of artifacts whose design is based on mathematics, such when we ride a bike (see Collins, 2010). The practice of creating and using mathematical models has been included among the US Common Core State Standards for Mathematical Practice (Meyer, 2015; NGA, 2010). There has also been much interest among mathematics educators in the teaching and learning of mathematical modeling, with emphasis usually placed on using mathematics to understand the real world (see Blum & Ferri, 2009; Blum, Galbraith, Henn, & Niss, 2007; Kaiser & Sriraman, 2006). But,

[1] We refer to our participants as *teens* and sometimes as *youth* to eschew the institutional role called forth with words like *students* or *learners*, as the first round of task design that we report here was done outside of school and without the expectation that any particular knowledge had to be learned.

as Hanna and Jahnke (2007, see also Halverscheid, 2008) have shown, the practice of mathematically modeling the real world also serves to understand mathematical ideas and processes. Along those lines, Herbst, Fujita, Halverscheid, and Weiss (2017) propose that the study of geometry in secondary schools could be organized from a modeling perspective: The study of geometry as engagement in the practice of mathematically modeling students' experiences with shape and space in spaces of different scale. They say,

> [T]he development of geometric knowledge at the secondary level consists of the progressive sophistication of students' intellectual means to model, predict, and control geometric representations—that is, the progressive sophistication in students' ways of organizing those artifacts (words, diagrams, and others) so that they can be reliably used in making and transacting meanings. (Herbst et al., 2017, p. 3)

We propose that the work that students do in geometry instruction needs to involve them in the *multimodal modeling* of space at various scales. By *modeling* we mean the representation of their experiences with space and shape using a semiotic system that affords its own mechanisms for making inferences. For example, when algebraic notation is used to represent an aspect of experience, the formal calculations that can be done with symbols (or, as Kaput, 1999, p. 141, put it, the "meaningful operations on opaque symbols") permit the making of inferences that could point to new possible meanings of the experience originally represented. Thus, symbolic representation using algebraic symbols, inasmuch as one can calculate with those symbols, is a form of modeling. That said, in the geometric modeling of one's experience with space and shape, we need to be open to the possibility, indeed the likelihood, that modeling is not only symbolic. To understand this contention it is helpful to consider that pre-algebraic reasoning is often mediated by representations that also have their own mechanisms for inference-making even though they are not symbolic (Kaput, 1991). Cuisenaire rods, for example, not only allow one to represent fractions, but also to juxtapose rods of same length and compare strings of juxtaposed rods; the latter is a mechanism for inference-making, one that permits one to decide which fractions are equivalent (Behr & Post, 1992). Such a system of representation can be considered a model even if the signs mobilizing the representations are not symbolic. By suggesting that geometric modeling may be *multimodal*, we mean that it might be done with a variety of semiotic resources, including not only symbols such as the notations for figures, but also diagrams and other artifacts (e.g., stick figures), which are taken as signs (Otte, 2006). Diagrams are quite important in our argument for the use of three-dimensional geometry as a context for students' engagement in mathematical modeling, as we argue that geometric diagrams play the role of representation system in the practice of modeling. In this chapter, we demonstrate this point using, as an example, students modeling what Berthelot and Salin (1998) have called the *mesospace*.

Experiences at Different Scales of Space and Shape and Conceptions of Figure

The psychological literature has distinguished between large- and small-scale spaces (Acredolo, 1981; Battista, 2007; Hegarty, Montello, Richardson, Ishikawa, & Lovelace, 2006; Siegel, 1981). Building on that distinction, Berthelot and Salin (1998) distinguished among three spaces—the microspace, the mesospace, and the macrospace. *Microspace* refers to space of scale smaller than that of the human body, such as the scale of drawings in notebooks or screens, or hand manipulation of tangram pieces or Lego blocks. *Macrospace* refers to the space of scale much larger than that of the human body, such as that of buildings, athletic fields, landscapes, and seascapes. *Mesospace* refers to the space of scale of the same order of magnitude as the human body including everyday objects such as household appliances, furniture, and sports equipment (e.g., exercise balls, treadmills).

Herbst et al. (2017) argue that students come to secondary schools with four conceptions of geometric figure that enable them to organize and manage their experiences in those three spaces: A conception of figure as *navigation* of the macrospace, a conception of figure as *capture* of objects in the mesospace, a conception of figure as *construction* of shapes in the microspace, and a conception of figure as *description* of shapes in the microspace. By *conception of figure* we mean, in general, any one of the relatively independent systems of practices in which children engage with space and that an observer might describe in reference to the geometric concept of figure. We draw this epistemological sense of conception-as-practice from Balacheff and Gaudin (2010), who also provide guidance as to how to account for those conceptions in terms of *problems*, *operators*, *representations*, and *controls*.

In saying that children come to secondary school with conceptions of figure associated with the three spaces described above, we mean that there are systems of practices they have engaged in, inside or outside of school, related to those spaces in ways that an observer could describe using the mathematical concept of figure. Those practices vary in the extent to which students reflectively use school-sanctioned geometric representations (e.g., diagrams, geometric vocabulary) or operators (e.g., measuring, constructing, proving) even if their practices reveal dexterity or skill.[2] Crucially, as Berthelot and Salin (1998) illustrate, there are differences across practices in the theorems-in-action (Vergnaud, 1996) that act as controls of correctness in solving problems.

The management of experience in the macrospace, mesospace, and microspace does not require the agent to engage explicitly in geometric modeling; but professional practices in architecture, industrial and graphic design, and other professions show that the geometric modeling of those spaces supports more

[2]For example, a player's management of positions in the soccer field can be quite dexterous and such practices might be describable by an observer using geometry, but that geometry may not be used reflectively by the player in enacting those practices.

efficient management of such experience (see Müller, Wonka, Haegler, Ulmer, & Van Gool, 2006). The literature on the use of mathematics in the workplace also suggests that, even if the mathematical conceptions apparent in the practices of workplace specialists refer to the same geometric concepts as the microspace conceptions usually practiced in school, the differences in scale often call for distinct techniques (Masingila, 1994; Millroy, 1991). This suggests that, while attention to the needs of practice in spaces of different scale may benefit from geometric knowledge, the specifics of practicing geometric modeling in a space of a particular scale might bring their own challenges, possibly favoring the use of elements of geometric knowledge different from those used at another scale. Hence, not only may the learning of geometric concepts support improved performance in spatial tasks in spaces of different scale, but doing such learning in the context of spaces of a given scale may also help the learner understand the constraints on geometric knowledge brought in by scale, or commonalities across spaces of different scale that they previously did not recognize.

Modeling One's Experience in the Macrospace and with the Mesospace

Geometry instruction usually supports the building of representational registers for microspace conceptions of geometric figure better than it supports the building of representational registers for macrospace or mesospace conceptions: Microspace conceptions of figure (construction of figure and description of figure) are supported by some geometric knowledge from elementary school. This support is mostly provided by registers of representation, including vocabulary and iconography (e.g., *rectangle*, ○), but also through the operators associated to the use of tools (e.g., how to use a compass or a ruler) and through the controls provided by some statements of definitions or properties (e.g., *all squares are rectangles*). Herbst et al. (2017) describe a modeling approach to the teaching and learning of geometry as one in which conceptions of figure from the microspace are brought to make sense of experiences with space and shape in the mesospace and macrospace. They argue that this modeling approach could support the evolution of conceptions of figure in these spaces, in particular making the use of geometric knowledge more explicit and modifying this knowledge to be reflectively adapted to the constraints of scale. A modeling approach would seek the evolution of these conceptions of figure in the macrospace and mesospace; this evolution could include the imposition of structure to those spaces—the expectation that objects and transformations in those spaces may have similar properties as objects and transformations in the microspace.

The mesospace and macrospace are not theorized in school in the same way that the microspace is. In the microspace, some artifacts (e.g., drawings) are "seen-as" (Coliva, 2012) signs or representations of geometric objects. They are described by definitions and axioms, and their properties are calculated or proved (see Dimmel &

Herbst, 2015, for a semiotic analysis of how drawing choices support geometric representation). Material objects and behaviors in the macrospace and mesospace are not, by default, seen as signs of geometric concepts. But, like the drawings and small objects encountered in the microspace, their behavior and properties could be construed as geometric. That is, they could be described, explained, and predicted using geometric representations that relate to each other by way of geometric theory, rather than merely by encountering them in experience. A question education designers need to ask is: In what task spaces might adolescents use geometric theory to frame their interactions with the macrospace or mesospace? Or, alternatively, how can task spaces afford opportunity for teens to develop the preference to anticipate and/or reflect on their engagement with such spaces using geometry?

A poor example of an answer to that question is readily available in the form of the so-called real-world application problems in geometry textbooks. For example, in Boyd et al. (1998, p. 24), a "ranching" problem reads,

> A rancher is adding a corral to his barn so that the barn opens directly into the corral. He has 195 feet of fencing left over from another project. Find the greatest possible area for his corral using this length of fence.

We consider the problem poor because, in addition to the statement, the textbook provides a diagram that shows the corral as a small rectangle sharing a side with the barn. We see the choice to include that diagram as unfortunate, as it removes what we see as the problem's greatest potential—the need for the student to model the real world context and attempt to use their model to solve the problem. Because of the inclusion of the diagram, the ranching problem also barely (if at all) affords students opportunities to cope with and understand the affordances of geometry in the context of the challenges of scale.[3] In contrast with that poor example, arts and crafts, social studies, and physical education present interesting contexts in which students might engage in the management of macrospace or mesospace activities, but these activities are rarely embraced as opportunities to do mathematics in school (just as mathematics is rarely incorporated in artistic, geographical, or physical problem-solving). Yet, the mere doing of those activities does not entail any explicit geometric modeling.

How can we produce in students the disposition to create and use geometric models of their experiences in the macrospace and mesospace? We might be able to assume that secondary school students are competent navigators of the macrospace and competent capturers of objects in the mesospace (referring to two of the four conceptions of figure mentioned earlier); but, while we might see conceptions of figure "frozen" (Gerdes, 1986) in their activities,[4] we'd like to develop

[3]Dan Meyer's 2010 TED talk (https://www.ted.com/talks/dan_meyer_math_curriculum_makeover) includes a similar example of a task about a water tank. In that talk, he demonstrates how the task found in the textbook could be revised in order to involve students in more of the modeling.

[4]Gerdes (1986, p. 12) says that "[t]here exists 'hidden' or 'frozen' mathematics" in the work of

in students the disposition to bring forth their more explicit knowledge of geometry (available in the conceptions of construction and description of figure in the microspace) to model their experiences in the macrospace and mesospace. Therefore, we ask what characteristics a context for mathematical work might have, if we were to be able to use it to devolve to teens the choice to use the geometric theory involved in microspace conceptions of figure to understand the macrospace or the mesospace. These characteristics are examples of what we will hereforth refer to as *didactical variables*—an important concept in Brousseau's (1997) approach to task design, defined as a feature of a task that can be manipulated by the teacher and where different values of the variable affect the chances for the student to engage in different strategies (see Grugnetti & Jaquet, 2005). With the verb *devolve*, we are alluding to Brousseau's (1997) notion of *devolution*—that the subject is afforded the choice to do something, or that they are afforded the opportunity to see it as a choice they can make (or choose not to make), on account of their analysis of the milieu[5] and their efficiency negotiating its demands.[6] We come back to this theme after we describe the specifics of our task design and consider the question of what sorts of problems a teacher may still expect to run into when they use this type of task.[7]

When designing our first approximation of a context in which students were engaged in the modeling of their experience with shape and space (i.e., the moving task that we describe in the next section), we were interested in understanding the structure of the learning milieu, or task space, and the developmental trajectory inscribed or assumed in it. As the question needs to be asked separately for the macrospace and mesospace, we dedicate this chapter to the mesospace.

"the artisan who imitates a known production technique." He calls it *hidden* or *frozen* because "the artisan is generally not doing mathematics" though "the artisan(s) who discovered the technique, did mathematics, developed mathematics, was (were) thinking mathematically." Gerdes (1986) describes the work of ethnomathematicians as one of uncovering the mathematical thinking that may have produced the mathematics frozen in cultural practices such as the patterns made in basket weaving.

[5] The *milieu,* for Brousseau (1997), is the system counterpart to the cognitive agent in a task. The everyday meaning of *milieu* alludes to the environment. But in didactique of mathematics, *milieu* points specifically to those elements of the environment that inform the agent's actions in a task (e.g., consequences of those actions, constraints on those actions).

[6] We also note that such assumptions about the subject are initial simplifications that address characteristics of the cognitive subject but not all of what it means to be a student. As successive design cycles take the task into institutionalized spaces like classrooms, these assumptions need to be complemented by understanding of the relationships among teacher, student, and content (i.e., the didactical contract) that envelope the cognitive subject and that are especially important in school (see Brousseau & Warfield, 1999).

[7] Some of the issues that we discuss are particular to the moving task, while others are issues that we expect teachers will have to struggle with whenever they create contexts for mathematical work in which they devolve to teens the choice to use the geometric theory involved in microspace conceptions of figure to understand the macrospace or the mesospace.

The Task of Moving Furniture and its Didactical Variables

In this design paper, we seek to identify the didactical variables that might be used to create a task space in which students would have to confront the decision to model three-dimensional space. The viability of the task rests on the assumption that teenagers already have some knowledge of plane geometry, learned in elementary and middle school, and that they are able to use it in the microspace to describe and construct shapes (see Herbst et al., 2017). The knowledge at stake in the task we propose consists of endowing the mesospace with the geometry of the microspace then confronting some of the epistemological constraints that the mesospace poses to the organization of knowledge about microspace objects. Specifically, while the handling of conceptions like *parallelism* or *congruence* in the microspace avails itself of operators such as visualization and manipulation, conceptions of parallelism or congruence, once expected in the mesospace, might still need to be mobilized there with different operators. Thus, some other properties of geometric figures and ways of handling them may become interesting and useful. In this paper, we focus on task design and on the representations of mathematical ideas that the teens' work on the task afforded us as observers. The design features and the possibility to observe such mathematics in student activity undoubtedly rely on psychological processes of the teens (e.g., visualization) that we do not attend to, but expect others will.

Moving Furniture as a Context for Modeling Experiences in the Mesospace

Starting to think about our own experiences using geometry in the mesospace, we found some hospitable grounds in the activity of moving furniture, one in which the second author had extensive experience. In reflecting on this experience, he recounted:

> While it is uncommon for residential movers to use "scholarly mathematics" (d'Ambrosio, 1985) to do their work, they do use some. The typical case is when they need to move an object through a relatively small opening, at which point they might take and compare measurements of the object and of the opening. The probability that measurements will be taken increases when the object is heavy. Objects that are large enough to prompt questioning whether they can fit through a given opening usually can be approximated by rectangular prisms (e.g., couches, fridges, dishwashers). If movers measure these objects, they tend to only measure their length, width, or height, as well as the dimensions of the openings through which they might move the object. That said, as someone with a reasonable amount of experience working as a residential mover, yet an excessive background in scholarly mathematics for that type of work, I have often

found myself taking both more and different measurements than my colleagues. Reflecting back on these experiences, I realized that what both I and many other movers were doing was, to varying degrees, precisely what we might want students to do—reducing mesospace objects to what they deemed to be their essential elements (i.e., their dimensions) in a given context (e.g., moving it through a doorway) and using those elements to consider whether and how to move the objects.

The second author's experience suggested to us that moving objects through spaces, especially heavy objects through spaces that the objects would only fit through if held in particular ways, might be an activity that would engage students in creating and using models of mesospace objects. Yet, we wondered whether high school geometry teachers would see such a task as worth assigning to their students, given that the only school mathematics that either the second author or his former colleagues did was to measure the dimensions of three-dimensional objects and the openings they hoped to move them through, and compare those numbers—mathematics likely to have been learned in elementary school. We therefore asked ourselves: could we design a moving task in a way that would require teens to do some of the geometry that they would only learn in high school?

It was important that the moving furniture task could be designed in such a way that the teens' performance could be read as an instance of modeling with geometric ideas. Reciprocally, because teens would be using those ideas to model objects in the mesospace, we wanted to construct the task around household objects, such as tables and mattresses that they could think of having to move. We asked ourselves how we could create the conditions for youth to see the mathematization of the moving of furniture as a choice they might want to make, on account of intellectual needs (in contrast to doing so on account of instructional norms, such as being told to do it by an adult). The notion that, in negotiating the moving of solid objects through constrained spaces, it is helpful to understand sections of a solid, suggested to us that for moving furniture to require modeling, we would need a constrained space. A quarter-turn, winder staircase from the basement to the main floor of a family home and large furniture (e.g., a boxspring) provided an interesting set of constraints.

Conjecturing Needed Didactical Variables

Before imposing additional constraints on the moving task, we asked ourselves what mathematics high school geometry students could encounter in this context, if they had the opportunity to work on a related problem. We imagined that, like residential movers, students could measure objects and spaces and compare those numbers. But we also saw the possibility that, with appropriate choices of task variables, students might be compelled to draw microspace diagrams to represent mesospace objects and perform measurements and calculations on those diagrams,

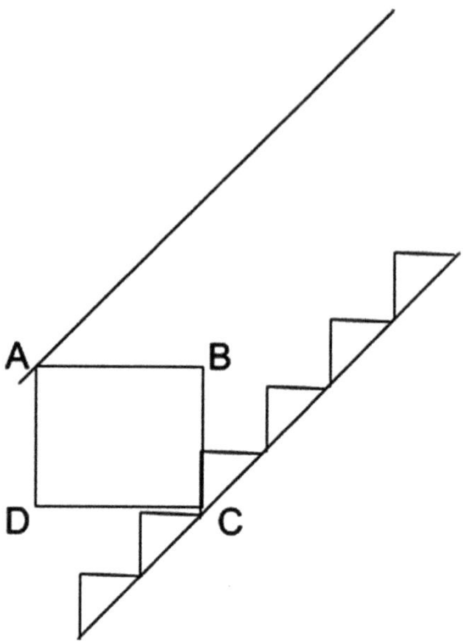

Fig. 13.1 Moving a boxspring with its width perpendicular to the steps of the staircase

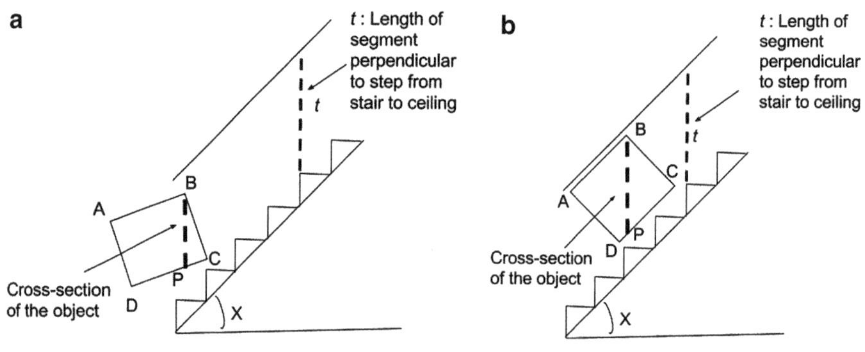

Fig. 13.2 (**a**) Tilting a boxspring to enter a staircase. (**b**) Carrying the boxspring with its length parallel to the ceiling

then use those to make inferences about and direct action with mesospace objects. We asked ourselves what topics commonly taught in high school geometry courses could be used to describe how to move a mesospace household object through a relatively small space, for example, how to move a queen-sized boxspring up an indoor staircase. One thing that movers would have to agree upon when doing this is how to hold the object, as it may not fit up the stairs when held in certain ways. For example, as most staircases are too narrow to hold a boxspring with its top and bottom parallel to the top of each step (i.e., perpendicular to the walls of the staircase), we expect that movers would try to carry the boxspring with its top and bottom parallel to the walls (as shown in Figs. 13.1 and 13.2). But they would still need to decide at what angle of inclination (i.e., the angle formed by the length of the boxspring and the top of any step under it) to hold it. One option would be to hold the width of the boxspring perpendicular to the steps of the staircase (as shown in Fig. 13.1). Yet, given the length and width of the boxpsring relative to the staircase, movers might realize that the boxspring could hit the stairway's ceiling, if held in this way.

Consequently, we expect that movers would tilt the boxspring, so that its longer edge ($\overline{AB}$) would become parallel to the stairway's ceiling (see Fig. 13.2b). There is mathematics frozen in those actions. In particular, to determine if a given angle of inclination would be a viable one at which to hold the object while the movers ascend the stairs (see Fig. 13.2a), the movers could usefully employ some trigonometry: For a given angle of inclination (note[8] that this will be equal to $m\angle CBP$ in Fig. 13.2a, b), they would need to consider the two-dimensional vertical cross-section of the three-dimensional object above the tip of each step (e.g., $\overline{BP}$, Fig. 13.2a, b) and determine whether the length of this cross-section is less than the *stairway overhead clearance,* or the length *t* of the segment between any point on the edge of that step and the point on the ceiling directly above it (see Fig. 13.2a, b). If the length of the cross-section is longer than *t*, the object will not fit when held at that angle. Therefore, to consider whether the object could be held with its length parallel to the staircase's ceiling, a mover could use the following procedure (see Fig. 13.2b):

1. Measure the length *t* (i.e., the stairway overhead clearance) of the segment between a point on the edge of any step and the point where the perpendicular to that step through that point meets the ceiling of the staircase.
2. Determine the measure *X* of the angle of inclination of the staircase,[9] for example, by measuring the rise and the depth of a step, then calculating the measure *X* of that angle, where $X = \text{arc tan}\left(\frac{\text{rise}}{\text{depth}}\right)$. After that two alternate courses of action arise.
3. A measurement option: Extend a tape measure across one of the boxspring's surfaces from what will be its top corner (*B*; see Fig. 13.2a) and making an angle of measure *X* (measured with a protractor) with the short side of the boxspring ($\overline{BC}$), identifying point *P* as the intersection of that ray with side $\overline{CD}$ of the boxspring. Then measure length *BP*. Of course, a mover would likely find it inconvenient to have to measure an angle and draw a ray on a boxspring. The other option avoids such inconvenience.

[8] The angle $\overline{AB}$ makes with the horizontal equals the angle $\overline{BC}$ makes with the vertical ($\overline{BP}$) because $\angle ABC$ is a right angle.

[9] Note that the angle that the staircase makes with the floor will be congruent to $\angle CBP$ in Figure 2b, when the length of the boxspring is parallel to the stairs. In that position, cross-section $\overline{BP}$ will reach its maximum length among all the possible cross-sections used while the boxspring is being tilted from originally being horizontal (Figure 13.1), to an initial angle to take on the stairs (Figure 13.2a) to being parallel to the stairs (Figure 13.2b).

4. A calculation option: The mover could calculate the maximal length of BP using trigonometry: This maximal length is equal to the length BC of the boxspring divided by the cosine of X (i.e., $BP = \frac{BC}{\cos X}$).
5. In either case, it would be feasible to ascend the stairs, while holding the object parallel to the stairs' ceiling, if and only if the length BP is less than t.

Keeping the boxspring upright while moving it up the staircase (consider, again, Fig. 13.1) also requires analogous considerations of its length AB and of the difference between BP and t at each measure of $\angle CBP$. We leave this investigation to the reader.[10] These anticipations of the mathematics potentially involved in modeling the problem of moving mesospace objects contrast with the second author's experience moving household objects without doing any such calculations explicitly, suggesting that, if a related task was going to engage teens in geometric modeling, we would need to design the task in such a way that teens would be likely to produce diagrams.

To encourage the teens to create and reason with diagrams, we decided to frame the task as a communication challenge: To prompt them to draw or calculate, some teens (hereafter, *the planners*) would be asked to write instructions for other people who would implement the instructions (hereafter, *the movers*). This suggested several potential variables in the design: (1) who the movers would be (in terms of what mathematical knowledge they should have), (2) whether the planners would be allowed to talk with the movers during the move, (3) whether the planners would be able to observe the movers implementing their instructions (or, instead, whether they would have to rely on the movers' feedback after attempting to implement them), and (4) which objects they would move. We note first that it is the planners who may, in this task, be modeling objects in the mesospace, enacting and increasing their geometric knowledge by adapting to the feedback of the milieu, while the movers are elements of the communication milieu, who interpret the instructions and provide feedback based on the success of the instructions. Thus, the first didactical variable, who the movers are, matters a lot: What geometry do they know? But, as the task is one of geometric modeling, the communication code they will use is at stake at the same time as the mathematical knowledge is: A transitional language is expected to be used–a language that includes everyday words like *stairs* and *couch* and everyday icons and indices (e.g., drawings of stair steps, arrows) along with some incipient and possibly metaphorical use of geometric representations, including words like *parallel* and *angle*, diagrams of rectangles that might not look rectangular, and possibly geometric indices (e.g., labels for points). For the task to provide an opportunity to learn geometric modeling, it would need to allow the movers to negotiate how exacting to be with the instructions: If movers interpreted

[10] Indeed, advancing up the stairs requires the movers to advance up using $\angle CBP < X$ for an initial horizontal distance before rotating the boxspring to reach $\angle CBP = X$. The difference $t - BP$ will allow for some rotations that increase $\angle CBP$ and for some translation up with a vector that makes that angle $\angle CBP$ with the horizontal.

the instructions of the planners generously, they could repair practically any conceptual mistakes in the instructions thus not provide formative feedback, and waive the need to improve the instructions, hence closing off the opportunity to learn. But, if movers interpreted the instructions very strictly, communication failures could make the activity go very slowly. We therefore wondered how we could enforce that instructions be taken literally, while at the same time create a task environment in which the planners could gradually learn to write better instructions, by having to adapt to successive sources of feedback, as opposed to having to do all that learning at once, by adapting to ungraded feedback.

As we imagined that planners would also do more mathematics if they believed that their instructions needed to be very detailed, we decided to tell the planners that they would not be able to talk with the movers while the movers implemented the instructions, that the movers would have to follow the instructions strictly (i.e., implement them without improvisation), and that the goal of the task was to write instructions that would allow the movers to move the object without bumping the staircase's walls or ceiling. If the movers bumped a surface, they would have to return the object to its starting position, debrief with the planner, ask the planner to revise the instructions, then attempt to implement the revised instructions. As we imagined that such discussions could lead the planners to make their instructions more precise (and, therefore, do more mathematics), we decided that the movers should also debrief with the planners after each time they successfully moved an object from its starting location to its final destination. During these debriefs, the planners would be allowed to ask the movers questions.

While we felt confident that these didactical variables would increase the probability that students would create and use mathematical models, there was still a fair amount of uncertainty on our part as to how the task would play out. For example, would the framing of the task as a communication challenge be strong enough to enable the students to use the mathematical knowledge and skills that they had learned in high school (e.g., trigonometry)? And what mathematical knowledge and skills would they use both in planning the move using microspace models (such as those in Figs. 13.1 and 13.2a, b) and in executing the move when following the instructions while handling mesospace objects? Answers to those questions would be important to enable high school geometry teachers to see this task as a viable way both to teach students about the mathematical concepts, skills, and practices prescribed by the curriculum and to buy into the idea that providing students with opportunities to model mesospace objects is important. But, we decided that it would be fair to learn a bit more about the task from a first round of design in which we would try it outside of school, knowing part of what we wanted to learn was how we might have to tweak such a task or support teachers in implementing it.

Trying Out a First Design

In order to explore our speculations about the mathematics that high school geometry students could engage in when completing a task with the characteristics described above,[11] we recruited a convenience sample of three local teens and played the role of teacher ourselves. All three teens had taken high school geometry the year prior. We had originally planned to have two movers and two planners. The reason for having two movers was practical: The objects needed to be large enough to make it worthwhile taking measurements and creating models, and it would take two people to move such large objects safely. We planned to have two planners so that there would be a natural stimulus for them to think aloud and so that they would have to negotiate what they thought were the constraints of the task.[12] In the end, however, one of the teens had to cancel participation, so we could only have one planner, which was likely a liability, inasmuch as the remaining planner lost a crucial source of advanced feedback for the instructions he produced.

We asked the teens to move three long objects up an indoor quarter-turn (90°), winder staircase with a ceiling parallel to the stairs (as illustrated in Fig. 13.1) and approximately 8 ft of stairway overhead clearance in the two straight portions. We asked them to move three objects available in the home: a table tennis conversion top, a twin-sized boxspring, and a heavy couch (although we did not tell them what the third object was until after they had moved the first two). The reason for having two light objects (the conversion top and boxspring) initially was that we wanted objects that could be moved safely. As mentioned above, the reason for having the third object be heavy was that we did not want the teens to attempt to move it; instead, we wanted them to use what they had learned from planning and moving the previous two objects to write detailed instructions that would enable putative professional movers to move the object, without causing any damage to the house, at a later date.

The conversion top interested us because it was large enough to require care in moving it up the narrow staircase, but also very thin, so we imagined that students might model it as a two-dimensional object (a rectangle), making its cross-sections one-dimensional. In contrast, the boxspring was thick enough that modeling its width could help plan how to move it around the corner in the staircase. We were interested in the couch not only because it was heavy enough that they would not want to learn how to move it by trial and error, but also because there were different

[11] Note that we say "a task with the characteristics described above," because we have not yet specified what objects would be moved or what space they would be moved through, even though we imagined a staircase. This was intentional: We imagine that we have assigned what would be the essential characteristics of the type of task — that the objects and space should be chosen based on the mathematics that one would like students to engage in and could expect students to engage in, if assigned the task by their teacher — and that other researchers or teachers could imagine variations of it that achieve the same goal (having students model mesospace objects).

[12] Having students work on tasks in pairs is common in mathematics education research, in order to make some of their thinking visible; see e.g., Lochhead and Whimbey (1987).

ways in which it could be modeled that might be consequential (e.g., as a rectangular prism, or as an L-shaped prism). The following are the exact instructions that we read to the teens:

We will ask you to move three large objects from the basement to upstairs—a table tennis conversion top, a boxspring, and a third object to be determined. One of you (the planner) will plan the move of each object using any tools that you require, while the other two (the movers) will carry out the instructions (i.e., move the objects). In particular, we would like the planner to plan the move of the conversion top and boxspring while the movers stay upstairs. Then, the planner will give the movers the written instructions. The movers will come downstairs and, using the instructions, will move the objects.

It is important that the instructions be very precise so that objects do not bump against the walls or the ceiling. If they do, the movers will have to restart. The goal of the task is to move each object up the stairs with as few restarts as possible.

After each time that the movers have to restart, they will report to the planner what went wrong and the planner will have an opportunity to revise the instructions. After the movers successfully move the conversion top, the movers will debrief the planner on how things went. After they successfully move the conversion top, the planner will have an opportunity to revise his plan for the boxspring. Then, the movers will implement those instructions and, again, debrief the planner on how that went, after each time they restart and/or after they successfully move the boxspring. Last, the movers and the planner will plan how to move the unknown object together, but they will not actually move it; somebody else will move it, following your instructions.

The planner will be allowed to take as long as he wants to plan the move of each of the objects and to revise the plans for the move of the boxsprings after the movers report back on their move of the conversion top. After moving the first object, the planner can also take as long as he wants to plan how to move the second object. After moving the second object the planner and movers may also take as long as they want to plan how to move the third object. They may use anything they see a need for, and should feel free to ask.

The following materials were available to be used, if needed: Tape measure, paper, pencil, calculator, ruler, protractor, compass, and some handyman tools to measure angles and lengths. To record the planner's writing of the instructions, as well as the implementation by the movers (i.e., the moving of each object), we used two video cameras. One camera was set up at the top of the staircase and another was held by a member of our team who followed the movers as they attempted to walk objects up the stairs and focused on the planner as he wrote instructions. We also gave a notebook to the planner and asked that they use it to write their instructions for the movers. We took notes during the planning, debriefs, and attempts to move the objects. After the first two objects were successfully moved and the teens were done planning the move of the third object, we also held a brief conversation with them, which we recorded and during which we took notes.

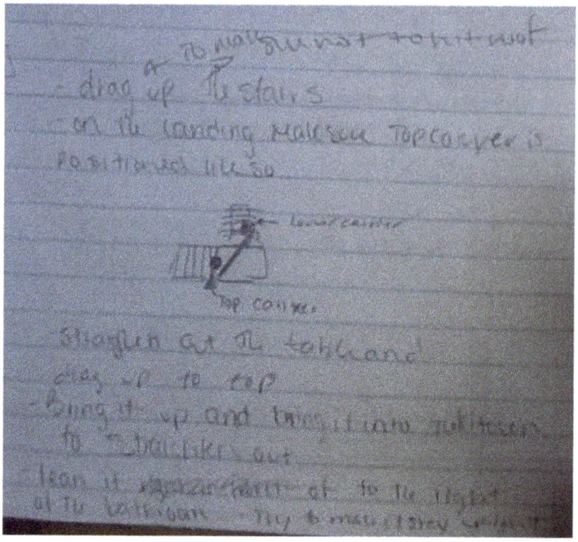

Fig. 13.3 The planner's revised instructions for how to move the conversion top

What We Saw Happening as Teens Tried out the Task

Moving the Conversion Top

As the planner was familiar with the home and the conversion top (a characteristic of this first instantiation of the task, which, in hindsight, we would not recommend for later iterations), he did not look at the space or the conversion top when planning how to move it; neither did he take any measurements or perform any calculations. However, his first draft of the instructions proved insufficient for the movers to complete the task: The instructions did not specify how to move the object around the corner in the staircase and so, implementing them faithfully, the movers dragged the object up the stairs until one of them said that he would hit the wall if he walked any further. Following our instructions, they therefore brought the object back down the stairs and went to debrief with the planner.

The movers told the planner what had happened. The planner was annoyed by the level of detail the movers suggested they needed, but adjusted the plans for them. With these revised instructions (see Fig. 13.3)[13], the movers were able to successfully move the conversion top up the staircase.

One aspect of these instructions that we thought was noteworthy was how vaguely described step 3 ("straighten out...") through step 7 ("Try to make it stay

[13] Transcription: (1) "drag up the stairs to make sure not to hit the roof"; (2) "on the landing make sure top corner is positioned like so…"; (3) 2D top view of staircase with line segment, representing the conversion top, spanning the landing, with the "top corner" and "lower corner" labeled; (4) "straighten at the table and drag it up to top;" (5) "bring it up and bring it into kitchen to straighten out;" (6) "lean it [illegible] of to the right of the bathroom; (7) "try to make it stay upright" (numbers added by authors for reference, corresponding with the bullets used by the planner).

upright") were, given how much inference and improvisation this would have required of the movers. That vagueness could have come up in the debrief and prompted the planner to add more detail to the instructions, but it did not, and we did not realize that the instructions were so vague at the time, so we also missed our opportunity to ask whether and how the movers had to improvise (it is partly for this reason we don't analyze the actions of the movers). Another aspect of the instructions that we found interesting was the fact that the planner told the movers to drag the object along the stairs (something that we had not disallowed). This was interesting to us as we imagined that this was the planner's way of, on the one hand, dealing with the fact that the conversion top was wider than the staircase and so could not be transported lying horizontally and, on the other hand, ensuring that the top of the object would remain parallel to the ceiling, thereby, avoiding bumping it against the ceiling, as the stairway overhead clearance was not much greater than any vertical cross-section of the conversion top (see Fig. 13.2a, b). We imagine that a high school geometry teacher could have added a condition to the problem that the object should not touch the stairs to see if the students would refer to the concept of parallelism to specify the angle at which the object would have to be held in order to avoid bumping the ceiling. This is particularly important since the conversion top (or any rectangular object whose height is close to the stairway overhead clearance) could not initially be positioned parallel to the staircase when starting from a flat floor; rather the parallelism would have to be achieved gradually, as the movers moved into the staircase, starting from a position in which a cross-section of the conversion top is shorter than the stairway overhead clearance (see Fig. 13.2a, b). This might also apply if the conversion top had to be carried making a non-right dihedral angle with the plane of the staircase ceiling—in that case, the vertical projection of a cross-section would have to be shorter than the stairway overhead clearance.

To explore the conjecture that the planner was trying to have the students hold the object with its length parallel to the steps, we asked him how he would determine at what angle the movers should hold the object, if we added the constraint that they would have to restart if the object touched the floor, to the ceiling, or walls. In response, he said that the important thing would be to keep the object parallel to the stairs and that if the ceiling were a little higher, they would not have dragged the object, but would have kept it approximately parallel to the stairs.

Fig. 13.3 provides evidence that the planner expected the movers to perform what an observer could describe as two rotations about the z-axis (see Fig. 13.4, where the black rectangle represents successive positions of the conversion top, as seen from above, first getting to the landing, then rotating to undertake the second straight portion of the staircase[14]).

[14] Note that the black rectangle in Figure 13.4 represents the projection on the xy-plane (floor) of the conversion top; its side projection (say on the xz-plane in the left-most image in Figure 13.4) would show a rectangle with sides parallel to the ceiling, making an angle equal to the inclination of the stairs with the xy-plane. It is likely that the z-axis rotations made to negotiate the landing would also be accompanied by slight rotations on the x and y axes, and translations, however the instructions do not provide any evidence of this.

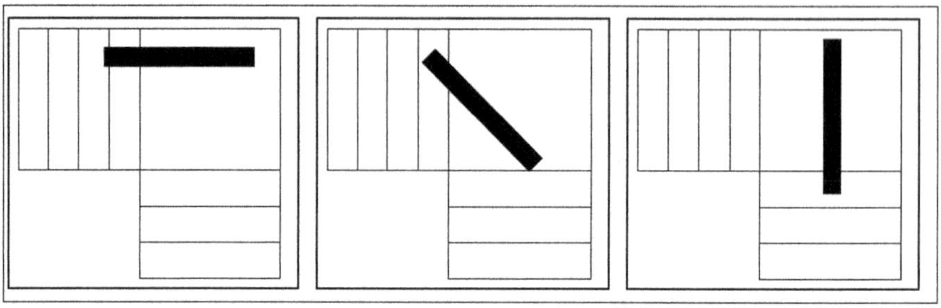

Fig. 13.4 Rotating the conversion top to negotiate the landing

In hindsight, we realize that the task could be improved by finding a way to get the planner to identify where those rotations needed to be made, as the locations of the axes of rotation are quite consequential to avoid bumping the conversion top to the walls. We should have also asked the teens what they could have measured in advance to be able to gauge whether or not the conversion top would fit without dragging it, if we wanted him to think about a cross-section, such as the one discussed above and depicted in Fig. 13.2. This question and the additional condition that the object should not touch the stairs are two more examples of didactical variables that we imagine would make the task more attractive to a high school geometry teacher, as they would engage the students in more of the topics in the high school geometry curriculum.

Moving the Boxspring

Despite the discussion we had just had with the planner, and possibly encouraged by the success of his earlier instructions that had allowed the movers to take the conversion top upstairs without needing to measure the object or the space, the planner did not measure the boxspring or any of the space, as he planned its move. Figure 13.5[15] contains the planner's instructions for how he expected the movers to move the boxspring. What seems noteworthy about these instructions is the attention paid to the position of the boxspring, where the planner writes "when roof of stairs jump up." In particular, the note to "angle the top of the frame up a little bit" suggests awareness that the boxspring was too long to merely rotate it around the corner in the stairs without changing its angle of inclination, as they had done with the conversion top. "Angle the top…" would alter the parallelism of the boxspring to the stairs to permit a rotation to the right without touching the walls.

While these instructions were still vague, they were sufficient for the movers to successfully move the boxspring up the stairs (without bumping the walls or ceiling).

[15]Transcript: (1) "drag up initial flight of stairs"; (2) "when roof of stairs jumps up, angle the top of the frame up a little bit"

Fig. 13.5 The planner's instructions for how to move the boxspring

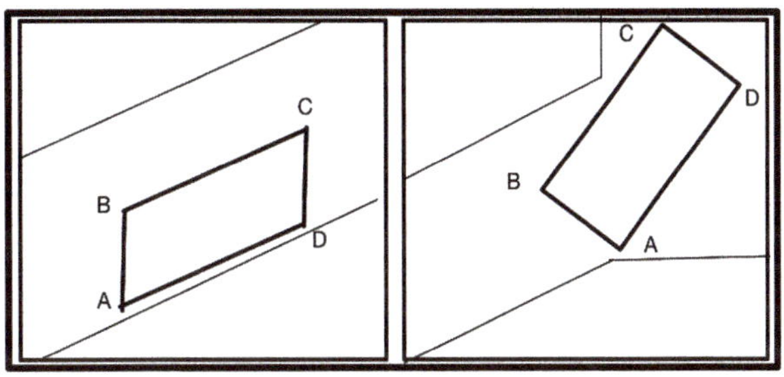

Fig. 13.6 Change in the boxspring model "when the roof of the stairs jumps up"

Moreover, while the instructions include no measurement, they do make use of the resources of microspace conceptions to model objects in the mesospace. Figures 13.5 and 13.6 (in which we reproduce the detail we note in Fig. 13.5) show two consecutive states of the boxspring being moved as two quadrilaterals separated by an arrow (viz., in different frames in Fig. 13.6). The quadrilaterals (*ABCD*, in Fig. 13.6) model the boxspring as seen from the side: In the first position (left frame in Fig. 13.6), two opposite sides of the quadrilateral ($\overline{AD}$ and $\overline{BC}$) are parallel to the line that represents the ceiling; the other sides of the quadrilateral ($\overline{AB}$ and $\overline{CD}$) are vertical, seemingly parallel to the movers' bodies (see the movers' bodies in Fig. 13.5; this verticality had also been shown in the case of the conversion top). The design of a boxspring is a rectangular prism, hence all its faces are, by design, rectangles. But, the first model of the boxspring (left frame in Fig. 13.6), a non-rectangular parallelogram, suggests either that the properties of the rectangle have not all been taken into consideration yet or that the planner believed that, for the sake of communicating the initial steps in the instructions, a non-rectangular parallelogram would be a good-enough model of the boxspring.[16] In

[16] Why the planner might have drawn the first shape as a non-rectangular parallelogram is not known to us. As he was drawing quickly, it could be that it was easier to draw vertical line segments than line segments that are perpendicular to a non-horizontal line segment (e.g., $\overline{BC}$), and as the angle of inclination of the object was small, the object still looked close to a rectangle. What is interesting, however, is not why the rectangle was distorted in the first model of the boxspring, but

the second position of the boxspring (right frame in Fig. 13.6), the ceiling has "jumped up" and the quadrilateral model of the boxspring looks different than the model of the boxspring in the first position. On the one hand, the representation of the boxspring has been "angle[d] … up," and in the new position, the sides $\overline{AD}$ and $\overline{BC}$ that had been parallel to the staircase's ceiling are no longer parallel to it: The boxspring has been rotated around a line on the floor passing through point A, so that $\overline{BC}$ and the staircase's ceiling make an acute angle. But it is interesting to note that the other sides of the quadrilateral ($\overline{AB}$ and $\overline{CD}$) have also changed their angles with respect to the ceiling, no longer showing them in a vertical position, but depicting segments that are clearly not parallel to the bodies of the movers. Instead, sides $\overline{AB}$ and $\overline{CD}$ appear to make right angles with the sides $\overline{AD}$ and $\overline{BC}$ referred to before; quadrilateral $ABCD$ in the figure on the right is not only a rotation of the original quadrilateral, but now it also appears to be a rectangle. This illustrates how the conception of figure as construction in the microspace serves as a modeling language for the mesospace: A rectangle, with the property that consecutive sides are perpendicular, is useful to communicate to the movers that the angles the boxspring make with the vertical orientation of the movers' bodies have changed so much that they may need to grasp the boxspring differently (e.g., while the mover carrying the boxspring around point A might be able to put both hands under $\overline{AD}$ and rest AB on their torso, when the mover increases the angle of inclination of the boxspring, this mover would have to either move his hands to be under $\overline{AB}$ and push or place one hand on $\overline{AB}$ and the other one on $\overline{AD}$ and pull).

Curious as to how much of the planner's not taking any measurements or doing any calculations in order to write his instructions for moving the boxspring had to do with his familiarity with the space or object, we asked the planner how he would have completed that same task if he was unfamiliar with them. He said that he would measure the object, specifically its "diagonal lengths" (which might refer to its cross-sections), so that they would not have to do as much "trial and error." To us, this suggested that the task might enable more explicit mathematical modeling if the objects involved in moving were not initially familiar to the teens.

Moving the Couch

The third object to be moved was a couch that might be described as an L-based right prism. When it came time to move the couch, as mentioned earlier, we asked the planner and the two movers to plan together, so that they could speak across their experience writing and implementing instructions. The two previous objects (conversion top and boxspring) had been assigned to prepare them for modeling the complexity of moving the couch, but they had also been simpler, inasmuch as trial

rather that the rectangle is corrected in the second model to accurately represent perpendicularity of the sides.

and error could be a source of feedback: In the case of the couch, the students would have to engage in making inferences by reasoning from the model.

They coped with the need for feedback by including a first phase of modeling objects in the mesospace: They reduced the couch to what they thought was its essential information, but maintained aspects of its mesospace scale to preserve their access to the controls (Balacheff & Gaudin, 2010) issued from the space of the staircase. Specifically, the teens started by measuring the length of the couch with a tape measure and walking the extended tape measure (what could be considered a one-dimensional model of the couch) up the stairs. They ostensibly did this to see how the length of the couch would factor into their design of how movers would have to carry the couch: Could the couch be held with its length parallel to the steps for the whole first straight portion of the staircase, as they had done with the conversion top, or would they have to increase the angle of inclination of the object to have it fit around the corner in the staircase, as they had with the boxspring? Next, they measured the depth of the couch, along its base, and built a two-dimensional model of the couch by holding two tape measures perpendicular to each other and walking that model up the stairs. Realizing that the depth of the couch was only slightly smaller than the width of the staircase, the teens wondered if there was a third dimension that would make the couch easier to fit around the corner. The planner then measured the length of an imaginary segment on a side of the couch, that extended from the lower back vertex of that side[17] (*A*; in Fig. 13.7) to a point somewhere in the middle of the couch's side (*M*). After some questioning by both us and the movers (now, co-planners), we understood this point to be an expected intersection with a segment that extended from the lower front vertex (*C*) to the upper back vertex (*B*) of the couch's side, and formed a right angle (at *M*; see Fig. 13.7 for our representation of this measurement). At this stage of their modeling work, we would contend the teens were producing a model inasmuch as the artifact they were using to make their inferences represented the couch, but was not the couch; rather, it was a physical assemblage of measuring tapes extended to cover similar space as the couch. For this modeling, measurement was essential.

Figure 13.8[18] contains the teens' instructions for how to move the couch. There are several elements of these instructions that warrant comment. For one, despite having taken many measurements of the couch and the fact that a 2D model with specific lengths had been walked up the stairs, none of those measurements were included in the instructions. This is reasonable, given that the purpose of the earlier model had simply been to determine whether and

[17] Figure 13.7 represents our depiction of the measures the teens took of the couch. To ease the reader's understanding of what the teens measured, we have labeled the extremes of the couch and use those labels *A, B, C, and M*, though the teens did not label any drawing they did of the couch and in fact did not draw the triangle shown in the Figure: The triangle is our reconstruction of the directions on which they trained the measuring tapes.

[18] Transcription: "When first person gets to landing gradually begin to increase the angle of elevation; When the second person gets to the landing straighten the couch and increase the angle until it is as straight up as possible but still not hitting the doorway at the top; When you get to the top width (?) of the stairs the couch should be completely upright."

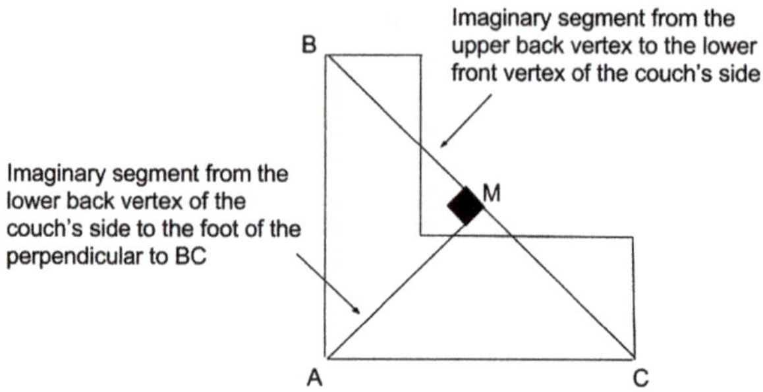

Fig. 13.7 Authors' two-dimensional model of student's imagined line segments from the side-view of the couch

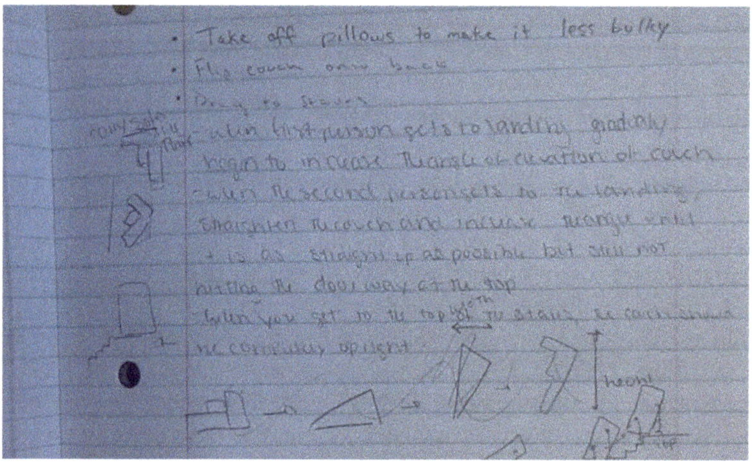

Fig. 13.8 The teens' instructions for how to move the couch

how the object would fit. Yet, the ways in which they hypothesized the object would need to be held and moved did not require the measurements to be explicitly reported. It is worth noting that in the first sub-step of step 3 ("when first person gets to landing…"), the teens did not provide the movers with any idea of what the "angle of elevation" should be, and the details of how to get the couch around the corner are only specified in so far as the instructions specify that the object should be as upright as possible.

Note, again, how the microspace model of the couch included in the instructions evolved from being a rectangle to being a parallelogram, then to being a rectangle, as the information of how the changes in height of the object and the rotation to undertake the second straight portion of the staircase need to be negotiated. To be clear, this is not a critique of the teens; these are simply remarks on what mathematics we expect such a task might have students do. We consider these critiques in order to ask ourselves whether the task could be engineered differently so as to get more of what a high school geometry teacher would likely want from the task (more use of mathematical concepts or practices relevant to the high school geometry curriculum). As researchers, we also consider to what extent and in what ways the task engaged students in the modeling of mesospace objects (as we mentioned earlier, something uncommon in the American high school geometry curriculum) and, hence, in developing understanding of three-dimensional geometry, particularly transformations (such as rotations and translations).

Discussion

This paper describes a first iteration in a design research project that aims at exploring whether communication tasks related to the experience of moving mesospace objects (e.g., large household items) can support the creation of a milieu for students to engage in the geometric modeling of objects in the mesospace. This first iteration of the design was done out of school and with teens who had already finished their high school geometry course. It is clear that those simplifications need to be addressed at later stages of the design process.

We have not yet made the case that motivated the paper. Yet, we have made some strides in that direction, and it is worth summarizing what we have learned through this first iteration that paves the way for making that case. The exploratory work we shared informs three kinds of issues: (1) didactical variables for the design of a communication task related to mesospace objects, (2) what three-dimensional geometry we can expect to be at stake in this kind of task, and (3) what could be learned through this kind of engagement in modeling as a way to learn geometry (see Herbst et al., 2017). This exploratory work also has some important limitations in its conceptualization: Undertaking this kind of geometric modeling of experience in school would require not only improvement of the task space, but also better conceptualization of the actors involved in such tasks. We end this discussion with a brief elaboration of what that means.

Didactical Variables

A first set of gleanings from this exploration concerns the identification of task features that can be described as didactical variables in that they are choices in the design of the task that may make a difference in regard to the conceptions they summon. Within this set we include four considerations.

First, the design of a communication milieu to initially include actual movers that interpreted (as opposed to merely executed) instructions from the planner and that indicated when the instructions were failing was useful to support the evolution (and increasingly mathematical nature) of the planner's messages. This was also useful to devolve to the teens the choice to make messages effective and more precise, even when they would not have to finish the task (as attested by their construction and actual move of a model of the couch).

Second, the task setting of moving household objects up a narrow staircase with a quarter turn and limited overhead clearance presented reasonable opportunities for making mathematical considerations. Our gradual increase of the dimensions of the object to be moved provided natural scaffolding to the activity: From a virtually two-dimensional object (the conversion top), to a three-dimensional object (the boxspring), to a three-dimensional object that was too heavy to manipulate (the couch). Considerations of size and shape within each case provided enough difficulty to make intuitive solutions unsuccessful and require considerations that could be described as mathematical, even if some of them were still embodied and tacit. And this didactical variable (variations in the dimensions and sizes of the objects to be moved) could be manipulated even further (e.g., a queen-sized boxspring would have combined the constraints imposed by the conversion top and twin boxspring).

Third, the comparison of the teens' plans for the movement of the couch with their plans for the movement of the other objects, noting in particular how much mathematical discussion there was around moving the couch, suggests that our initial idea to have a pair of planners, as opposed to a single planner, had potential. It was the teen who had been the planner earlier who came up with the idea of measuring the "diagonal width" of the couch; the immediate lack of clarity his peers saw in that expression led to him showing more concretely what he meant. We expect that ambiguous messages might receive similar scrutiny if the planning was assigned to a team, rather than to an individual, and suggest that having a planning team should be enforced in future trials. This consideration of team configuration along with the earlier suggestion to try the task with objects of more extreme sizes come together to suggest that it might be worth making a second trial of this task in a similar out of school setting.

Fourth, while it is clear that the teens used microspace conceptions of figure to represent what they were doing in the mesospace, it is less clear that all aspects of modeling with geometry were involved: The teens reduced the three-dimensional objects to their key projections onto two-dimensional paper and were able to interpret those, but (as we show in Figs. 13.2a, b, 13.4, and 13.6) the task could have also led to important inferences about how to handle mesospace objects based on microspace diagrams. While we suspect some of that may have happened tacitly, the evidence is not conclusive, as we did not examine the movers' reasoning. It would be important to improve this task with additional requests that elicit the reading of the diagrammatic messages before the plans are enacted. The production of more conditional (if-then) assertions might be stimulated by a follow-up task where competing groups of students each have to produce a general set of instructions for moving a set of unknown objects, all of a given shape but of unknown dimensions, up a staircase of a known type, but also with unknown dimensions, that movers would later try to use to move a set of actual objects of that shape, and varying dimensions. An immersive three-dimensional simulation environment could even be designed in which these variables could be explored.

Three-Dimensional Geometry

A second set of gleanings corresponds to what knowledge from the high school geometry course was elicited by the task. By implementing the task outside of school, we were able to mute an important constraint of mathematics instruction in classrooms: The notion that the work that students are asked to do needs to have an exchange value in terms of the knowledge at stake in the course of studies in which the class is engaged (Herbst, 2006). Our muting of that constraint enabled us to observe the affordances of the task without it being encumbered by the possible frustration of a teacher who might not be seeing enough evidence of school mathematics[19] in the students' work and might therefore change the task into one where the mathematical ideas are suggested to them (Baxter & Williams, 2010). In that context, it is important to identify the mathematics that could be afforded by the task and the moves that a teacher might legitimately make and that might support the summoning by the students of those mathematical ideas. Along this line, the task summons discussions of angle and parallelism, rotations, basic trigonometric ideas, and plane projections of solids. It is notable also that the task stimulated teens to deal with three-dimensional space by reducing it to two-dimensional models, so the exchange alluded to above does not benefit from resemblance between the ideas that students use in problem solving and what the problem solving entails: We and the teacher can say that, in order to complete a task like the one described, students must use their knowledge of three-dimensional Euclidean geometry (i.e., solid geometry). But, while that is the mathematical meaning that can be ascribed to their work, it is not the mathematics that students might explicitly see in their work.[20] This is possibly a place where the task could use some modifications: Would the request to investigate how to move a complicated objects, such as the couch, using scale models possibly built with Legos or constructed in applications like SketchUp (www.sketchup.com), support the use of concepts of solid geometry or maintain the problem solving within the realm of the concrete manipulation? Similarly, if planners could provide their instructions through a video,[21] would their use of

[19] We note here, even if belatedly, that the mathematics at stake in this task could be seen from at least three perspectives. From school mathematics, we could wonder what among the ideas in the high school geometry curriculum are used in and through doing this task. From the discipline of mathematics, we might add that there are aspects of doing mathematics, such as the practice of modeling, that are at stake. A third stakeholder is the ethnomathematics of household moving—which might be described as a systematic elaboration of the concepts and propositions that undergird the practice of competent movers, pretty much in the way that other ethnomathematics researchers and cultural psychologists have inspected other practices, such as carpet-laying or carpentry (e.g., Masingila, 1994; Millroy, 1991).

[20] This is similar to early algebra, where students may be involved in solving problems like $5 + 7 = [\ ] + 3$. Their work may amount to understanding operations algebraically, while students may only see explicitly actual numbers and operations among them (Carraher & Schliemann, 2007).

[21] We thank Mike Battista for the interesting suggestion that messages could be conveyed in video.

gesture, body position, and body movement in creating the video support making more explicit the transformations (rotations, translations) described above?

A teacher could use these variations of the task to bring explicit attention to the three-dimensional nature of the space and the objects involved: Specifically attending to the location of the planes onto which a two-dimensional model of three-dimensional space is referring to, and attending to the relative positions of different axes of rotation. To provide such attention might require the teacher to manipulate didactical variables, along the lines of what was said above. For example, upon reading line (2) in Fig. 13.3 ("on the landing make sure top corner is positioned like so…"), the teacher could ask how they could achieve that position as they walk up the stairs, possibly requesting a play-by-play interpretation, followed by a revision of the instructions. The goal of such a question would be to elicit from students the identification of the need for, and location of, a rotation in space. This could be followed by considerations of what could happen if the rotation was applied to a rectangular prism whose length and width were the same as that of the conversion top but whose thickness was much more significant.

Consider Fig. 13.9, which depicts the projection on a horizontal plane (parallel to the xy-plane, overhead view) of the move of such a rectangular prism: The prism has moved up the first straight portion of the staircase with its longer dimension parallel to the stairway (position 1), and rotates around the z-axis to take on the second straight portion of the staircase (position 2, then 3) having no room to complete the rotation. A rotation around an axis parallel to the xy-plane (position 4) results in a reduction of the size of the projection of the prism onto the xy-plane, which enables more efficient negotiation of the space in the staircase's landing to finish the desired rotation on the z-axis (position 5). The rotation on the axis parallel to the xy-plane is undone (in position 6) to optimize taking on the second straight portion of the staircase. The axes of rotation and planes of projection would need to be elicited through ad-hoc questioning that might easily serve the purpose of ensuring, say, that the object would not bump against walls or the ceiling. At the same time, such questioning could make more apparent the elements of three-dimensional geometry that could be learned through engagement in this task.

Mathematical Modeling

A third set of gleanings relates to the notion of mathematical modeling and what could be learned about it through this activity. The work of the teens illustrates variously the meanings associated with *representation* and *model* and their resources for inference. A language representation such as *turn* or *rotation*, for example, comes with the preposition *around* and associated grammatical forms that beget more geometry—e.g., *rotate around [X]* calls for identifying *X*. Arguably, this serves to infuse some structure to the mesospace, as concepts like projection planes, axes of rotation, and so on, could become resources for spatial visualization later on. Likewise, two-dimensional diagrams of three-dimensional objects are representations that reduce the

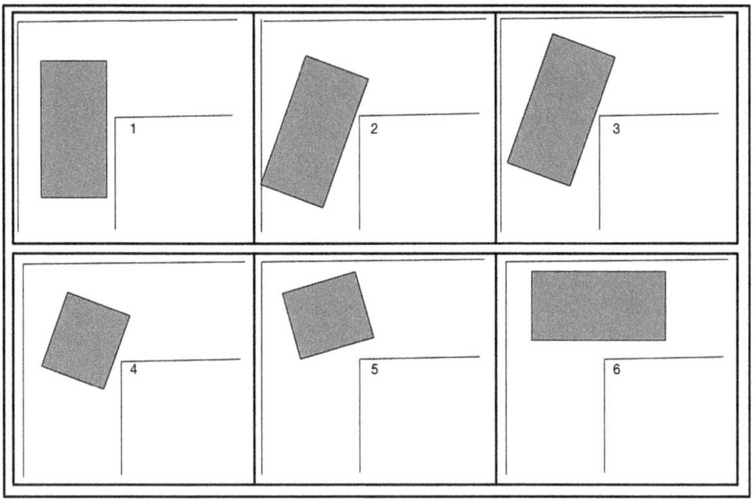

Fig. 13.9 Moving a large prism up a staircase (*xy*-plane projection)

complexity of the object to something simpler that can be manipulated better than the actual object (for example, by drawing successive stages in a move next to each other). And once those diagrams are drawn they also claim other aspects as being represented, some of which may need to be assessed as to their truth if they are going to be built on. The most patent case in point is the boxspring representation in Figs. 13.5 and 13.6. While we can't be sure that the teens were thinking in this way, the correction of the interior angles of the quadrilaterals used to represent the boxspring in successive stages of its move—first not right angles, then right angles—is concomitant with a change in the relation between the boxspring and the body of the first mover, whose head is positioned differently in relation to the boxspring. This change in position would be made intuitively by a mover, but it is the communication game that brings forth the need for geometry, not only to represent, but also to calculate (i.e., to infer that there will be a change in body position). The teacher could ask the planners what they meant to communicate when they drew the first mover in the way they did, or could ask the movers to read from the diagram how they are expected to grasp the boxspring.

In their modeling approach to the teaching and learning of geometry, Herbst et al. (2017) provide the diagram shown in Fig. 13.10 to represent the relation between experiences with shape and space and geometries as the theories which are sources of models. If we think of the work of residential movers, the staircase, and the boxspring as what goes in the cloud "Real world objects and activities" and words, stick figures, and broken lines as representations of those real world objects and activities, then the box named "geometric models of representations of real world objects" identifies the set of correspondences proposed through such identifications. As when the boxspring is associated with a set of strokes that we can take as the diagram of a quadrilateral and then this is described to be parallel to the

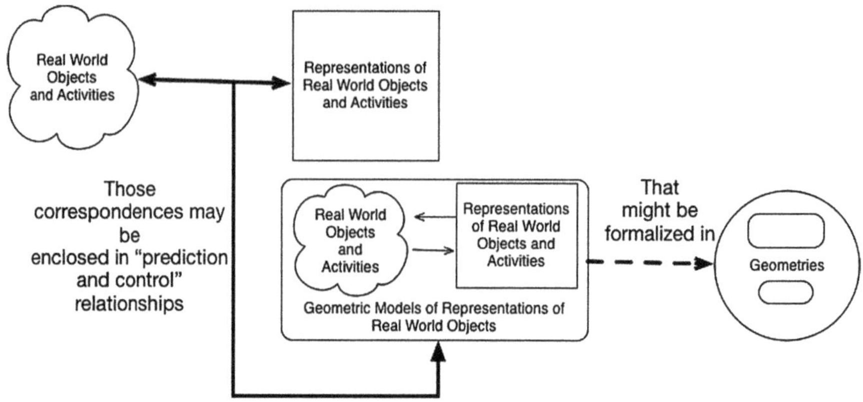

Fig. 13.10 A modeling approach in geometry (from Herbst et al., 2017, p. 4)

stairs, the selection and concentration of such a set of correspondences starts getting more and more accountable to geometry (rather than to the realism of the representation) as the source of rationality (Dennis, 1970; Simon, 2013). Geometries, such as synthetic or coordinate Euclidean geometry, provide the theoretical basis for some of those inferences. For example, the properties of isometries guarantee that the angles of a rectangle will continue to be right after a rotation, which in turn justifies that the first mover, whose location remains unaffected by the rotation of the boxspring, will have to grasp the boxspring differently after the rotation. The general properties of geometric objects entailed by the postulates of Euclidean geometry justify diagrammatic statements made in the messages that describe how the boxspring is to be moved. The diagram is a model in the sense that its capacity to support anticipations of what will happen is powered by inferences derived from geometric theory. This more general point can be made in many occasions with activities like the one presented, but the specific inferences can help both anticipate specific claims about action in the world and highlight specific properties of geometric figures. For example, the location of the axis of rotation as the object reaches the landing might be initially made by trial and error, but both the actual object and its diagrammatic representation contain enough resources to bring in the properties of the circle (or the cylinder) to separate viable from impossible locations for the axis. A consideration of what space will be swept by the object as it rotates not only summons earlier understandings of the circle, but also could afford the teacher an opportunity to re-introduce the concept of cylinder,[22] this time along with its properties. The paper suggests that these inferences are afforded by the task space, but it would take more attention to the way individual students reason when

[22] We are assuming here that students may have encountered holistic descriptions of the cylinder in middle school, but they could be reintroduced to it using its definition as the locus of as a set of points in space.

engaging in these tasks as compared to others that possibly already modeled the three-dimensional space using geometry to ascertain that this is true.

Institutional Considerations

An important limitation of this design paper lies on our sole attention to the task space or our lack of attention to the institutional conditions in which a task like this one would eventually need to be deployed in the teaching and learning of high school geometry. In thinking of the learner merely as somebody who knows, reasons, and communicates logical and experiential meanings, we've eschewed consideration of how schooling mediates relationships with adults and peers. We kept that simplification in mind as we spoke of our participants as teens rather than students; but it's worth noting that there is a need to add the complexity of being a student into the mix when *teen* is replaced by *student*. Students do what the teacher asks them to do because it is the teacher who asks; they cooperate (or not) with others not only in response to their sense making of what the task demands but also to manage enduring social and power relationships. Both the didactical contract that relates students to the teacher and the subject of studies (Brousseau, 1997; Herbst, 2002), and other institutionalized identities and relationships, need to be considered as we examine what we can expect students to do when attempting to complete a task that is presented in school and as we explain what they eventually do. While we leave for another paper the specifics of how the a priori analysis of this task might change depending on those issues, it is important to highlight the bottom line of such analysis: As students respond to a task like the one presented here, we need to be prepared for the possibility that they will not always pursue efficiency or the economic maximization of benefits: They will also second-guess what the teacher wants and use the social interactions prescribed in the task to pursue social relationships of broader scope (e.g., friendship, rivalry).

Similarly, the teacher has been conceptualized rather lightly as someone who assigns a task and observes what students do with it. But issues related to the institutional position of the teacher need to be considered when a version of this task is studied in school context. We discuss them briefly in terms of the professional obligations of mathematics teaching—a teacher's obligations to the discipline of mathematics, to the needs of their individual students, to the institutions within which they work, and to the social group of the class (Chazan, Herbst, & Clark, 2016; Herbst & Chazan, 2012): The teacher's obligation to the discipline of mathematics has arguably been addressed centrally in the motivation for implementing a modeling approach. We can expect this obligation would justify moves by the teacher to encourage students to anticipate outcomes and to delay *try-and-see* approaches, to promote precise reasoning and not be happy with lucky guesses. The teacher's obligation to individual students was considered, at least in part, when the design distinguished among the objects that teens could be expected to lift and objects they should not lift (e.g., the couch). We can expect this obligation would also justify moves to support multiple solutions and students' engagement in diverse physical and social activity. The obligation to the school institution has so far been present only in our discussions of the curriculum, but we can also expect it to justify consideration of what spaces and objects in a school could be used to enact a task like this one, without incurring liabilities or creating problems for other professionals. The interpersonal obligation to the social group of the class has been absent from consideration thus far, but we could expect it to justify attention to logistical questions such as what would other students do when planners design how to move objects.

Conclusion

This paper illustrates how a modeling perspective on the design of geometry tasks can engage adolescents in using what they know about geometry to make sense of a problem about moving furniture. The paper identifies task variables that can be manipulated to summon the engagement of different conceptions of figure, illustrates how the geometric representation of everyday objects can give rise to mathematical modeling, and how such a design can target the teaching and learning of ideas that are at stake in the high school geometry curriculum. The discussion section provides suggestions for future iterations of the same task in an out-of-school context and for reconceptualization of the work for its exploration in school context.

A modeling perspective emphasizes the progressive development of geometric tools to understand the world. The mesospace (i.e., the space of objects of scale comparable to that of the human body) is used to illustrate how considerations of scale are important in such a modeling perspective, not only as a context where one can apply more sophisticated conceptions of figure (such as those developed in the microspace of paper and pencil constructions), but also to elicit particular geometric ideas (e.g., projection planes, rotation axes) whose relevance is especially apparent at this scale.

References

Acredolo, L. (1981). Small- and large-scale spatial concepts in infancy and childhood. In L. Liben, A. Patterson, & N. Newcombe (Eds.), *Spatial representation and behavior across the life span: Theory and application* (pp. 63–81). New York: Academic.

Balacheff, N., & Gaudin, N. (2010). Modeling students' conceptions: The case of function. *CBMS Issues in Mathematics Education, 16*, 207–234.

Battista, M. T. (2007). The development of geometric and spatial thinking. In F. Lester (Ed.), *Second handbook of research on mathematics teaching and learning* (pp. 843–908). Charlotte, NC: NCTM/Information Age Publishing.

Baxter, J. A., & Williams, S. (2010). Social and analytic scaffolding in middle school mathematics: Managing the dilemma of telling. *Journal of Mathematics Teacher Education, 13*(1), 7–26.

Behr, M., & Post, T. (1992). Teaching rational number and decimal concepts. In T. Post (Ed.), *Teaching mathematics in grades K-8: Research-based methods* (2nd ed., pp. 201–248). Boston: Allyn and Bacon.

Berthelot, R., & Salin, M. H. (1998). The role of pupils' spatial knowledge in the elementary teaching of geometry. In C. Mammana & V. Villani (Eds.), *Perspectives on the teaching of geometry for the 21st century: An ICMI study* (pp. 71–77). Dordrecht, The Netherlands: Kluwer Academic.

Blum, W., & Ferri, R. B. (2009). Mathematical modelling: Can it be taught and learnt? *Journal of Mathematical Modelling and Application, 1*(1), 45–58.

Blum, W., Galbraith, P. L., Henn, H. W., & Niss, M. (2007). *Modelling and applications in mathematics education*. New York: Springer.

Boyd, C., Burrill, G., Cummins, J., Kanold, T., & Malloy, C. (1998). *Geometry: Integration, applications, connections*. New York: Glencoe-McGraw Hill.

Brousseau, G. (1997). *Theory of didactical situations in mathematics: Didactique des mathématiques 1970–1990* (N. Balacheff, M. Cooper, R. Sutherland, & V. Warfield, Eds. and Trans.). Dordrecht, The Netherlands: Kluwer.

Brousseau, G., & Warfield, V. M. (1999). The case of Gaël. *The Journal of Mathematical Behavior, 18*(1), 7–52.

Carraher, D., & Schliemann, A. (2007). Early algebra and algebraic reasoning. In F. Lester (Ed.), *Second handbook for research on mathematics teaching and learning* (pp. 669–705). Charlotte, NC: Information Age.

Chávez, Ó., Papick, I., Ross, D. J., & Grouws, D. A. (2011). Developing fair tests for mathematics curriculum comparison studies: The role of content analyses. *Mathematics Education Research Journal, 23*(4), 397–416.

Chazan, D., Herbst, P., & Clark, L. (2016). Research on the teaching of mathematics: A call to theorize the role of society and schooling in mathematics. In D. Gitomer & C. Bell (Eds.), *Handbook of research on teaching* (5th ed., pp. 1039–1097). Washington, DC: AERA.

Coliva, A. (2012). Human diagrammatic reasoning and seeing-as. *Synthese, 186*(1), 121–148.

Collins, H. (2010). *Tacit and explicit knowledge*. Chicago: University of Chicago Press.

d'Ambrosio, U. (1985). Ethnomathematics and its place in the history and pedagogy of mathematics. *For the Learning of Mathematics, 5*(1), 44–48.

Dennis, J. (1970). Mondrian and the object art of the sixties. *Art Journal, 29*(3), 297–302.

Dimmel, J. K., & Herbst, P. G. (2015). The semiotic structure of geometry diagrams: How textbook diagrams convey meaning. *Journal for Research in Mathematics Education, 46*(2), 147–195.

Gerdes, P. (1986). How to recognize hidden geometrical thinking: A contribution to the development of anthropological mathematics. *For the Learning of Mathematics, 6*(2), 10–12.

González, G., & Herbst, P. (2006). Competing arguments for the geometry course: Why were American high school students supposed to study geometry in the twentieth century? *International Journal for the History of Mathematics Education, 1*(1), 7–33.

Grugnetti, L., & Jaquet, F. (2005). A mathematical competition as a problem solving and a mathematical education experience. *Journal of Mathematical Behavior, 24*(3–4), 373–384. https://doi.org/10.1016/j.jmathb.2005.09.012

Halverscheid, S. (2008). Building a local conceptual framework for epistemic actions in a modelling environment with experiments. *ZDM Mathematics Education, 40*(2), 225–234.

Hanna, G., & Jahnke, H. N. (2007). Proving and modelling. In W. Blum et al. (Eds.), *Modelling and applications in mathematics education* (pp. 145–152). New York: Springer.

Hegarty, M., Montello, D. R., Richardson, A. E., Ishikawa, T., & Lovelace, K. (2006). Spatial abilities at different scales: Individual differences in aptitude-test performance and spatial-layout learning. *Intelligence, 34*(2), 151–176.

Herbst, P. (2006). Teaching geometry with problems: Negotiating instructional situations and mathematical tasks. *Journal for Research in Mathematics Education, 37*(4), 313–347.

Herbst, P., & Chazan, D. (2012). On the instructional triangle and sources of justification for actions in mathematics teaching. *ZDM Mathematics Education, 44*(5), 601–612.

Herbst, P. G. (2002). Engaging students in proving: A double bind on the teacher. *Journal for Research in Mathematics Education, 33*(3), 176–203.
Herbst, P. G., Fujita, T., Halverscheid, S., & Weiss, M. (2017). *The learning and teaching of geometry in secondary schools: A modeling perspective*. New York: Routledge.
Kaiser, G., & Sriraman, B. (2006). A global survey of international perspectives on modelling in mathematics education. *ZDM Mathematics Education, 38*(3), 302–310.
Kaput, J. (1991). Notations and representations as mediators of constructive processes. In E. von Glasersfeld (Ed.), *Radical constructivism in mathematics education* (pp. 53–74). Dordrecht, The Netherlands: Kluwer.
Kaput, J. J. (1999). Teaching and learning a new algebra. In E. Fennema & T. A. Romberg (Eds.), *Mathematics classrooms that promote understanding* (pp. 133–155). Mahwah, NJ: Erlbaum.
Lochhead, J., & Whimbey, A. (1987). Teaching analytical reasoning through thinking aloud pair problem solving. In J. E. Stice (Ed.), *New directions in teaching and learning* (pp. 73–92). San Francisco: Jossey-Bass.
Masingila, J. O. (1994). Mathematics practice in carpet laying. *Anthropology and Education Quarterly, 25*(4), 430–462.
Meyer, D. (2015). Missing the promise of mathematical modeling. *The Mathematics Teacher, 108*(8), 578–583.
Millroy, W. L. (1991). An ethnographic study of the mathematical ideas of a group of carpenters. *Learning and Individual Differences, 3*(1), 1–25.
Müller, P., Wonka, P., Haegler, S., Ulmer, A., & Van Gool, L. (2006). Procedural modeling of buildings. *ACM Transactions on Graphics, 25*(3), 614–623.
National Governors Association (NGA) Center for Best Practices & Council of Chief State School Officers. (2010). *Common core state standards for mathematics*. Washington, DC: Authors.
Otte, M. (2006). Mathematical epistemology from a Peircean semiotic point of view. *Educational Studies in Mathematics, 61*, 11–38.
Siegel, A. W. (1981). The externalization of cognitive maps by children and adults: In search of ways to ask better questions. In L. S. Liben, A. H. Patterson & N. Newcombe (Eds.), *Spatial representation and behavior across the lifespan* (pp. 167–194). New York: Academic Press.
Simon, B. (2013). Mondrian's search for geometric purity: Creativity and fixation. *American Imago, 70*(3), 515–555.
Skovsmose, O. (2000). Aporism and critical mathematics education. *For the Learning of Mathematics, 20*(1), 2–8.
Vergnaud, G. (1996). The theory of conceptual fields. In Steffe et al. (Eds.), *Theories of mathematical learning* (pp. 219–239). Mahwah, NJ: Erlbaum.

Chapter 14
Visualization Abilities and Complexity of Reasoning in Mathematically Gifted Students' Collaborative Solutions to a Visualization Task: A Networked Analysis

A. Gutiérrez, R. Ramírez, C. Benedicto, M. J. Beltrán-Meneu, and A. Jaime

Abstract We analyze the solutions given by secondary school mathematically gifted students to a collaborative task designed to promote the development of students' competence of visualization. Each student was provided with two different orthogonal projections of a set of buildings made of cubes and other verbal data, and they were asked to place the buildings on a squared grid. We analyze students' use of visualization abilities and the complexity of their reasoning. Results show that there is a relation between the objective of students' actions and the kind of visualization abilities used, and, also, between students' strategies of solution and the cognitive demand necessary to fulfill them. Finally, we network both analyses to gain insight and look for global conclusions.

Keywords Visualization abilities · Cognitive demand · Mathematical giftedness · Networked theories · Problem solving · Secondary school · Visualization

A. Gutiérrez (✉) · C. Benedicto · A. Jaime
Departamento de Didáctica de la Matemática, University of Valencia, Valencia, Spain
e-mail: angel.gutierrez@uv.es; adela.jaime@uv.es

R. Ramírez
Departamento de Didáctica de la Matemática, University of Granada, Granada, Spain
e-mail: rramirez@ugr.es

M. J. Beltrán-Meneu
Departamento de Educación, Jaume I University, Castellón, Spain
e-mail: mmeneu@uji.es

K. S. S. Mix, M. T. Battista (eds.), *Visualizing Mathematics*, Research in Mathematics Education, https://doi.org/10.1007/978-3-319-98767-5_14

Introduction

Teaching mathematics is more effective when it includes diagrams, pictures, drawings, etc. visually representing concepts and relationships. To take advantage of this teaching methodology, students should develop visualization abilities and know effective ways of using visualization as part of their mathematical reasoning. The use of visualization in mathematics classrooms is considered an important object of research in mathematics education (Battista, 2007; Presmeg, 2006; Rivera, 2011). A significant open research question on visualization is the need to identify aspects of classroom cultures which promote the use of visualization in mathematics (Presmeg, 2006). To answer this question, we investigated the promotion of visualization in cultures of collaboration between mathematically gifted students (m-gifted students hereafter). Collaborative learning has proved to be beneficial (Davis, Rimm, & Siegle, 2014) for such students, but research on ways to deepen its effects is needed.

Solving mathematical tasks dealing with visualization requires the use of two kinds of elements (Gutiérrez, 1996): external data, mainly objects (e.g., pictures or real models of geometrical figures) and verbal information (e.g., written statements or oral information), and internal elements, mainly visual elements (e.g., mental images; Presmeg, 1986), visual thinking (to manage visual information) and mathematical reasoning. Two visualization processes—interpretation of figural information (IFI) and visual processing (VP) (Bishop, 1983)—and several visualization abilities (Del Grande, 1990) control the intrapersonal communication between external and internal data and the interpersonal communication between different subjects. We designed a workshop aimed at promoting the development of m-gifted students' use of visualization abilities and collaborative learning. To capitalize on the benefits of collaborative learning for m-gifted students, the workshop encouraged the interpersonal communication to increase the use of visualization abilities by providing each student with only a part of the data for the tasks (Fig. 14.1), so they needed to share information, and they had to verbally communicate visual information efficiently to solve the problem.

In this chapter, we present a networked analysis (Bikner-Ahsbahs & Prediger, 2010) of three pairs of m-gifted students' use of visualization abilities and the complexity of their reasoning while solving the problems. This introduction presents an overview of recent research on the constructs that conform the theoretical background of our research, describes the problems posed in the workshop, and states the research objectives.

Visualization in Mathematics Education

Researchers in psychology, mathematics, and mathematics education possess diverse interpretations of terms such as visualization, visual reasoning, spatial ability, and so on. Gutiérrez (1996) presented a model integrating partial results from diverse areas, which characterize the different visualization components and that is relevant for mathematics education research (Presmeg, 2006).

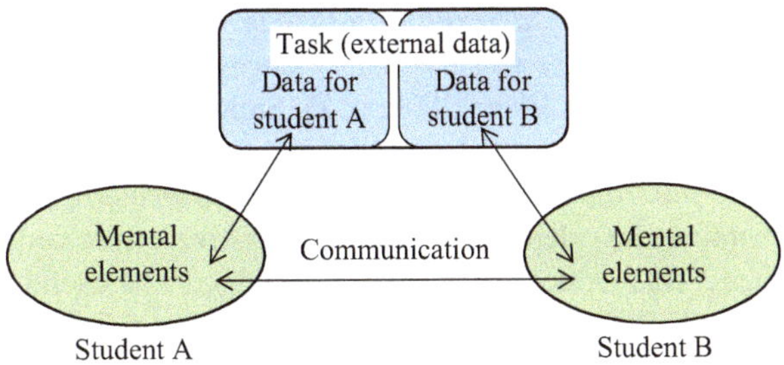

Fig. 14.1 Components of the virtual workshop

Among the different definitions of visualization pertinent to mathematics education found in the literature, we highlight those comprising the types of images, processes, and abilities necessary to produce, analyze, transform, and communicate visual information related to objects, models, and geometric concepts (Arcavi, 2003; Gutiérrez, 1996). Visualization consists of four main elements (Gutiérrez, 1996, p. 10): *Mental images* are "any kind of cognitive representation of a mathematical concept or property by means of visual or spatial elements." *External representations* are "any kind of verbal or graphical representation of concepts or properties including pictures, drawings, diagrams, etc. that helps to create or transform mental images and to do visual reasoning." A *process of visualization* is "a mental or physical action where mental images are involved." *Visualization abilities* are stable capacities of the subject which are necessary for effective learning of geometry (Bishop, 1980). In general, different visualization abilities have to be mastered "to perform the necessary processes with specific mental images for a given problem" when solving mathematical tasks (Gutiérrez, 1996, p. 10). Del Grande (1990) compiled several visualization abilities with great relevance for the development of mathematics students.

Several authors have emphasized the importance of visualization in mathematization (Arcavi, 2003; Clements & Battista, 1992) and problem solving (Ozdemir, Ayvaz-Reis, & Karadag, 2012), but results of research do not show a unified position on the relation between visualization and mathematical giftedness (Lean & Clements, 1981; Ryu, Chong, & Song, 2007; Van Garderen, 2006), although other recent research has shown significant evidence of the relation between visual perception and mathematical ability (Ramírez, 2012; Rivera, 2011).

The Complexity of Mathematical Reasoning

The tasks that teachers pose to their pupils are an important element to promote m-gifted students' learning of mathematics. There are different criteria to assess their suitability for students. A relevant criterion is the cognitive complexity of their solutions. Felmer, Pehkonen, and Kilpatrick (2016) argued that it is necessary to pose cognitively demanding tasks to make students engage in higher order thinking and improve the quality of their learning of mathematics. We agree with this criterion, since the problem solving experiment we present here was aimed to make students struggle to solve an unusual challenging task.

The *cognitive demand* of a task is "the kind and level of thinking required of students in order to successfully engage with and solve the task" (Stein, Smith, Henningsen, & Silver, 2009, p. 1). Smith and Stein (1998) elaborated the *Levels of Cognitive Demand*, which organize mathematical tasks in four levels (*memorization*, *procedures without connections*, *procedures with connections*, and *doing mathematics*) depending on the cognitive effort necessary for students to solve them. This model has been acknowledged as a useful tool for teachers to promote students' higher order thinking (NCTM, 2014; Schoenfeld, 2014). We present this model in detail in Section "Theoretical Background".

The levels of cognitive demand have been used mainly to train teachers in identifying the levels of the tasks they select for their classes and maintaining their intended level during the classes (Smith & Stein, 1998). All studies we have read assigned levels of cognitive demand to tasks by analyzing their statement and the solution considered as correct by teachers. This procedure does not acknowledge that most mathematics tasks may be solved correctly in several ways, requiring from students different degrees of cognitive effort. Furthermore, there are not studies about classification of tasks that attend to the needs of m-gifted students. To overcome these issues, we have adapted the levels of cognitive demand in an innovative way to the characteristics of the visualization tasks, to analyze students' outcomes during the solution of problems (Benedicto, Gutiérrez, & Jaime, 2017). On the other hand, the characteristics of the levels, as presented in Smith and Stein (1998), are generic and a bit ambiguous, not sufficiently precise to be applied to the visualization tasks nor to the m-gifted students' answers we have analyzed, so we have also particularized the definitions of the levels of cognitive demand to the specific context of spatial visualization and the type of tasks we deal with in this chapter. This way of using the levels has proved to be a reliable framework to identify tasks adequate to students with diverse mathematical capabilities, in particular to m-gifted students (Benedicto, Acosta, Gutiérrez, Hoyos, & Jaime, 2015).

Networking Theories in Mathematics Education

In mathematics education research, several theories live together to contribute, from different approaches, to provide complementary analysis or solutions to a specific mathematics education issue. Researchers usually adopt one theoretical framework to carry out their research, but there is a growing interest in establishing links between different theories, to take advantage of the most useful components of each one by making interwoven analyses of data. Bikner-Ahsbahs and Prediger (2010) considered that:

> … *networking strategies* are those connecting strategies that respect on the one hand the pluralism and/or modularity of autonomous theoretical approaches but are on the other hand concerned with reducing the unconnected multiplicity of theories and theoretical approaches in the scientific discipline. (p. 492, italics added)

There are different ways of networking theoretical approaches depending on the objectives aimed and the strategies used for finding connections (Bikner-Ahsbahs &

Prediger, 2014). We are interested in the networking strategy of *combining*, since we have combined the theories of visualization abilities and the levels of cognitive demand to analyze the outcomes of m-gifted students solving some visualization tasks. We do not intend to merge both theories, but to use them as complementary analytical tools to gain insight into the data of the experiment.

A Collaborative Visualization Task

The experiment that we present was based on a set of collaborative visualization tasks that were designed to be solved by a pair of students linked by videoconference. The objective of the tasks is to place a set of colored buildings on a squared grid. Buildings are made of equal interlocking cubes, with all buildings in the same/different color having the same/different height. The data are the four side orthogonal projections (north, south, east, and west views hereafter) and other data like the number of buildings of each color and some restrictions in the positions of the buildings. Each student is provided with only part of the data, which is not sufficient to solve the task. Therefore, students have to gather together their data to succeed in solving the task, with the restriction that they cannot share graphical information (pictures, drawings, etc.), although they can describe it verbally.

The buildings tasks consist of two parts. The first part is an introduction for students who do not know orthogonal projections; it presents a perspective representation of a city built on a squared grid (Fig. 14.2) and students are asked to make the buildings with cubes and place them on a paper grid. Then, students are guided by the teacher to compare their view of the buildings with the orthogonal projections provided (Fig. 14.2).

For the second part, each student is provided with a set of interlocking colored cubes, a 2 cm. squared grid oriented with the cardinal points and a coordinate system (Fig. 14.3), two views of another set of buildings, information about the number of buildings and their colors, and some restrictions to the position of the buildings (see an example in Table 14.5). Students are asked to write the coordinate numbers in the marks near the axes based on the information provided in the views. Finally, they are asked to

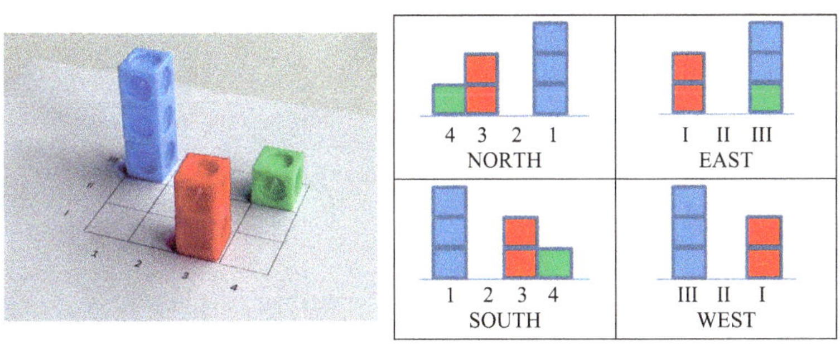

Fig. 14.2 An example of the information provided in the first part of the buildings tasks

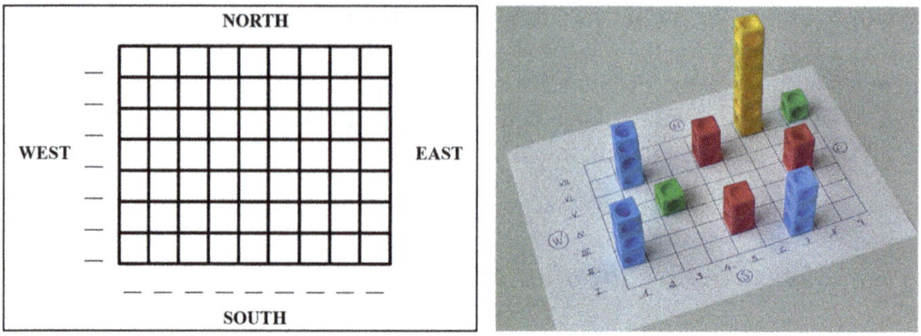

Fig. 14.3 The grid provided in the second part of the buildings tasks and a solution to the task

place the buildings (made with the cubes) on the grid (Fig. 14.3 shows a solution). Several variables may modify the difficulty of the tasks, such as the number of buildings in each line of the grid, buildings hidden in some views, or the number of solutions.

Research Objectives

To solve this kind of task successfully, students have to make extensive use of visualization. Furthermore, as the communication between the students is only verbal, they have to use their visualization competence to convert visual information into a meaningful verbal explanation, and vice versa. In the tasks we present, we will concentrate on looking at the visualization abilities. Our research objective is to analyze the use of visualization made by pairs of students during their interactions. Such objective is made operative by the following specific objectives:

1. Analyze the use of visualization abilities by pairs of students while solving the buildings task, looking for trends and relationships between abilities used and students' aims at that moment.
2. Analyze the variations in the cognitive effort made by pairs of students while solving the task, looking for relationships between levels of cognitive demand and kinds of actions made.
3. Relate the results from objectives 1 and 2 into a networked analysis of students' behavior, looking for relationships between use of visualization abilities and levels of cognitive demand.

Theoretical Background

We devote this section to present in detail the three theoretical components grounding the analysis of data. First, we characterize the visualization abilities as used in the context of the buildings tasks. Then, we characterize the levels of cognitive demand particularized to the specificities of the tasks. Finally, we discuss the networking of both theoretical models.

Visualization in Mathematics Education

We consider visualization as "the set of types of images, processes, and skills necessary for students of geometry to produce, analyze, transform, and communicate visual information related to objects, models, and geometric concepts" (Gutiérrez, 1996, p. 9). We analyze the presence of visualization in students' outcomes by identifying their use of visualization abilities, which constitute one of the four main elements of visualization described in the Introduction section. We do not analyze the processes of visualization because they are ever present throughout the solution of the tasks, so they do not provide relevant information on students' behavior, and students' mental images because we did not have a reliable tool to identify them.

Del Grande (1990) characterized a set of abilities necessary for a fruitful use of visualization in mathematics. For an accurate identification of the abilities used by students in this experiment, it is necessary to make a particular characterization of each ability narrowly related to the tasks being solved. Table 14.1 presents those Del Grande's (1990) visualization abilities used by our students characterized in the specific context of the buildings tasks. Section "Visualization Abilities in Students' Answers" includes examples of students' answers showing the different abilities.

The Model of Cognitive Demand in Visualization Tasks

The *model of Cognitive Demand* consists of four levels which allow classify tasks *and* solutions according to the cognitive effort necessary for students to solve them, and allow teachers and researchers understand the complexity of the mathematical knowledge and reasoning used by students in their solutions. Smith and Stein (1998) defined each level by a set of characteristics to be used to assign levels to tasks. We offer below a detailed characterization of the levels of cognitive demand specific for the solutions to the buildings tasks (Tables 14.2, 14.3, and 14.4). This analytical framework, focused on students' use of visualization abilities, is an original contribution because, as far as we know, the model of cognitive demand has never been used to analyze visualization tasks or their answers.

We focus on the levels pertinent to our research, so we omit the level of memorization. The characteristics stated in the tables are organized in several *categories* which refer to different components of the solution to a mathematical task: the process of solution, the learning objective, the cognitive effort necessary to solve the task, the mathematical content implicit in the statement, the kind of explanations asked of students, and the systems of representation of information used by students.

Table 14.1 Characterization of the visualization abilities used to solve the buildings tasks

Abilities	Characterization of the abilities for the tasks
Figure-ground perception (FG)	– Recognize that isolated squares in the views are part of a particular building – Discriminate one or several buildings in a view – Recognize that different colors correspond to different buildings
Perceptual constancy (PC)	– Recognize that the position of a building on the grid is invariant even when it is not seen in a view – Recognize that the coordinates of a building are invariant no matter the observer's position – Recognize that buildings that are apart on the grid continue being apart although they are seen together in a view
Positions in space (PS)	– Identify the positions of buildings by using coordinates and/or cardinal points – Imagine a view corresponding to another student's position – Relate two views to determine the position of a building – Relate several buildings on the grid by using terms like "in the same street," behind, hidden, diagonally, etc.
Spatial relationships (SR)	– Identify a relation between the positions of two or more buildings on the grid, without depending on the observer's point of view or their coordinates, by using terms like "they touch/do not touch each other," "they are apart," etc. – Mention the heights of two buildings, e.g., to justify that one hides the other in a view
Visual discrimination (VD)	– Compare an orthogonal projection of the buildings on the grid with the corresponding view given in the data or with another student's projection – Compare the locations of buildings on the two students' grids – Compare the buildings placed on a grid with verbal data

Table 14.2 Characteristics of the level of procedures without connections

Categories	Characteristics of solutions
Process of solution	Are based only on the observation and interpretation of simple explicit relationships between data available in student's part of the statement (e.g., a student places a building just by coordinating her two views of the building)
Objective	Place buildings correctly without needing to coordinate the four views or logical-deductive reasoning to understand the relationships between buildings (e.g., it is not necessary to relate the views of both students)
Cognitive effort	A successful solution requires limited cognitive effort. Little ambiguity exists about what needs to be done and how to do it, since the views available to a student clearly show how to place the building
Implicit content	Students do not need to be aware of the implicit connections between the four views and other data and the buildings to be placed. They can be placed by using only the data of a student
Explanations	Are focused only on describing the procedure used. It is not necessary to identify relationships between the other student's views and the building
Representation of solution	Students use the manipulative representation to show the solution, but they might also use a graphical representation (e.g., by making some marks on the grid to indicate cells that can or cannot be locations of one or more buildings)

Table 14.3 Characteristics of the level of procedures with connections

Categories	Characteristics of solutions
Process of solution	Consist of following a sequence of steps based on implicit complex relationships. Students should consider different possibilities and make logical-deductive decisions about which data to combine and how to combine them (e.g., coordination of the four views to decide where to place a building)
Objective	Understand the underlying relationships between the different data, and make logical-deductive reasoning to select or reject cells in the grid for a building based on the information available (e.g., after having identified the buildings which are on a street, students analyze the data to reject or select cells to place the buildings)
Cognitive effort	Requires some degree of cognitive effort, to logically connect different elements of the task and deduce which procedure of solution should be followed
Implicit content	To solve the tasks, students need to consider explicitly the relationships underlying their different elements, like the four views, buildings already placed, verbal data, etc.
Explanations	Requires explanations that include deductive justifications for the decisions made (e.g., about choosing or rejecting cells to place a building), based on combination of information from the views, buildings already placed and still not placed, etc.
Representation of solution	Students use the manipulative representation of the solution, but they might also use a graphical representation (e.g., by marking in different colors cells where building can or cannot be placed)

Table 14.4 Characteristics of the level of doing mathematics

Categories	Characteristics of solutions
Process of solution	Students analyze the data in detail and coordinate the information. They identify buildings having more than one possible location, and get all possible solutions to the task (e.g., two buildings can be placed in different cells fitting the four views)
Objective	Explore and combine the information provided by the task and use logical-deductive reasoning to realize the existence of several feasible locations for some buildings, and get all possible solutions
Cognitive effort	Solutions require a considerable cognitive effort, since there may be several solutions, so students have to be aware of this fact and take decisions, based on the data, about the possible locations of each building
Implicit content	Students identify that there is more than one possible solution. They solve the tasks by relating the four views and other data, analyzing the different possibilities, and getting logical deductions
Explanations	Justify the existence of several solutions, as well as rejected cells and chosen locations, based on the available information
Representation of solution	As there are several solutions, students combine manipulative and graphical representation to mark cells that can or cannot be locations of buildings (e.g., students mark the cells around a placed building as not available for other buildings)

The Level of Procedures Without Connections

Students' solutions in the level of *procedures without connections* consist of performing in a routine manner an algorithmic process already known, without the need of being aware of connections to mathematical contents underlying the tasks. These tasks are focused on getting correct answers but not on producing mathematical understanding of the underlying contents. The characteristics of this level are particularized for solutions to buildings tasks in Table 14.2.

The Level of Procedures With Connections

Students' solutions in the level of *procedures with connections* consist of solving the task by following a solution process that is procedural but not routine, since it presents some ambiguity on how to carry it out, and students need to be aware of certain connections to mathematical contents underlying the tasks to decide on their way to the answer. These tasks are focused on discovering the underlying contents and gaining mathematical understanding of them. Table 14.3 shows the characteristics of solutions to buildings tasks in this level of cognitive demand.

The Level of Doing Mathematics

Students' solutions in the level of *doing mathematics* require complex and non-algorithmic thinking, because there is not a predictable approach to solve them. Students have to understand the underlying mathematical contents and their relationships to make appropriate use of them while working through the tasks. Table 14.4 presents the characteristics of solutions to buildings tasks in the level of doing mathematics.

Networking Theories of Visualization and Cognitive Demand

As mentioned in section "Networking Theories in Mathematics Education," we consider the networking strategy of *combining* as the most interesting for our purposes in this chapter, because it is useful to make a networked analysis of empirical experiments like ours, by looking at the same data produced by the experiment from two theoretical perspectives.

Based on this analytical tool, we will analyze the pairs of students' solutions by looking at the use of visualization abilities and at their levels of cognitive demand, by means of the theoretical constructs presented in the two last subsections. Then, we will complete the networking by comparing and contrasting the results of both analyses to get global conclusions. Previous examples of this kind of networking may be found in the ZDM special issue in volume 40(2), 2008.

Methodology

In this section, we describe the specific task posed to students, the characteristics of the m-gifted students whose solutions will be analyzed, and the two research methodologies applied to analyze the presence of visualization abilities in students' outcomes and the cognitive effort required from students to solve the task.

Description of the Experiment

The experiment consisted of posing a buildings task to several pairs of m-gifted students, the same task to all them. One student in each pair was living in Valencia (Spain) and the other one in Granada (Spain). They were linked by a group videoconference with each other and with the researchers. The researchers only intervened in students' dialog when it was evident that students had misunderstood some instruction or to answer their questions. We describe the solutions to the same task produced by three pairs of m-gifted students (we name them A1-B1, A2-B2, and A3-B3). They were aged 14–16 and studied grades 9 or 10 (lower secondary school), and were recruited from an out-of-school workshop for mathematical enrichment of m-gifted students. The pairs of students were provided with the materials mentioned in the Introduction and the information shown in Table 14.5.

This task has several elements of complexity: It has two solutions (Fig. 14.4). There are two red buildings in street 5, which cannot be discriminated from north and south views, so it is necessary to consider other buildings and the verbal conditions (Table 14.5) to get a solution. Every view shows two blue buildings in streets 1 and 2 or I and II, which might induce students to believe that they are the same buildings. Placing the red and blue buildings is only possible by coordinating information from both students.

The task may be solved by using several strategies. One is based on determining the cells in the grid where the buildings of each color may be placed by observing the views and analyzing the feasibility of the different combinations of cells. Another strategy is based on careful recursive trial and error, placing buildings on the grid and checking whether they fit or not the four views and the other data, and then making adjustments.

Analysis of Students' Solutions to the Buildings Task

The main sources of information were the video recordings of computer screens and students' dialogs, which were transcribed. To analyze a pair of students' solution, we first divided the protocol into fragments corresponding to the different actions. Then, each students' outcome was analyzed twice, to identify the visualization abilities exhibited during their actions and reasoning and to characterize the levels of

Table 14.5 Information provided to the pairs of students for the second part of the buildings task

Student A	Student B
• North-south direction: Streets are numbered with Arabic numbers 1–9, from west to east • There are buildings of four colors. One is yellow, two are green and three are red • The buildings with the same height have the same color • The buildings are placed on the squares of the grid. The buildings cannot touch each other • South and west orthogonal projections are:	• West-east direction: Streets are numbered with Roman numbers I to VII, from south to north • There are nine buildings • Blue buildings are three floors high • North and east orthogonal projections are:
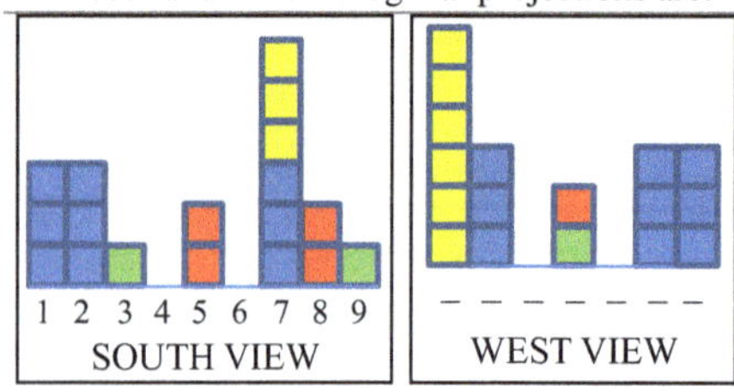 	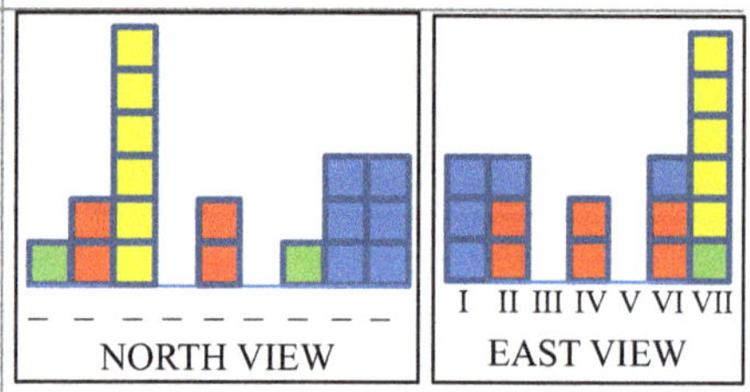

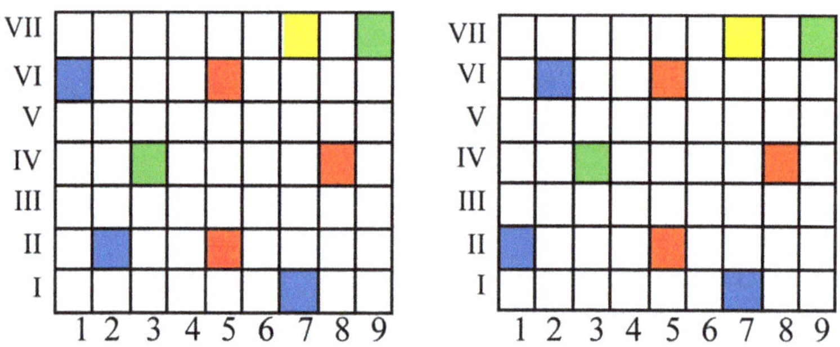

Fig. 14.4 The two solutions to the buildings task analyzed

cognitive demand associated with their reasoning. Our objective in the networked analysis is to identify trends in and relationships between the use of abilities and levels of cognitive demand.

A global observation of students' solutions to the buildings task evidenced several *phases* in the solutions, devoted to different types of actions performed by the students characterized by their operational aims:

- *Placement* (of buildings): Students try to place buildings on the grid. These were the most frequent and time-consuming actions.
- *Checking*: Students compare the buildings placed on the grid with the views and verbal data to check whether the buildings' positions and colors are correct or not.
- *Correction* (of errors): Students realize that some buildings are misplaced on the grid, and they try to identify their correct positions.

- *Request of information*: A student asks the other student to provide him with data from his views, positions of buildings, verbal data in the statement, etc. in a quite systematic way.
- *Recapitulation*: A pair of students share the positions of the buildings placed on their grids to verify whether both grids match or not.

To identify the visualization abilities put to work by students, we looked carefully into each student's outcome, since most of them showed one or more abilities. To complete this analysis, we counted the number of appearances of each ability in every phase of the solution.

To identify the levels of cognitive demand exhibited in students' reasoning while solving the buildings task, we looked globally at each phase of solution, since the cognitive effort made by students cannot be reliably identified in a single outcome, but instead it is necessary to consider the whole students' dialogue along each phase. To complete this analysis, we put together the levels of cognitive demand of the consecutive phases of the solution.

As an example, we analyze below a short fragment of the dialog between A1 and B1. They had already placed the yellow building in (7,VII) and a green building in (9,VII) (refer to Fig. 14.4 and Table 14.5). Now they began trying to place the red buildings:

B1: (12:41) *But the red* [building] *is not hidden by the yellow one. The red, in fact, is in street VI in my east view.*
A1: *In street VI?*
B1: *In the north-south street, street 6* [B1 really meant east-west street VI].
A1: *Ok. I don't see the red in street VI, I have a blue building.*
B1: *You don't see the red! Ah!*
A1: *I don't see the red.*
B1: *Of course, I only see a blue square* [in street VI from east view], *which should be... the one behind it. Therefore, it should be in street 7 and street VI.*

We identify occurrences of the ability of positions in space when the students used coordinates to talk about possible positions of the red building (e.g., *The red, in fact, is in street VI in my east view*). We also identify the ability of figure-ground perception when students identified the blue building behind the red one (*I only see a blue square, which should be... the one behind it*).

Students showed reasoning in the level of procedures without connections, since B1 worked in identifying the position of a red building by using only his own data, taking and applying simple explicit relationships found in his two views. Even though B1 got information about A1's data and views, he did not use it to place a red building in (7,VI). B1's objective was not to gain a global understanding of the set of red buildings, but only to place one of them. B1's explanations were only descriptions of what he observed in his views.

Analysis of Students' Use of Visualization Abilities

We present the analysis of students' use of visualization abilities during the solution of the buildings task. We first present examples of use of the different visualization abilities. To complete this section, we provide an overview of the three pairs of students by comparing their ways of solution.

Visualization Abilities in Students' Answers

We present examples of students' use of different Del Grande's (1990) visualization abilities. Refer to Table 14.5 for the orthogonal views and other data provided to each student, and to Fig. 14.4 for the solutions to the task.

Ability of figure-ground perception: Students put to work this ability to isolate buildings or parts of buildings from their context with different aims:

- To recognize that isolated squares in the views are part of buildings partially hidden:

B2: (16:55) *I believe that there is a three-floor building… I see it in the east view, in* [street] *VI there is a three-floor blue building.*

- To recognize that squares in different colors represent buildings of different heights:

B1: (20:28) *I have green, red, and yellow squares. I believe that a building cannot have a part in a color and the other part in another color.*

Ability of perceptual constancy: It is necessary when students have to recognize that:

- The position of a building on the grid does not change when it is hidden behind another building:

B1: (11:30) *In the south view, then, in street 7 north-south, there is some blue building that I cannot see* [from the north view].
A1: *Ok. Do you see the yellow building complete* [from north view], *or is there another building that I cannot see* [from the south view]*?*

- Two buildings may be apart on the grid even when they are seen together in a view:

A3: (40:18) [referring to the blue buildings in streets I and II] *One in (II,7), for instance, and another one, for instance, in (3,I), and they would look like an entire building. Do you understand me?*

Ability of positions in space: This is the ability most frequently used by students, because the context of the problem is a grid with coordinate axes. This ability is present when:

- Students identify positions of buildings by mentioning the coordinates and/or cardinal points of cells in the grid:

A1: (37:22) *Wait. The blues* [blue buildings] *in (1,II) and (2,VI) can be placed in (1,VI) and (2,II).*

- A student imagines buildings in a view from the other student's data. B1 described what he thought should be seen in street II from A1's west view:

B1: (27:31) … *Furthermore,* [in my east view,] *the one* [blue building] *in street II is hidden by the red* [building] *in* [street] *II, which you see, in your west view, hidden by the blue one. In your west view, you only see one blue building in the* [street] *II, right?*

- Students coordinate two views to determine the position of a building on the grid. The two views may be from the same student or one view from each student:

B1: (30:05) *In my north view, I see two blue buildings in* [streets] *2 and 1, so this street, the north-south, must have a blue* [building]. *And, if you tell me that in your west view it hides the red* [building] *in VI, … Do you see the red building in your west view?*

- Students relate one building to others by using terms like "in the same street," behind, etc.:

A3: (33:10) [A red building] *Could be behind the tall building in (7,VII)* [the yellow building].

Ability of spatial relationships: This ability is used by students when they have to relate several buildings:

- To identify an internal relation between buildings. A characteristic of this use of the ability is to verbalize terms like touch, diagonally, they are apart, etc.:

A3: (29:17) [A3 and B3 had placed the buildings shown in Fig. 14.5] *Then, the* [blue] *one in… (7,VI) cannot be there, because it touches the yellow one. It should be in another place.*

- To compare or relate their heights:

B1: (39:00) … *I see that the red building hides the green one* [from the east view, in street IV] *but it does not hide the blue* [buildings] *because they are taller…*

Ability of visual discrimination: This ability is necessary to compare visual pieces of information, like buildings on the grid and in some view(s), looking for similarities or differences:

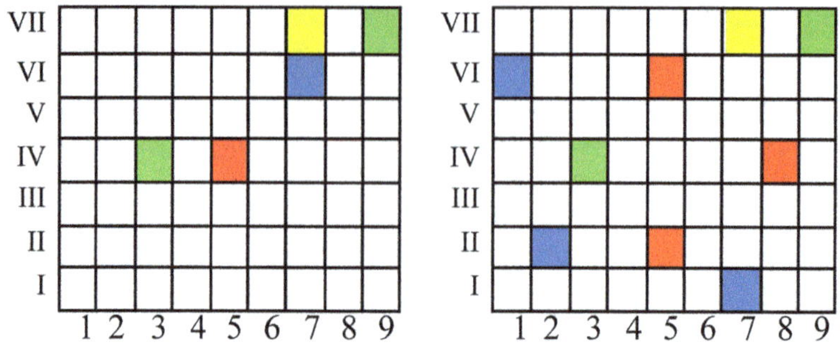

Fig. 14.5 Buildings placed on the grid by A3 and B3 (29:17) (left) and a solution (right)

- To compare a view of the buildings already placed on the grid to the same view in the data:

B2: (16:26) [B2 is inclined comparing his views with the buildings already placed on the grid] *The building in (V,3) does not fit* [the east view].

- To compare the positions of buildings on both students' grids, to check whether they match:

B1: (33:27) *I tell you, from right and top. Yellow (7,VII), green (9,VII) and (3,IV), red (5,II), …*

A1: *Yes. It's ok. Everything fits my views.*

- To check the positions of buildings and verbal data:

A1: (20:50) *Look, my clue says that there is one yellow, two greens and three reds. Three reds!?*

B1: *Yes, three reds. Because, in my east view, there are three reds in* [streets] *VI, IV, and II…*

A1's surprise was because his views showed one and two red buildings (Table 14.5).

Global Analysis of Students' Answers

This section offers a synthesis of the visualization abilities shown by the students in their solutions. The charts in Figs. 14.6, 14.7, 14.8, 14.9, 14.10, and 14.11 show the relation between the phases of solution of the task and the abilities used. The ability most used by all pairs of students was positions in space, since it was necessary to share or transmit information about placement of buildings or to refer to cells in the grid, mainly by means of coordinates or cardinal points.

The pair A1-B1 used 144 times the visualization abilities along their solution of the buildings task. Figures 14.6 and 14.7 show that A1 and B1, apart from the ability

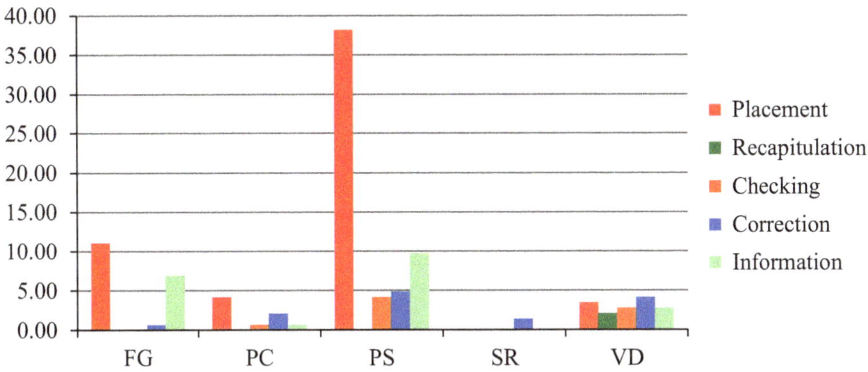

Fig. 14.6 Distribution (%) of A1 and B1's use of visualization abilities between the phases of solution

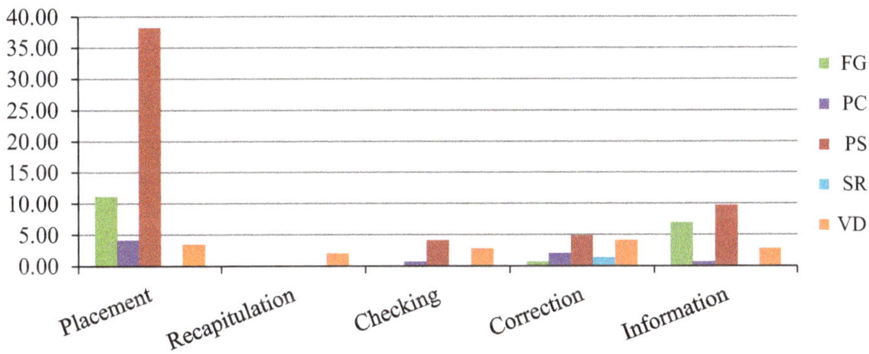

Fig. 14.7 Distribution (%) of A1 and B1's use of visualization abilities over each phase of solution

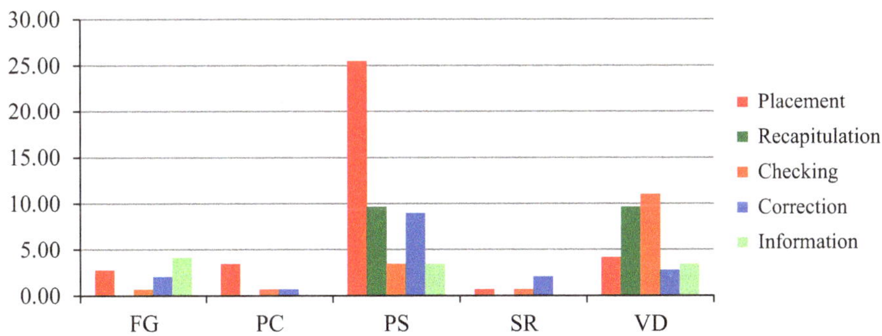

Fig. 14.8 Distribution (%) of A2 and B2's use of visualization abilities between the phases of solution

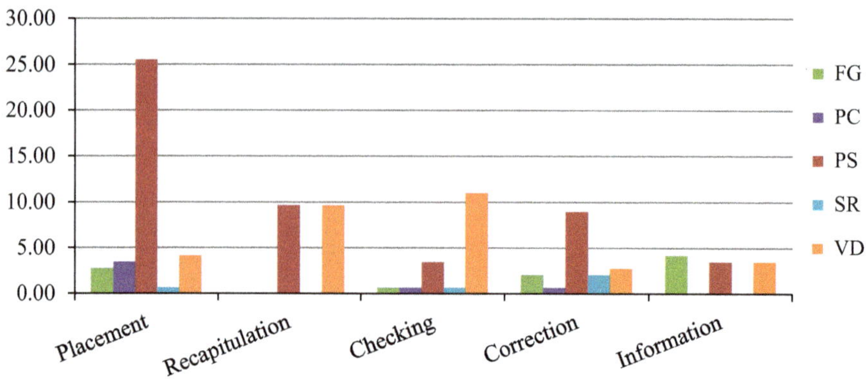

Fig. 14.9 Distribution (%) of A2 and B2's use of visualization abilities over each phase of solution

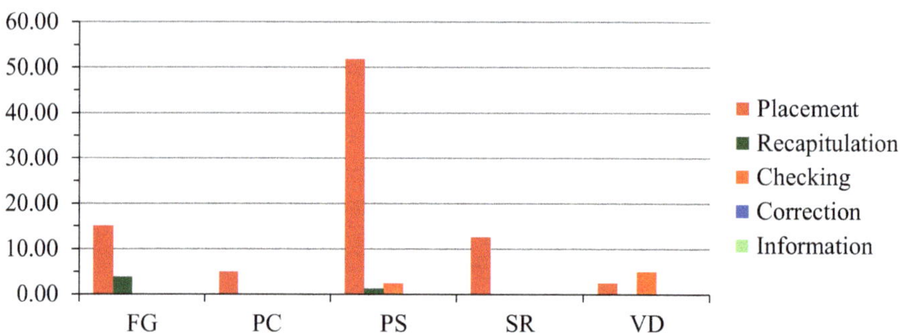

Fig. 14.10 Distribution (%) of A3 and B3's use of visualization abilities between the phases of solution

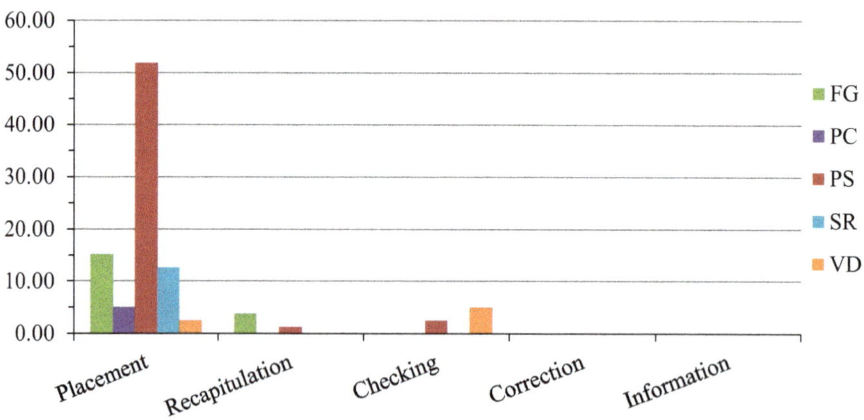

Fig. 14.11 Distribution (%) of A3 and B3's use of visualization abilities over each phase of solution

of positions in space (56.9% of all appearances of visualization abilities), also used quite frequently the abilities of figure-ground perception (18.8%) and visual discrimination (15.3%). They used the other abilities too, but sporadically and without any apparent pattern of use.

The pair A2-B2 used 145 times the visualization abilities along their solution. A2 and B2 took much more time than the other pairs to solve the buildings task, mainly because they used the abilities of recapitulation and checking many more times than the other pairs. According to Figs. 14.8 and 14.9, the most used ability was positions in space (51%), but its difference of use with respect to the other abilities was smaller than for the other pairs of students. The ability of visual discrimination (31%) was also used often by A2 and B2, mainly in the phases of recapitulation and checking.

The pair A3-B3 only used 79 times the visualization abilities along their solution. A3 and B3 made a very efficient solution, devoting most time to actions of placement of buildings, and they did not use the phases of correction and information (Figs. 14.10 and 14.11). As a consequence, the ability most used by A3 and B3 was positions in space (55.7%). They only used significantly figure-ground perception (19%) and spatial relationships (12.7%).

A global overview of the three pairs of students' solutions shows some patterns of behavior in their use of visualization abilities. All of them made scarce use of the abilities of perceptual constancy and spatial relationships, mainly because the task required little use of them. The students in the sample were aware of the perceptual constancy of the buildings on the grid, so they usually did not make explicit use of this ability. The spatial relationships ability was necessary mainly to identify whether two buildings touch each other or to locate several buildings in the same street (buildings that, in some views, are seen superposed or are hidden by a taller building).

Other patterns of behavior are specific to different levels of efficiency in solving the buildings task. Students A1-B1 and A3-B3 were the most efficient solving it. Some characteristics of their behavior were:

- The phase of placement of buildings occupied most time of solution (70–80% of the time).
- The ability of figure-ground perception was mostly used in the placement phase (80% of the times A3-B3 used this ability, 59.3% for A1-B1, and only 28.6% for A2-B2).

Students A2-B2 were the less efficient in solving the task. Some characteristics of these students that explain their difficulties, were:

- They used much time in the phases of solution different from placement: 36% of the time devoted to placement of buildings, 20% to checking, 18% to recapitulation, 16% to correction of errors, and 10% to requests of information.
- The ability of visual discrimination was quite used by A2-B2, mainly in the phases of recapitulation (31.1% of the times they used this ability) and checking (35.6%).

Analysis of the Cognitive Complexity of Students' Visual Reasoning

In this section, we analyze the cognitive effort required by the visual reasoning made by the m-gifted students when solving the buildings task. We have assigned levels of cognitive demand, as characterized in section "The Model of Cognitive Demand in Visualization Tasks", to students' outcomes. We first present a classification of the strategies which may be used by students to solve this kind of task, analyzing their levels of cognitive demand and presenting examples taken from our experiments. Then, we analyze the trajectory of each pair of students' levels of cognitive demand while solving the buildings task.

Classification of Strategies of Solutions According to Their Level of Cognitive Demand

The strategies used by m-gifted students to communicate information to each other and get the positions of buildings on the grid are the main source of information to understand why they expended more or less cognitive effort to solve the buildings tasks. We have identified six types of strategies that are present throughout their solutions of the buildings task and analyzed the cognitive demand required by each of these strategies.

Strategies Requiring the Level of Procedures Without Connections

Students do not connect the contents underlying the tasks (views, verbal data, buildings already placed, and relationships between them), since each student can manage his own information but they are not able to combine the information shared. Sharing pieces of information without combining them, or providing directions to help the other student to place a building are strategies typical of this level of cognitive demand. These strategies require only a limited cognitive effort, but they cannot be used to get the correct position of all buildings, since only the yellow and green ones can be placed by using just the data available to one student. We have identified two strategies in this level of cognitive demand:

Share and build: Students exchange pieces of information about a building, but they do not combine them operatively so, finally, a student gets a (maybe correct) location for the building by using only his own data. In the following excerpt, A3 and B3 did not make sense of the data they shared, so they were not able to combine their views to find a correct cell for a blue building. Then, A3, considering only his own views, located a blue building in (7,VI), which is a wrong position.

A3: (24:50) *In the south view, in* [street] *7, I have a three-floor blue* [building] *hiding half yellow building. Do you see it?*

…

B3: *I see three blues in I, II and VI* [east view].
A3: *I have them in I, II and VI too* [west view]. *Do you have them together in I and II?*
B3: *Yes.*
A3: *Then, it has to be in VI, hasn't it? Because it is detached. (VI,7), right?*
B3: *Yes, maybe.*

Build and direct: A student places a building on the grid by using only his views and then he guides the other student to place the building in the same cell. We see below that B2, after having (wrongly) deduced from his views that there are two blue buildings in (1,I) and (2,II), gave directions to A2 to help him place two blue and a red buildings.

B2: (10:53) *I've found the place of the blues* [buildings]. *It is in the south-west corner, the one with 3 floors, the first one. The second, diagonally... towards the north-east corner. I better use coordinates... The south-west corner is (1,I). There is a three-floor blue in the corner (1,I).* [A] *Three* [floor blue building] *in (2,II) and* [a] *two* [-floor red building] *in (8,II).*

Strategies Requiring the Level of Procedures with Connections

Students need to be aware of certain connections between contents underlying the task and be able to use them to decide on how to proceed to the answer, which requires some degree of cognitive effort. These strategies do not help students realize that there may be several correct positions for some buildings. We have identified three different strategies:

Study positions: Students discard cells where a building cannot be located based on the observation and application of implicit relationships between views and buildings already placed on the grid. Students might not be able to get the position of a building but they identify the possible positions.

The dialog below shows an example of this strategy. A3 and B3 combined their views to get the correct conclusion that there is a red building in street 8. Then, A3 identified as possible locations for the red building all non-empty streets in his west view, and B3 discarded the cells not having a red building in his east view.

A3: (21:59) [Fig. 14.12 shows the buildings already placed] *Let's see, in the* [street] *8, in both north view and south, we see the red building without hiding anything.*
B3: *Yes, it is the same.*

…

A3: *Then, I do not have it in the west view. It is in* [street] *I or II or IV, or VI or VII. Do you have something in…?*
B3: *I have that it may be in II, IV, or VI.*

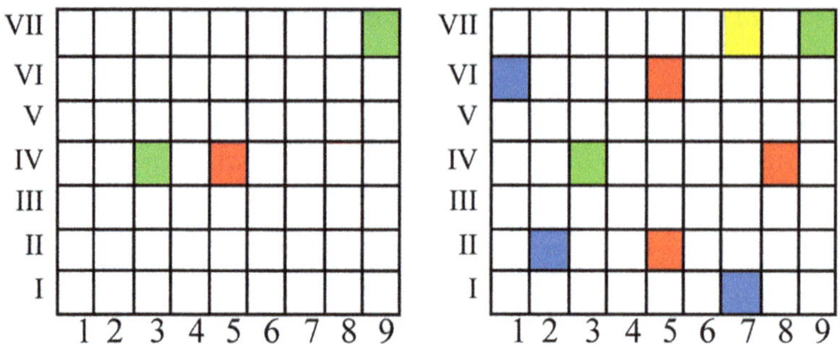

Fig. 14.12 Buildings placed by A3 and B3 (21:59) (left) and a solution (right)

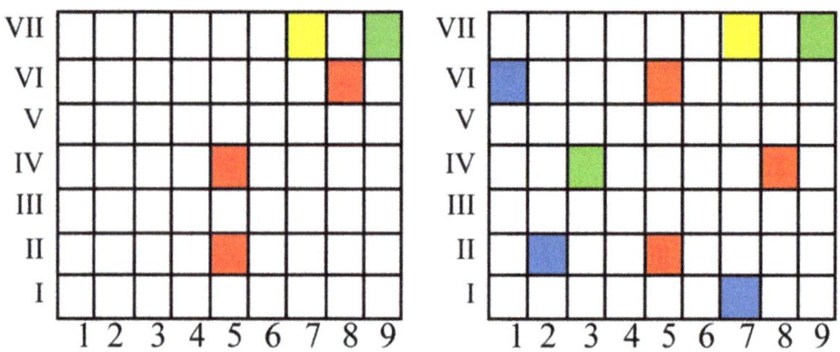

Fig. 14.13 Buildings placed by A1 and B1 (20:51) (left) and a solution (right)

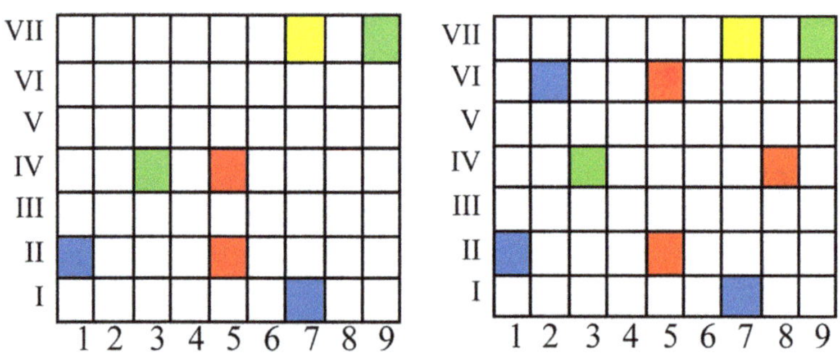

Fig. 14.14 Buildings placed by B1 (30:10) (left) and a solution (right)

Combine and build: Students exchange information and combine it operatively to correctly place a building. In the excerpt below, A1 and B1 looked for the position of a green building. Figure 14.13 shows the buildings already placed. They combined operatively data from both students and succeeded in finding the correct position of that green building.

B1: (20:51) *Ok. Have you placed a green* [building] *in 3 north-south, IV east-west?*
A1: *Yes, I have a green there or in (IV,9). I have as* [possible] *greens (IV,9) and (IV,3).*
B1: *I believe that it is in (IV,3) because, in my east view, I see the second green building you say in street (9,VII). You do not see it from your west view...*
A1: *Because the yellow hides it. Ok. So there is a green there. Ok... So, as you said, the other green is in (3,IV).*

Not all possible: Students do not note that two blue buildings may be placed correctly in another cell. In the following excerpt, B1 had already placed some buildings (Fig. 14.14). To try placing the blue buildings, the students shared information from all the views and decided to place a blue building in (2,VI) without realizing that the blue buildings in streets 1 and 2 could be correctly placed also in another position.

B1: (30:10) *As, in my north view, I see two blue buildings in* [streets] *2 and 1,* [then] *in street 2 north-south there must be a blue* [building]. *If you tell me that, in your west view,* [the blue building] *hides the red* [building] *in* [street] *VI,... Then, it* [the blue one] *would be in 2 north-south, VI west. Or in 1... In your west view, do you see the red building in* [street] *VI?*
A1: *No.*
B1: *Ok. I also have a red building in* [street] *VI in my west view and behind it* [I see] *a blue square. Then, there is a blue building hiding the red one* [from the east view].
A1: *But, in the north or south views, the* [red building] *which is in* [street] *VI east-west is in* [street] *1 or 2 north-south?*
B1: *It is in* [street] *2, because in my north view I see two blue buildings in* [streets] *2 and 1. Then, it* [the blue building] *has to be in* [street] *2, because in the 1 we had already placed one* [blue building] *which is the one hiding the red* [building] *in your west view.*

Strategies Requiring the Level of Doing Mathematics

These strategies require a complex non-algorithmic thinking, and considerable cognitive effort, to explore the implicit relationships between the data available and make appropriate operational use of them, allowing students deduce that some buildings may be correctly placed in several cells and find all possible solutions (two blue buildings in the task we are analyzing). We have identified one strategy in this level:

All possible: Students A3 and B3 had already correctly placed all buildings except the blue ones in streets 1 and 2 (Fig. 14.15). They explored all the possible positions of those blue buildings to find out the two solutions (Fig. 14.4).

A3: (44:27) *I don't know whether it* [a blue building] *is in (VI,1) or (VI,2).*
B3: *I see.*

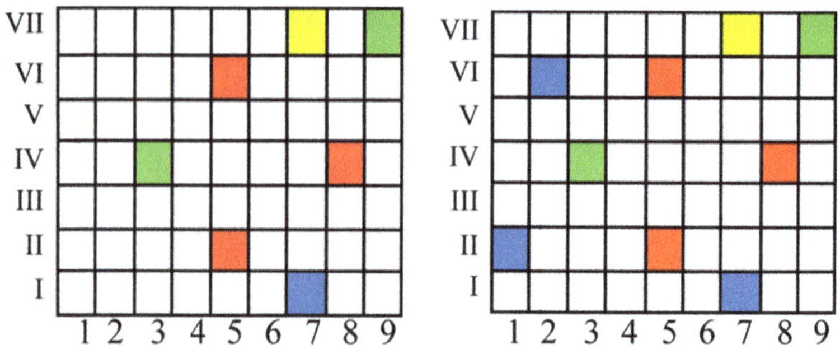

Fig. 14.15 Buildings placed by A3 and B3 (44:27) (left) and a solution (right)

A3: *If one is in (VI,2), then the other has to be in (II,1). And, if one is in (VI,1), then the other has to be in (II,2). It wouldn't matter, I think. Because there are not more buildings, are they?*
B3: *No, the nine are there* [placed in the grid].
A3: *Then, they could be in both cells.*

Trajectories of Cognitive Demand of Students' Solutions

In previous pages we have presented, exemplified, and analyzed the cognitive demand required by the different strategies used by the three pairs of students to solve the buildings task. To summarize our analysis of the complexity of those students' visualization reasoning, we present three graphs showing the trajectory of each pair of students' cognitive demand during the solution of the task (Figs. 14.16, 14.17, and 14.18); the horizontal axis represents the strategies used in the consecutive phases of the solution and the color of the marks corresponds to the buildings students were dealing with. We have analyzed only the phases of placement and correction of errors, since these are the only phases where students' actions might end up placing buildings on the grid.

A1 and B1 started (Fig. 14.16) placing the yellow building without needing to combine their information, so requiring a cognitive effort in the level of procedures without connections. Next, they tried independently (each student using only his views) to place the green buildings, in the same level of cognitive demand. They did not succeed, so they shifted to a combine-and-build strategy, requiring from them a higher level of cognitive effort to correctly place the green buildings. A1 and B1 made several partially successful attempts to place the red buildings without combining their pieces of information operatively, which required from them a reduced cognitive demand. When they used the strategy of combine-and-build, they were able to get the correct locations of all red buildings. Finally, A1 and B1 worked the same way to place the blue buildings, but they did not realize the existence of more than one possible solution (a not-all strategy).

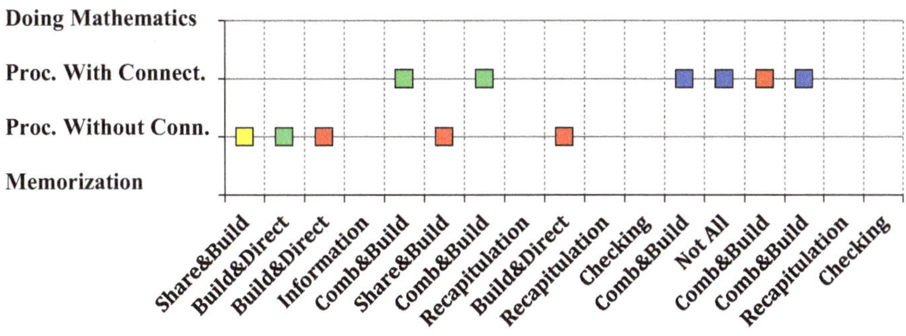

Fig. 14.16 Levels of cognitive demand of A1 and B1's strategies over the phases of solution

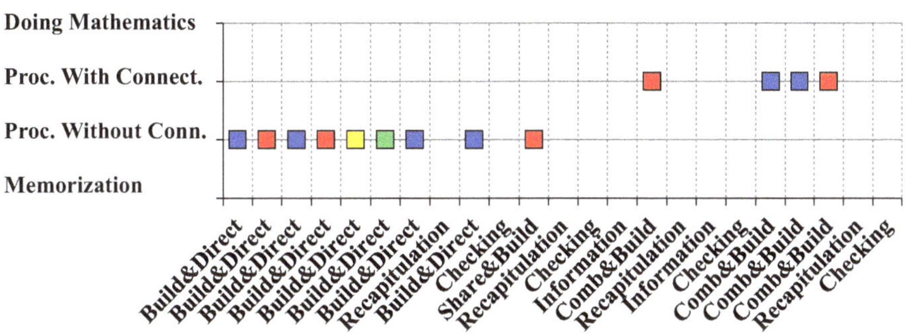

Fig. 14.17 Levels of cognitive demand of A2 and B2's strategies over the phases of solution

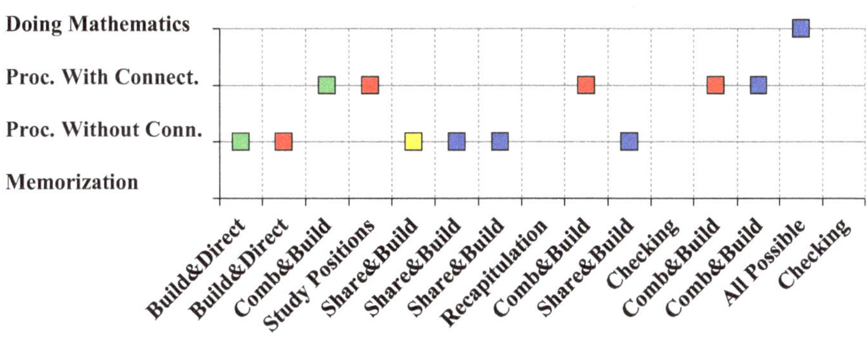

Fig. 14.18 Levels of cognitive demand of A3 and B3's strategies over the phases of solution

The graph in Fig. 14.17 shows that A2 and B2 solved the task in 23 phases and only 13 of them included location of some building. During the first part of their solution, students' level of cognitive demand was procedures without connections, which allowed them to correctly place the green buildings but not the red and blue ones. When A2 and B2 began doing real collaborative work, moving to the strategy combine-and-build, they were able to place correctly the red and blue buildings, although they did not realize the two possible solutions of blue buildings.

A3 and B3 were the most efficient students. Figure 14.18 shows that 12 out of the 15 phases of their solution were devoted to place buildings. The students worked mostly collaboratively, except during the first two phases of the solution: A3 started placing green and red buildings based only on his views. Next, A3 and B3 combined information from their views, by using the strategy of combine-and-build, and they succeeded in placing the second green building and the red buildings. When they first tried to place the blue buildings, they had some difficulties because they could not combine their pieces of information operatively. When A3 and B3 succeeded in combining operatively their information, by means of strategies combine-and-build, they correctly placed all the buildings and even identified the two possible solutions, showing a cognitive effort in the level of doing mathematics.

Networked Analysis of Students' Visualization Behavior

We have presented in sections "Analysis of Students' Use of Visualization Abilities" and "Analysis of the Cognitive Complexity of Students' Visual Reasoning" two parallel analyses of three pairs of students' solutions to a visualization task, taking into consideration the use of visualization abilities and the levels of cognitive demand posed to students by the task and their strategies of solution. Those analyses focus on different aspects of students' solutions, so some relationships and concordance between them should be expected, although it has never been explored. In this section, we make an interwoven analysis trying to relate both points of view. Our sample is only three pairs of students, so we do not pretend to get any generalizable conclusion, and we have found that students' behaviors to be quite different. A3-B3 were the most efficient and successful solvers of the buildings task, since they needed the least number of phases (12) to solve it and found the two solutions. A1-B1 were also very efficient solving the task, needing a few more phases (16) than A3-B3 but they only found one solution. In contrast, A2-B2 had more difficulties, needed almost twice as many phases (23) as A3-B3 and they required more help.

The aim of this networked analysis is to explore a possible relation between the use of the visualization abilities and students' levels of higher cognitive demand (procedures with connections and doing mathematics). To do it, we have focused on the phases of placement of buildings and correction of errors, since these are the only phases in which the levels of cognitive demand can be evaluated. Table 14.6 presents a synthesis of the quantitative data (absolute and percentage) describing the solutions. For instance, the pair A1-B1 used 55 times the ability of positions in space in the phases of placement, which represents 38.2% of the 144 times they used any visualization ability throughout their solution. And A1-B1 showed high levels of cognitive demand in 28 out of the 55 times they used the ability of positions in space in the phases of placement (50.9% of them).

The ability of positions in space is the only one that has been consistently and extensively used by students throughout all phases during their solutions; this result is reasonable given the characteristics of the buildings tasks. The ability of visual

Table 14.6 Use of visualization abilities related to levels of cognitive demand

		Number of occurrences of each ability			
		Placement phases		Correction phases	
Abilities	Students	Total[a]	With high cognitive demand[b]	Total[a]	With high cognitive demand[b]
Positions in space	A1-B1	55 (38.2%)	28 (50.9%)	7 (4.9%)	0 (0.0%)
	A2-B2	37 (25.5%)	21 (56.8%)	13 (9.0%)	12 (92.3%)
	A3-B3	41 (51.9%)	29 (70.7%)	0 (0.0%)	0 (0.0%)
Visual discrimination	A1-B1	5 (3.5%)	2 (40.0%)	6 (4.2%)	6 (100%)
	A2-B2	6 (4.1%)	4 (66.7%)	4 (2.8%)	2 (50.0%)
	A3-B3	2 (2.5%)	0 (0.0%)	0 (0.0%)	0 (0.0%)

[a]Percentages with respect to the total number of uses of visualization abilities along the solution
[b]Percentages with respect to the number of occurrences of the ability in the phases

discrimination is the other ability having a significant presence in students' outcomes, but less consistently than the ability of positions in space. With respect to the use of the higher levels of cognitive demand, the data from our experiments do not show any clear trend or relation between the visualization abilities used by students and the higher levels of cognitive demand required from them to solve the buildings task. We could only raise a relation between the use of higher levels of cognitive demand and the ability of positions in space.

Conclusions

Mathematically gifted students need to be posed challenging problems and tasks that help them progress in the learning of mathematical content and the development of their mathematical capabilities, in particular their competence with mathematical visualization. In this chapter, we have presented a kind of challenging task, the buildings task, that is useful to improve students' visualization abilities while demanding from them a high level of cognitive activity.

The objective of this research was to analyze students' solutions to a buildings task (1) to identify their use of visualization abilities and (2) to evaluate the level of cognitive demand used by them to solve the task successfully. We have adopted a networking position to combine both analyses to gain a deeper knowledge of students' activity. Each pair of students solved the task in a different way, which allowed us to get some conclusions that, due to the small sample, we do not claim are generalizable.

A buildings task may be designed to have several solutions. To find all them, students have to work collaboratively, communicate efficiently, use visualization abilities, and reach the highest level of cognitive demand. When our students did not succeed in sharing and combining operatively information, they were unable to correctly place some buildings. The use of the visualization abilities was more necessary when the solution to the task required from students higher levels of cognitive demand.

All pairs of students progressed in learning to work collaboratively and using more demanding reasoning, to manage their visualization abilities, and to improve their communication with each other. Our analysis shows that m-gifted students can understand and learn quickly new, more efficient strategies of solution.

The research presented in this chapter is part of the R+D+I projects EDU2015-69731-R (Spanish Government/MinEco and ERDF) and GVPROMETEO2016-143 (Valencian Government).

References

Arcavi, A. (2003). The role of visual representations in the learning of mathematics. *Educational Studies in Mathematics, 52*(3), 215–241.

Battista, M. (2007). The development of geometric and spatial thinking. In F. Lester (Ed.), *Second handbook of research on mathematics teaching and learning* (Vol. 2, pp. 843–908). Charlotte, NC: Information Age Publishing.

Benedicto, C., Acosta, C., Gutiérrez, A., Hoyos, E., & Jaime, A. (2015). Improvement of gifted students' visualization abilities in a 3d computer environment. In N. Amado & S. Carreira (Eds.), *Proceedings of the 12th International Conference on Technology in Mathematics Teaching* (pp. 363–370). Faro, Portugal: University of Algarve.

Benedicto, C., Gutiérrez, A., & Jaime, A. (2017). When the theoretical model does not fit our data: a process of adaptation of the cognitive demand model. In T. Dooley & G. Gueudet (Eds.), *Proceedings of the 10th Congress of the European Society for Research in Mathematics Education (CERME10)* (pp. 2791–2798). Dublin, Ireland: ERME.

Bikner-Ahsbahs, A., & Prediger, S. (2010). Networking of theories - An approach for exploiting the diversity of theoretical approaches. In B. Sriraman & L. English (Eds.), *Theories of mathematics education* (pp. 483–506). Dordrecht, The Netherlands: Springer.

Bikner-Ahsbahs, A., & Prediger, S. (Eds.). (2014). *Networking of theories as a research practice in mathematics education*. Dordrecht, The Netherlands: Springer.

Bishop, A. J. (1980). Spatial Abilities and Mathematics Education: A Review. *Educational Studies in Mathematics, 11*(3), 257–269.

Bishop, A. J. (1983). Spatial abilities and mathematical thinking. In M. Zweng et al. (Eds.), *Proceedings of the 4th ICME* (pp. 176–178). Boston, MA: Birkhäuser.

Clements, D. H., & Battista, M. T. (1992). Geometry and spatial reasoning. In D. A. Grouws (Ed.), *Handbook of research on mathematics teaching and learning* (pp. 420–464). New York: Macmillan.

Davis, G. A., Rimm, S. B., & Siegle, D. (2014). *Education of the gifted and talented*. Boston, MA: Pearson.

Del Grande, J. J. (1990). Spatial sense. *Arithmetic Teacher, 37*(6), 14–20.

Felmer, P., Pehkonen, E., & Kilpatrick, J. (2016). *Posing and solving mathematical problems. Advances and new perspectives*. New York: Springer.

Gutiérrez, A. (1996). Visualization in 3-dimensional geometry: In search of a framework. In L. Puig & A. Gutiérrez (Eds.), *Proceedings of the 20th PME Conference* (Vol. 1, pp. 3–19). Valencia, Spain: Universidad de Valencia.

Lean, G., & Clements, M. A. (1981). Spatial ability, visual imagery, and mathematical performance. *Educational Studies in Mathematics, 12*(3), 267–299.

National Council of Teachers of Mathematics (NCTM). (2014). *Principles to actions. Ensuring mathematical success for all*. Reston, VA: NCTM.

Ozdemir, S., Ayvaz-Reis, Z., & Karadag, Z. (2012). Exploring elementary mathematics pre-service teachers' perception to use multiple representations in problem solving. In Z. Karadag & Y. Devecioglu-Kaymakci (Eds.), *Proceedings of the 1st International Dynamic, Explorative, and Active Learning Conference*. Bayburt, Turkey: Bayburt University.

Presmeg, N. (2006). Research on visualization in learning and teaching mathematics. In A. Gutiérrez & P. Boero (Eds.), *Handbook of research on the psychology of mathematics education* (pp. 205–235). Rotterdam, The Netherlands: Sense Publishers.

Presmeg, N. C. (1986). Visualization in high school mathematics. *For the Learning of Mathematics, 6*(3), 42–46.

Ramírez, R. (2012). *Habilidades de visualización de los alumnos con talento matemático* [visualization abilities of mathematically talented students] (unpublished PhD). Granada, Spain: Universidad de Granada. Retrieved from http://fqm193.ugr.es/produccion-cientifica/tesis/ver_detalles/7461/.

Rivera, F. D. (2011). *Towards a visually-oriented school mathematics curriculum*. New York: Springer.

Ryu, H., Chong, Y., & Song, S. (2007). Mathematically gifted students' spatial visualization ability of solid figures. In J. H. Wo, H. C. Lew, K. S. Park, & D. Y. Seo (Eds.), *Proceedings of the 31st Conference of the International Group for the Psychology of Mathematics Education* (Vol. 4, pp. 137–144). Seoul, South Korea: PME.

Schoenfeld, A. H. (2014). What makes for powerful classrooms, and how can we support teachers in creating them? A story of research and practice, productively intertwined. *Educational Researcher, 43*(8), 404–412.

Smith, M., & Stein, M. (1998). Selecting and creating mathematical tasks: from research to practice. *Mathematics Teaching in the Middle School, 3*(5), 344–350.

Stein, M., Smith, M., Henningsen, M., & Silver, E. (2009). *Implementing standards-based mathematics instruction: a casebook for professional development*. New York: Teachers College Press.

Van Garderen, D. (2006). Spatial visualization, visual imagery, and mathematical problem solving of students with varying abilities. *Journal of Learning Disabilities, 39*(6), 496–506.

Part IV
Commentaries

Chapter 15
Part II Commentary 1: Mathematics Educators' Perspectives on Spatial Visualization and Mathematical Reasoning

Beth M. Casey

Researchers generally agree that spatial problem solving skills involve the ability to generate mental images as a strategy for solving mathematics problems, often in conjunction with maintaining and manipulating those images. Further, translating these mental images into physical representations/graphics through drawings or diagrams is advantageous for many mathematics problems. The chapters by Sinclair, Moss, Hawes, and Stephenson (this volume) and Lowrie and Logan (this volume), point out Polya's (1965) recommendation to "draw a diagram" as one of the first steps in understanding a mathematics problem. Students who use this heuristic may be more successful on a wide range of problems across mathematics content areas. Ho and Lowrie (2014) report that Singapore students are taught to use the model method, which is a visual problem-solving heuristic prevalently used in Singapore classrooms, and Murata (2008) reports on the use of the tape diagram approach as visual-spatial tool used to solve many types of mathematics problems in Japanese classrooms—both countries that score highly on standardized testing.

Generating Diagrams for Solving Mathematics Word Problems

One beneficial effect of applying spatial reasoning to mathematics problems is the ability to utilize spatial imagery to solve problems under circumstances that do not obviously require their use for problem solving. Thus, this spatial representation approach may be particularly beneficial when no graphic is available for children to depend upon. One clear example of this is the application of spatial skills to mathematics word problems. In recent reviews, researchers have investigated the benefits of using spatially based schematic representations to solve word problems and have found that it can be quite effective (see review by Kingsdorf & Krawec, 2016;

B. M. Casey (✉)
Lynch School of Education, Boston College, Boston, MA, USA
e-mail: caseyb@bc.edu

K. S. S. Mix, M. T. Battista (eds.), *Visualizing Mathematics*, Research in Mathematics Education, https://doi.org/10.1007/978-3-319-98767-5_15

Jitendra, Nelson, Pulles, Kiss, & Houseworth, 2016). Typically, this approach involves the use of diagrams to represent the mathematics problem. It also often incorporates the representation of connections between the different problem parts in order to link the different steps in the problem-solving process (Gonsalves & Krawec, 2014). Hegarty, Mayer, and Monk (1995) proposed that there is a spatial component to word problems when students construct a mental-model of a problem and plan the solution based on that model. Hegarty and Kozhevnikov (1999) found evidence that sixth grade students who used schematic spatial representations such as diagrams had better mathematical problem solving success than students using other approaches. Use of schematic representations was also shown to significantly correlate with spatial skills.

Boonen and associates (Boonen, van der Schoot, van Wesel, DeVries, & Jolles, 2013) found that a substantial proportion of the association between spatial skills and numerical word problem solving (21%) was explained through the indirect effects of strategies involving visual-schematic representations. For numerically-based mathematics reasoning problems, spatial reasoning may facilitate the ability to translate complex verbal and number problems into an appropriate spatial array or diagram representing the problem solution. Thus, research suggests that spatial problem solving can be a useful tool when solving mathematics problems unrelated to either geometry and measurement, and even under conditions in which no decoding of graphics is required.

As the research on use of spatial representations progresses, Lowrie and Logan (this volume) point out an important consideration: Do educators introduce spatial representations and problem-solving approaches through heuristic models such as the Singapore approach—involving teaching practices where spatial heuristics are explicitly taught and practiced involving "draw a diagram"—or do they use an approach in which students are exposed to a diverse variety of mathematics representations and are encouraged to use their own personal strategies to solve these tasks, as is more typical of Western educational systems? Lowrie and colleagues (Lowrie, Logan, & Ramful, 2016) compared sixth grade students from Singapore to students from Australia in terms of their use of spatial and non-spatial problem solving approaches to numerical word problems. These researchers found that "… the Singapore students are able to use these foundational approaches and skills in quite flexible ways. Consequently, the restricted development of problem solving strategies actually enhances their capacity to solve unfamiliar tasks…It may be the case that too much variety in strategy development and not enough explicit teaching does not equip Australian students with a sufficient skill set when faced with unfamiliar or challenging tasks. The demands of the Australian school system place great importance on an inquiry approach. However, to maximize student learning potential, intentional teaching still needs to take place, as is the case in Singapore." (p. 107). These instructional issues will have to be addressed as we move to greater integrations of spatial approaches to solving mathematics problems across the curriculum.

Early Introduction of Spatial Reasoning Approaches to Arithmetic Problems

We have shown strong longitudinal support for a spatial-numerical association in a recent study in which we examined spatial skills in first grade girls as predictors of two types of mathematics reasoning skills 4 years later in fifth grade (Casey et al., 2015). The results showed that spatial skills, assessed as early as first grade functioned as key long-term predictors for numeric/algebraic mathematics reasoning skills in fifth grade, as well as for geometry/measurement mathematics-reasoning skills (even when controlling for early verbal skills and arithmetic accuracy). In a follow-up study on the same students, we found a strong pathway leading from spatial skills at the outset of first grade to use of advanced decomposition strategies by the end of first grade, and then leading to higher level numeric and algebraic mathematics reasoning skills in fifth grade (Casey, Lombardi, Pollock, Fineman, & Pezaris, 2017). Though correlational, this pattern of associations suggests the possibility that levels of spatial reasoning may impact arithmetic strategy choices at the outset of arithmetic learning, which in turn may have long-term effects on later mathematics reasoning.

A recent study by Frick (2018) further reinforces the importance of emphasizing spatial approaches to arithmetic instruction starting at early ages. Using structural equation modeling, Frick found that mental rotation and spatial scaling in kindergarten showed their strongest relation to the component of the mathematics test tapping arithmetic operations in second grade, whereas mental transformations and cross-sectioning were more strongly related to geometry and magnitude estimation. Thus, a future goal of spatial-mathematics research should be to examine in greater depths how different types of spatial skills impact different types of mathematics skills when applying spatial reasoning strategies to mathematics problems.

Spatial Skills as Predictors of Geometry/Measurement Versus Numerical/Algebraic Mathematics

To further argue for greater emphasis on the importance of spatial skills extending to a wider range of mathematics content, I would like to present data from a recent study that made it possible to directly examine spatial skills—both as predictors of geometry and measurement reasoning problems that involved the use of graphics—and as predictors of numerical/algebraic problems in which no graphics were provided. We examined spatial skills, consisting of the Vandenberg Mental Rotation task (Peters et al., 1995) and the Water Levels Task (Piaget & Inhelder, 1956) at the beginning of seventh grade as predictors of two types of mathematics reasoning skills at the end of seventh grade. The mathematics assessment tools were designed to maximize the number of geometry and measurement items that addressed spatial mathematics reasoning and the number of numerical and algebra items that addressed analytical reasoning. We conducted regression analyses to determine the extent to which spatial skills predicted these two types of mathematics items. The specific goal was to examine the strength of these associations on the two types of mathematics problems—one type that would seem to maximize the association with spatial skills, while the other type might be expected to be less likely associated with spatial skills.

For the geometry/measurement items with graphics, the standardized coefficient for the composite spatial measure was 0.53. For the numeric/algebra problems with no graphics, the standardized coefficient for the composite spatial measure was 0.51. Thus, the spatial skills-mathematics associations for both types of mathematics problems were very similar—and substantial. Next, we controlled for students' mathematics fact fluency and verbal skills, because these skills might account for substantially greater variance in predicting numeric/algebra performance than for geometry/measurement. When these additional measures were included in the regression analyses, spatial skill still significantly contributed to both the geometry/measurement items (standardized coefficient = 0.42) and the numerical/algebraic items (standardized coefficient = 0.33). Although the standardized coefficient for spatial skills as a predictor dropped more for the numeric/algebraic items than for the geometry/measurement, the association between spatial skills and numeric-based mathematics performance were still the strongest predictors in the regression analyses for both types of mathematics items.

In conclusion, in my commentary I have made the argument that more research should be done by mathematics educators to identify specific strategies for teaching students how to approach a much wider range of mathematics content areas utilizing their spatial reasoning processes. Findings from many research studies suggest a greater potential role for applying spatial problem solving approaches across mathematics content than is typically applied in practice within schools in US and other Western countries (Mix & Cheng, 2012). Now our task is to conduct intervention research in order to figure out explicit ways of helping teachers incorporate spatial thinking successfully throughout these mathematics content areas.

References

Boonen, A. J. H., van der Schoot, M., van Wesel, F., DeVries, M. H., & Jolles, J. (2013). What underlies successful word problem solving? A path analysis in sixth grade students. *Contemporary Educational Psychology, 38*, 271–279. https://doi.org/10.1016/j.cedpsych.2013.05.001

Casey, B. M., Lombardi, C., Pollock, A., Fineman, B., & Pezaris, E. (2017). Girls' spatial skills and arithmetic strategies in first grade as predictors of fifth grade analytical mathematics reasoning. *Journal of Cognition and Development, 18*(5), 530–555. https://doi.org/10.1080/15248372.2017.1363044

Casey, B. M., Pezaris, E., Fineman, B., Pollock, A., Demers, L., & Dearing, E. (2015). A longitudinal analysis of early spatial skills compared to arithmetic and verbal skills as predictors of fifth-grade girls' mathematics reasoning. *Learning and Individual Differences, 40*, 90–100. https://doi.org/10.1016/j.lindif.2015.03.028

Frick, A. (2018). Spatial transformation abilities and their relation to later mathematics. performance. *Psychological Research*. Published online: April 10, 2018. https://doi.org/10.1007/s00426-018-1008-5

Gonsalves, N., & Krawec, J. (2014). Using number lines to solve mathematics word problems: A strategy for students with learning disabilities. *Learning Disabilities Research and Practice, 29*(4), 160–170. https://doi.org/10.1111/ldrp.12042

Hegarty, M., & Kozhevnikov, M. (1999). Types of visual-spatial representations and mathematical problem solving. *Journal of Educational Psychology, 91*(4), 684–689. https://doi.org/10.1037/0022-0663.91.4.684

Hegarty, M., Mayer, R. E., & Monk, C. A. (1995). Comprehension of arithmetic word problems: A comparison of successful and unsuccessful problem solvers. *Journal of Educational Psychology, 87*(1), 18–32. https://doi.org/10.1037/0022/0663.87.1.18

Ho, S. Y., & Lowrie, T. (2014). The model method: Students' performance and its effectiveness. *The Journal of Mathematical Behavior, 35*, 87–100. https://doi.org/10.1016/j.jmathb.2014.06.002

Jitendra, A. K., Nelson, G., Pulles, S. M., Kiss, A. J., & Houseworth, J. (2016). Is mathematical representation of problems an evidence-based strategy for students with mathematical difficulties? *Exceptional Children, 83*(1), 8–25. https://doi.org/10.1177/0014402915625062

Kingsdorf, S., & Krawec, J. (2016). A broad look at the literature on mathematics word problem-solving interventions for third graders. *Cogent Education, 3*(1), 1–12. https://doi.org/10.1080/2331186X.2015.1135770

Lowrie, T., Logan, T., & Ramful, A. (2016). Cross cultural comparison of grade 6 students' performance and strategy use on graphic and non-graphic tasks. *Learning and Individual Differences, 52*, 97–108. https://doi.org/10.1016/j.lindif.2016.10.005

Lowrie, T., & Logan, T. (this volume). The interaction between spatial reasoning constructs and mathematics understandings in elementary classrooms. In K. S. Mix & M. T. Battista (Eds.), *Visualizing mathematics: The role of spatial reasoning in mathematical thought*. New York: Springer.

Mix, K. S., & Cheng, U. L. (2012). The relation between space and mathematics: Developmental and educational implications. In J. B. Benson (Ed.), *Advances in child development and behavior* (Vol. 42, pp. 199–243). Waltham, MA: Academic.

Murata, A. (2008). Mathematics teaching and learning as a mediating process: The case of tape diagrams. *Mathematical Thinking and Learning, 10*(4), 374406. https://doi.org/10.1080/10986060802291642

Peters, M., Laeng, B., Latham, K., Jackson, M., Zaiyouna, R., & Richardson, C. (1995). A redrawn Vandenberg and Kuse mental rotations test: Different versions and factors that affect performance. *Brain and Cognition, 28*(1), 39–58. https://doi.org/10.1006/brcg.1995.1032

Piaget, J., & Inhelder, B. (1956). *The child's conception of space*. London: Routledge & Kegan Paul.

Polya, G. (1965). *Mathematical discovery; on understanding and teaching problem solving*. Hoboken, NJ: Wiley.

Sinclair, N., Moss, J., Hawes, Z., & Stephenson, C. (this volume). Learning with and from drawing in early years geometry. In K. S. Mix & M. T. Battista (Eds.), *Visualizing mathematics: The role of spatial reasoning in mathematical thought*. New York: Springer.

Chapter 16
Part II Commentary 2: Disparities and Opportunities: Plotting a New Course for Research on Spatial Visualization and Mathematics

Kelly S. S. Mix and Susan C. Levine

Although the chapters contained in this volume focus on the singular topic of spatial visualization as it relates to mathematics, they span two distinct fields of study with different literatures and different scholarly approaches. In many ways, despite their common goals, the two sets of chapters seem worlds apart. We are reminded of Susan Carey's classic developmental psychology book, *Conceptual Change in Childhood* (1985), that discussed incommensurate ideas in science and the ways children reconcile structurally disparate conceptual systems as they grow and learn. The gist was that when one conceptual structure lacks isomorphism with another conceptual structure, it is a significant challenge. We believe the fields represented in this volume face a similar challenge. Yet, these disciplinary asymmetries can also define and stimulate fruitful new research questions, as the advances made in one discipline raise new questions for the other. In this commentary, we aim to identify such asymmetries and consider what new research directions they suggest. We organize our comments around three major questions that cut across research from both fields:

1. What is spatial visualization?
2. How does spatial visualization relate to mathematics?
3. How are these relations reflected in development and learning?

K. S. S. Mix (✉)
Department of Human Development and Quantitative Methodology, University of Maryland, College Park, MD, USA
e-mail: kmix@umd.edu

S. C. Levine
Departments of Psychology, and Comparative Human Development and Committee on Education, University of Chicago, Chicago, IL, USA

K. S. S. Mix, M. T. Battista (eds.), *Visualizing Mathematics*, Research in Mathematics Education, https://doi.org/10.1007/978-3-319-98767-5_16

What Is Spatial Visualization?

All of the chapters grappled with the nature of spatial visualization, but the two fields approached this issue somewhat differently. Some of the education chapters seemed willing to commit to a specific underlying process model as a starting assumption. For example, Battista invoked the construct of mental models (i.e., Johnson Laird, 1980), which are schematic representations arranged in space but without necessarily having a visual component. Gutiérrez et al. adopted Gutiérrez's (1996) mental imagery framework, which rests on the assumption that spatial processing uses visual images. In both cases, the definition of spatial visualization was presented early in the chapter as a way of contextualizing the research to follow.

In contrast, the psychology chapters tended to see the nature of spatial visualization as an open question. Although spatial visualization has certainly been defined in the psychology literature, the definitions are based on observable behaviors that seem to require similar processing, rather than a commitment to any particular representational format. For example, a common definition of spatial skill is the ability to mentally manipulate objects. Such manipulation may involve mental models or visual imagery, but need not. Indeed, pinning down the underlying structure of spatial skill has been a preoccupation of psychologists for decades (see Mix & Cheng, 2012, for a review) and this focus is clearly reflected in the psychology chapters included here. For example, Young, Levine, and Mix (this volume) focus on ways to model the underlying structure of spatial thought and how to interpret the results of different modeling approaches.

Because psychologists continue to work toward a generally accepted process model for spatial thought, it may be premature for related literatures to make strong claims regarding the underlying representational format of spatial visualization. For example, it may seem uncontroversial to claim that spatial representations are visual images (particularly because we call it "spatial visualization"), but there have been challenges to this view in the psychology literature. Research has shown that although spatial development is delayed and more error-prone in blind versus sighted children, blind children can perform tasks that require spatial visualization (Bigelow, 1996; Landau, Spelke, & Gleitman, 1984). A long debate in the psychological literature also centered on whether ordered syllogisms are solved via mental imagery or linguistic information, admitting the possibility that even tasks that seem likely to require spatial visualization may not (e.g., Clark, 1973; Huttenlocher & Higgins, 1971; Trabasso & Riley, 1975; Sternberg, 1980). Finally, there has been discussion about the level of visual detail needed for mental models to be useful. Research suggests that sparse, schematic spatial representations are better for mathematics problem solving than detailed, pictorial images (e.g., Hegarty & Koszhenikov, 1999). As Huttenlocher, Jordan and Levine (1994) pointed out, mental models may resemble physical models in some ways, but these representations need only preserve relevant critical features for problem solving situations when used for mathematics. When solving a mathematics problem involving number of pieces of fruit, for example, it is not necessary to accurately represent the color of the fruit, but only the number of pieces of fruit. If critical features are not preserved, and irrelevant features are, an erroneous answer may be obtained.

One new direction suggested by this contrast may be to use the paradigms offered by the mathematics education studies to more precisely determine the nature of the underlying representations. For example, the diagrams and visual supports used by Gutiérrez et al. or Herbst et al. could be manipulated to be more or less schematic. It might also be possible to see whether students' self-generated physical supports change over time, perhaps becoming less detailed and visual as they master a particular task. Another new direction would be to reframe some of the mathematics education work using current psychological theory. The mental models literature of the 1980s and 1990s provided part of the foundation for what has become the literature related to embodied cognition, or the idea that abstract thought is grounded in bodily movement, perception, and action (Barsalou, 2008; Clark, 1998, 2008; Glenberg, 2008, 2010; Lakoff & Nunez, 2000; Novack, Congdon, Hemani-Lopez, & Goldin-Meadow, 2014; Schöner & Spencer, 2015; Thelen & Smith, 1996). This literature may offer a stronger mechanistic explanation for phenomena such as those identified by Battista, and it would be interesting to see what new questions arise from a thorough meshing of the two. Sinclair et al.'s chapter provides a nice example of this (i.e., integrating mathematics education with theories of embodied cognition). Lowrie and Logan also provided a detailed and rich integration of current psychological theory regarding the nature of spatial thought and children's understanding of geometry. Perhaps extending this approach to other areas of mathematics content would be fruitful.

How Does Spatial Visualization Relate to Mathematics?

The central thesis of this book is that spatial visualization relates to mathematics, so all the chapters work from this premise. However, there are many potential characterizations of this relation and the chapters differed along several lines.

One salient distinction has to do with how inherent spatial processing is to mathematical thought. At one extreme, space can be seen as the representational medium for abstract thought and is thus inherently engaged whenever people perform mathematical tasks. This perspective is exemplified, to some extent, in all of the psychology chapters. For example, Jirout and Newcombe point to evidence that the meanings of numbers are represented as relative quantities using spatial scaling. Congdon et al. emphasize the role of measurement units in conceptualizing quantity across a range of tasks from counting whole objects to ordering fractions. On this view, individual differences in mathematics performance could be construed as individual differences in spatial processing. This strong view was not as evident in the mathematics education chapters. This is particularly surprising given that all focused on geometry—a mathematics topic for which, if any, spatial reasoning is arguably most inherent. Yet, only Battista's chapter made a strong argument along these lines. It was also interesting that none of the mathematics education chapters focused on the inherent nature of spatial processing in numerical thinking more broadly, though this was mentioned by Lowrie and Logan.

An alternative to this strong view is one in which spatial skill is not inherent to mathematical thought, but rather is an optional aid that may be recruited to ground concepts or support reasoning. This seems to be the view of Herbst et al. in that their spatial visualization activity is constructed to ground geometric problem solving in a real world context. Similarly, the explicit aim of Gutiérrez et al. is to develop spatial skills that may be recruited when children are reading diagrams used in mathematics, and Sinclair et al. demonstrate the benefit of figure drawing in children's understanding of geometry. In these studies, the idea seems to be that practice in a spatial skill will transfer back to the more spatial aspects of mathematics in a supportive way.

An interesting question following from this perspective is whether children spontaneously recognize how spatial reasoning can help them in mathematics and the extent to which teacher direction is needed to utilize spatial reasoning. The education chapters allude to the mapping between mental representations, real world, and symbols (including vocabulary) without taking advantage of the recent developments in structure mapping theory, relational learning, or perception-action/embodied cognition. These literatures may help in the design of instructional approaches that avoid the problems of lack of transfer or encapsulated learning. Indeed, by bringing the education and psychological literatures into closer alignment, it is likely that real advances in understanding the relation of spatial and mathematical learning could be made, with beneficial consequences for instructional approaches and children's learning outcomes.

Another contrast was in the level of multidimensionality acknowledged for either spatial skill, mathematics, or both. Nuances in spatial representation are very salient to psychologists, whose research centers more squarely on underlying cognitive processes. Attention to the multidimensionality of spatial thought permeated these chapters. There was also attention paid to the multidimensionality of mathematical thought but this was relatively impoverished and rough compared to what could be said about the multidimensionality of spatial thinking.

In contrast, the mathematics education chapters were striking in their fidelity to the mathematics underlying their phenomena of study, as well as the use of conceptual distinctions that arise purely from consideration of mathematics itself. Through careful analysis of the eventual learning outcomes and potential conceptual pitfalls along the way, these authors identified specific mathematical constructs or misconceptions that might benefit from spatial supports. This orientation is beautifully illustrated in Battista's chapter, in which he examined children's error patterns during geometry proofs. Interestingly, although Battista acknowledged the potential multidimensionality of spatial concepts as well (particularly in reference to Newcombe & Shipley's 2015 framework), this multidimensionality did not play a major role in the mechanism of change. The role of spatial grounding was clearly acknowledged, but the specific nature of that grounding seemed less important.

In terms of next steps, there seems to be great potential in a synthesis of these two approaches—that is, a rich, detailed account of underlying cognition married with a rich, detailed account of underlying mathematics. Moves in that direction are likely to reveal an entire host of new research questions and insights that emerge from drilling down to a deeper, more specific level of shared processing and bidirectional influence. Each of the mathematics education chapters offers a slice of mathematical development that is already fleshed out at this level. One approach may be to revisit these accounts with an eye toward achieving equally nuanced explanations of the same specific phenomena based on cognitive processing, and testing these explanations empirically.

How Are These Relations Reflected in Development and Learning?

A final question all of the chapters addressed was how relations between spatial skill and mathematics play out as children learn and change over developmental time. The chapters presented an interesting contrast between those describing the stages of development and those identifying the mechanisms that propel children through these stages. Historically, psychologists such as Jerome Bruner and Jean Piaget attempted to achieve both aims—to describe developmental stages and identify broad mechanisms of change. In the present volume, the two aims seemed to separate along the disciplinary lines.

The mathematics education chapters tended to offer detailed, carefully articulated stage theories. For example, though less rigid than a classical stage theory, Battista's learning trajectories seem very much like a Piagetian description of development, with movement from holistic to decomposed concepts, and from concrete to logic-based reasoning. In terms of learning mechanisms, the education chapters were more focused on the potential benefits of various spatial activities. For example, Gutiérrez et al. described the impact of training on a perspective-taking task. Herbst et al. sought to improve geometric reasoning via practice modeling three-dimensional space. Sinclair et al. discussed the role of drawing in understanding geometry. These are creative instructional approaches that show promise; however, they beg a host of questions related to the underlying processes. What process model can explain why they work? What is the active ingredient in these approaches that propels change?

For the psychologists' part, there was a strong emphasis on the mechanisms of change and less attention paid to typical developmental or learning trajectories. For example, Cipora et al. point to the correlations between spatial skill and mathematics performance, and raise the question of whether variation in spatial skill is driving the correlation, or perhaps the reverse (i.e., variation is arithmetic understanding leads to more sophisticated and accurate spatial representations). Increasing understanding of the mechanisms that drive the strong relation between spatial and mathematical thinking is an important goal for successfully incorporating spatially rich instructional strategies into the mathematics curriculum.

This is another dimension for which circling back and integrating the two literatures may be beneficial. For example, future research might focus on questions such as whether the mechanisms of change identified by psychologists can be manipulated experimentally so as to yield the various stages identified by the mathematics education researchers. Alternatively, a review of the detailed shifts identified by the mathematics education chapters may suggest new or revised mechanisms of change that have not been recognized previously. A bidirectional analysis and program of research such as this has the potential to yield exceptionally strong instructional approaches that may be missed by taking only one approach or the other.

In summary, the chapters in this volume make exciting strides toward understanding the relations between spatial skill and mathematics, but often do so in very different ways or from perspectives that are not easily aligned. By integrating these differing orientations, there is potential to increase our understanding and design more effective instructional interventions.

References

Barsalou, L. W. (2008). Grounded cognition. Annu. Rev. Psychol., 59, 617–645.

Bigelow, A. E. (1996). Blind and sighted children's spatial knowledge of their home environments. *International Journal of Behavioral Development, 19*(4), 797–816.

Carey, S. (1985). *Conceptual change in childhood.* Cambridge, MA: Bradford Books, MIT Press.

Clark, A. (1998). *Being there: Putting brain, body, and world together again.* Cambridge, MA: MIT press.

Clark, A. (2008). *Supersizing the mind: Embodiment, action, and cognitive extension.* New York: OUP.

Clark, H. H. (1973). Space, time, semantics and the child. In T. E. Moore (Ed.), *Cognitive development and the acquisition of language* (pp. 27–63). New York: Academic Press.

Glenberg, A. M. (2008). Embodiment for education. In P. Calvo & T. Gomila (Eds.), *Handbook of cognitive science: An embodied approach* (pp. 355–372). Amsterdam: Elsevier.

Glenberg, A. M. (2010). Embodiment as a unifying perspective for psychology. *Wiley Interdisciplinary Reviews: Cognitive Science, 1*(4), 586–596.

Gutiérrez, A. (1996). Visualization in 3-dimensional geometry: In search of a framework. In L. Puig & A. Gutiérrez (Eds.), *Proceedings of the 20th PME Conference* (Vol. 1, pp. 3–19). Valencia, Spain: Universidad de Valencia.

Hegarty, M., & Koszhenikov, M. (1999). Types of visuo-spatial representations and mathematics problem solving. *Journal of Educational Psychology, 91*(4), 684–689.

Huttenlocher, J., Higgins, E. T., & Clark, H. H. (1971). Adjectives, comparatives, and syllogisms. *Psychological Review, 78*(6), 487–504.

Huttenlocher, J., Jordan, N. C., & Levine, S. C. (1994). A mental model for early arithmetic. *Journal of Experimental Psychology: General, 123*(3), 284–296.

Johnson-Laird, P. N. (1980). Mental models in cognitive science. *Cognitive Science, 4*(1), 71–115.

Landau, B., Spelke, E., & Gleitman, H. (1984). Spatial knowledge in a young blind child. *Cognition, 16*, 225–260.

Lakoff, G., & Núñez, R. E. (2000). Where mathematics comes from: How the embodied mind brings mathematics into being. New York: Basic Books.

Mix, K. S. & Cheng, Y. L. (2012). The relation between space and math: Developmental and educational implications. In J. B. Benson (Ed.) Advances in Child Development and Behavior, (Vol. 42, pp. 197–243), New York: Elsevier.

Novack, M. A., Congdon, E. L., Hemani-Lopez, N., & Goldin-Meadow, S. (2014). From action to abstraction: Using the hands to learn mathematics. *Psychological Science, 25*(4), 903–910. https://doi.org/10.1177/0956797613518351

Newcombe, N. S., & Shipley, T. F. (2015). Thinking about spatial thinking: New typology, new assessments. In Studying visual and spatial reasoning for design creativity (pp. 179–192). Dordrecht: Springer.

Schöner, G., & Spencer, J. (2015). *Dynamic thinking: A primer on dynamic field theory*. New York: Oxford University Press.

Sternberg, R. J. (1980). Representation and process in linear syllogistic reasoning. *Journal of Experimental Psychology: General, 109*(2), 119–159.

Thelen, E., & Smith, L. B. (1996). *A dynamic systems approach to the development of cognition and action*. Cambridge, MA: MIT press.

Trabasso, T., & Riley, C. A. (1975). On the construction and use of representations involving linear order. In Robert L. Solso (Ed.), *Information Processing and Cognition: The Loyola Symposium* (pp. 381–410). Mahwah, NJ: Lawrence Erlbaum.

Young, C. J., Levine, S. C., & Mix, K. S. (this volume). What processes underlie the relation between spatial skill and mathematics? In I. K. S. Mix & M. T. Battista (Eds.), *Visualizing mathematics: The role of spatial reasoning in mathematical thought*. New York: Springer.

Chapter 17
Part II Commentary 3: Linking Spatial and Mathematical Thinking: The Search for Mechanism

Nora S. Newcombe

A multitude of factors affect mathematics learning: motivation, anxiety, gender stereotypes, working memory, teacher practices, teacher knowledge, and many more. Recently, spatial thinking has emerged as one of these factors. This conclusion was based initially on concurrent correlations but is now also supported by longitudinal studies with controls for other determinants, such as verbal intelligence and executive function (e.g., Frick, 2018; Verdine, Golinkoff, Hirsh-Pasek, & Newcombe, 2017; Zhang & Lin, 2017). Even stronger evidence is beginning to emerge from randomized control experiments that can evaluate true causal relations, i.e., whether interventions to support spatial thinking have downstream effects on mathematical achievement in comparison with an appropriate control group. Not all randomized-control studies show positive results, but enough of them do to encourage optimism (see review in Newcombe, 2017).

As the evidence base for a linkage strengthens, the crucial next step is to explore the mechanisms at work. Understanding mechanisms allows us to refine our interventions and understand why and when they work or fail. In this enterprise, it will be crucial to ensure cooperation between researchers in mathematical cognition and in mathematics education. Thus, this book is the beginning of a welcome exchange. In reading the chapters written by mathematics education researchers, I was struck by three points that seemed to warrant clarification from the cognitive psychology side of the exchange.

N. S. Newcombe (✉)
Department of Psychology, Temple University, Philadelphia, PA, USA
e-mail: newcombe@temple.edu

K. S. S. Mix, M. T. Battista (eds.), *Visualizing Mathematics*, Research in Mathematics Education, https://doi.org/10.1007/978-3-319-98767-5_17

Space-Mathematics Linkages Are Not Only, or Even Primarily, with Geometry

Many of the education chapters focus primarily or in substantial part on geometry. Geometry certainly seems very spatial, but it is not entirely spatial. Furthermore, other branches of mathematics are not non-spatial. Thus, research on space and mathematics should look widely at various branches of mathematics.

Let us look first at geometry. Geometric proofs clearly involve analytic and logical skills, as well as the acquisition and deployment of heuristics and examples. Proofs are not required until high school, and sometimes not even then, but early geometry is also more than just spatial. For example, learning shape names is deeply dependent on exposure to relevant linguistic input and well-constructed stimulus sets that involve atypical as well as typical shapes (Verdine et al., in press). Learning what the word "angle" means is similarly dependent on linguistic input paired with relevant environmental support (Gibson, Congdon, & Levine, 2015). Thus, understanding learning of geometry will require a broader approach than merely the spatial one.

Let us turn now to branches of mathematics other than geometry. Among preschoolers, Verdine et al. found that spatial skills predicted mathematics achievement at age 5 on a wide-ranging battery of mathematics learning at that age, not just shape learning. In elementary school, early spatial skills predict approximate calculation via enhancement of number line knowledge (Gunderson, Ramirez, Beilock, & Levine, 2012). The number line is a spatialization of children's conceptions of integers, and thus powerfully links the spatial and the mathematical domains. In addition, other studies have found that spatial skills relate concurrently and longitudinally to many types of mathematics beyond geometry in elementary school, including place value, word problems, fractions, and algebra (Frick, 2018; Mix et al., 2016). Even among expert mathematicians, engagement of areas of the brain identified with spatial thinking occurs when pondering mathematical problems of four different kinds, including analysis and algebra as well as topology and geometry (Amalric & Dehaene, 2016).

Thus, in thinking about why spatial thinking enhances mathematics learning—in pursuing the vital search for mechanism—we need to keep in mind that many non-spatial problems can be visualized and that thinking about numbers and equations can involve spatial manipulation either of the symbols themselves or of the underlying situation in the world for which they stand.

Spatial Ability Is Far from Unitary But How to Characterize It Is a Work in Progress

The term "spatial ability" is in widespread use, but it really is a misnomer. First, "ability" tends to connote a fixed trait, but spatial thinking is malleable (Uttal et al., 2013). The word "skill" is thus preferable to "ability" because for most people the word connotes an attribute that can be learned through practice. But the other problem with "spatial ability" is that it is singular. Spatial thinking is clearly multi-factorial—there are abilities, or (better) skills, in the plural. The same problem arises with the use of the singular terms "spatial cognition" and "spatial thinking."

The highest order distinction to make in thinking about spatial thinking is between navigation and object manipulation (Newcombe, 2018). These two sets of cognitive skills have different evolutionary roots and distinct neural bases. Navigation is a function shared with all other mobile species and involves coding location and movement tracking to maintain orientation to the external world. Navigation is what Uttal et al. (2013) had in mind in discussing extrinsic coding between and among objects. Object manipulation is relevant to the use and invention of tools, which involves the mental representation and transformation of the shapes of objects, i.e., intrinsic coding. Mental rotation is a much-studied example of intrinsic spatial transformation, but objects can also be transformed in other ways, e.g., by folding, bending, or breaking (Atit, Shipley, & Tikoff, 2013; Resnick & Shipley, 2013).

Of course, as a symbolic species, humans also spatialize thought in various symbolic ways, using tools such as language, metaphor, analogy, gesture, sketches, diagrams, graphs, maps, and mental images. We can use symbolic representations both for objects and for environments. That is, we can talk about both kinds of topics, either literally or metaphorically, make analogies for both kinds of information, sketch both, and imagine both. Thus, there is a large "third set" of spatial skills, corresponding to these symbolic tools.

The typology of spatial skills outlined in Uttal et al. (2013; see also Newcombe, 2018) is a work in progress. Several chapters in this book use it, but many authors do not seem to construe the extrinsic skills are navigational, but simply as between-objects on a small scale. However, between-object relations in small-scale space are quite different from remembering an environment at a larger scale. An important body of research on this theme comes from papers contrasting mental rotation and perspective taking. Many tests of mental rotation focus on single objects, including the classic Shephard-Metzler block task. However, it is also possible to study the mental rotation of tabletop arrays of separate objects. We can also change the relations between the objects by walking around the table. Walking around is often called perspective taking. It seems to be a fundamentally different process from mental rotation both behaviorally (in a line of research initiated by Huttenlocher & Presson, 1973) and neurally (Lambrey, Doeller, Berthoz, & Burgess, 2011). It is centrally involved in linking spaces encountered separately and integrating them into a flexible unified representation (Holmes, Newcombe, & Shipley, in press). Such skills are very relevant both to navigation (Nazareth, Weisberg, Margulis, & Newcombe, 2018) and to thinking in geography and geoscience (Nazareth, Newcombe, Shipley, Velazquez, & Weisberg, under review), but whether they are relevant to mathematics remains to be seen. I would guess not.

In sum, there are many spatial skills. Research needs to continue concerning how they are related and what typology to use. Minimally, mathematical cognition research needs to consider symbolic skills as well as perceptual and memory-based skills. In addition, it is possible that skills focused on the structure of a single object are distinct from skills involving small-scale relations between objects and those skills are in turn distinct from large-scale relations among objects.

Spatial Tools Are as Important as Spatial Skills

Researchers have often focused on spatial skills (i.e., attributes of the learner). However, as the previous section suggested, spatial tools provided in the classroom, curriculum, or learning environment are also important. One good example is encouraging students to sketch. This simple instruction can improve problem solving in mathematics (and in science) (Ainsworth, Prain, & Tytler, 2011; Miller-Cotto et al., under review; Sinclair, Moss, Hawes, & Stephenson, this volume). A second example is the use of gesture in teaching missing addend problems (Novack & Goldin-Meadow, 2015). A third useful technique is providing spatially aligned mathematical problems and gesturing between the two problems to highlight analogies (Begolli & Richland, 2016). Thus, "spatializing the curriculum" may be as important as enhancing spatial skills, and arguably even more important because it is more practical to implement. Of course, using spatial tools such as sketching, gesture, or spatial alignment can also enhance learners' spatial skills and hence spatializing the curriculum may both enhance learning in the moment and increase spatial skills for later learning—a "two for the price of one" intervention. What remains to be determined more exactly is whether a learner's level of spatial skills may affect what kinds of spatial tools work where and when. Many investigators believe in such aptitude by treatment interactions but the search for them has been elusive.

For me, the bottom line here is that a focus on the individual learner alone will reduce our view of how to utilize spatial thinking in the classroom. Researchers should integrate examination of skills with examination of how gesture, sketching, diagrams and graphs, and other such spatial tools, are best used to enhance mathematics learning, when and how, and possibly for whom.

References

Ainsworth, S., Prain, V., & Tytler, R. (2011). Drawing to learn in science. *Science, 333*(6046), 1096–1097.

Amalric, M., & Dehaene, S. (2016). Origins of the brain networks for advanced mathematics in expert mathematicians. *Proceedings of the National Academy of Sciences, 113*(18), 4909–4917.

Atit, K., Shipley, T. F., & Tikoff, B. (2013). Twisting space: Are rigid and non-rigid mental transformations separate spatial skills? *Cognitive Processing, 14*(2), 163–173.

Begolli, K. N., & Richland, L. E. (2016). Teaching mathematics by comparison: Analog visibility as a double-edged sword. *Journal of Educational Psychology, 108*(2), 194.

Frick, A. (2018). Spatial transformation abilities and their relation to later mathematics performance. *Psychological Research Psychologische Forschung*. https://doi.org/10.1007/s00426-018-1008-5

Gibson, D. J., Congdon, E. L., & Levine, S. C. (2015). The effects of word-learning biases on children's concept of angle. *Child Development, 86*(1), 319–326.

Gunderson, E. A., Ramirez, G., Beilock, S. L., & Levine, S. C. (2012). The relation between spatial skill and early number knowledge: The role of the linear number line. *Developmental Psychology, 48*(5), 1229.

Holmes, C., Newcombe, N. S., & Shipley, T. F. (in press). Move to learn: Integrating spatial information from multiple viewpoints. *Cognition*.

Huttenlocher, J., & Presson, C. C. (1973). Mental rotation and the perspective problem. *Cognitive Psychology, 4*(2), 277–299.

Lambrey, S., Doeller, C., Berthoz, A., & Burgess, N. (2011). Imagining being somewhere else: Neural basis of changing perspective in space. *Cerebral Cortex, 22*(1), 166–174.

Miller-Cotto, D., Booth, J. L., Chang, B. L., Cromley, J., Newcombe, N. S., & Williams, T. A. (under review). Sketching and verbal self-explanation: Do they help middle school children solve mathematics and science problems?

Mix, K. S., Levine, S. C., Cheng, Y. L., Young, C., Hambrick, D. Z., Ping, R., & Konstantopoulos, S. (2016). Separate but correlated: The latent structure of space and mathematics across development. *Journal of Experimental Psychology: General, 145*(9), 1206.

Nazareth, A., Newcombe, N. S., Shipley, T. F., Velazquez, M. & Weisberg, S. M. (under review). Beyond wayfinding: Navigation skills and geoscience education.

Nazareth, A., Weisberg, S. M., Margulis, K., & Newcombe, N. S. (2018). Charting the development of cognitive mapping. *Journal of Experimental Child Psychology, 170*, 86–106.

Newcombe, N. S. (2017). *Harnessing spatial thinking to support STEM learning* (OECD Education Working Papers, No. 161). OECD Publishing, Paris. https://doi.org/10.1787/7d5dcae6-en

Newcombe, N. S. (2018). Three kinds of spatial cognition. In J. Wixted (Ed.), *Stevens' handbook of experimental psychology and cognitive neuroscience* (4th Ed.), Wiley.

Novack, M., & Goldin-Meadow, S. (2015). Learning from gesture: How our hands change our minds. *Educational Psychology Review, 27*(3), 405–412.

Resnick, I., & Shipley, T. F. (2013). Breaking new ground in the mind: An initial study of mental brittle transformation and mental rigid rotation in science experts. *Cognitive Processing, 14*(2), 143–152.

Sinclair, N., Moss, J., Hawes, Z., & Stephenson, C. (this volume). Learning through and from drawing in early years geometry. In K. S. Mix & M. T. Battista (Eds.), *Visualizing mathematics: The role of spatial reasoning in mathematical thought*. New York: Springer.

Uttal, D. H., Meadow, N. G., Tipton, E., Hand, L. L., Alden, A. R., Warren, C., & Newcombe, N. S. (2013). The malleability of spatial skills: A meta-analysis of training studies. *Psychological Bulletin, 139*, 352–402.

Verdine, B. N., Golinkoff, R. M., Hirsh-Pasek, K., & Newcombe, N. S. (2017). Links between spatial and mathematical skills across the preschool years. *Monographs of the Society for Research in Child Development, 82*(1), 1–150.

Verdine, B. N., Zimmermann, L., Foster, L., Marzouk, M. A., Michnick, R., Golinkoff, K. H. P., & Newcombe, N. (in press). Effects of geometric toy design on parent-child interactions and spatial language. *Early Childhood Research Quarterly*. https://doi.org/10.1016/j.ecresq.2018.03.015

Zhang, X., & Lin, D. (2017). Does growth rate in spatial ability matter in predicting early arithmetic competence? *Learning and Instruction, 49*, 232–241.

Chapter 18
On the Multitude of Mathematics Skills: Spatial-Numerical Associations and Geometry Skill?

Krzysztof Cipora, Philipp A. Schroeder, and Hans-Christoph Nuerk

Battista et al. rightly point out the importance of the relation between spatial and geometric reasoning. They are also working towards more fine-grained analyses considering different aspects of both spatial and mathematical reasoning, and postulate that analyses should examine particular skills, rather than these very general constructs. The authors support their claim with the results of a series of well-designed and carefully conducted studies, using one-on-one interviews, one-on-one teaching experiments, and case studies. They conclude that the ability to visualize objects and build accurate mental models thereof is linked to property-based spatial reasoning.

Attempting to relate correlational psychological results to the perspective by Battista et al., we can identify important gaps in the literature and between disciplines. Of particular interest, the detailed skills required for geometric problem solving could partially substantiate explicit and implicit Spatial-Numerical Associations (SNAs) as introduced in our taxonomy, with the former also being

K. Cipora
Department of Psychology, University of Tuebingen, Tuebingen, Germany

LEAD Graduate School and Research Network, University of Tuebingen, Tuebingen, Germany
e-mail: krzysztof.cipora@uni-tuebingen.de

P. A. Schroeder
Department of Psychology, University of Tuebingen, Tuebingen, Germany

Department of Psychiatry and Psychotherapy, University of Tuebingen, Tuebingen, Germany

H.-C. Nuerk (✉)
Department of Psychology, University of Tuebingen, Tuebingen, Germany

Leibnitz-Institut für Wissensmedien, Tuebingen, Germany

LEAD Graduate School and Research Network, University of Tuebingen, Tuebingen, Germany
e-mail: hc.nuerk@uni-tuebingen.de

K. S. S. Mix, M. T. Battista (eds.), *Visualizing Mathematics*, Research in Mathematics Education, https://doi.org/10.1007/978-3-319-98767-5_18

reflected in the introspective reports of mental models by their participants. Overall, these observations invoke us to consider three distinctions:

1. variations in mathematics skill operationalization
2. variations in spatial dimensions (which can be combinations of extension and the different directionalities) and processing (e.g., spatial imagery, spatial reasoning)
3. the links between (1) and (2), which are in turn dependent on the choice of mathematics and space operationalization, but also perhaps on individual and instructional differences.

Variation in Mathematics Skill Operationalization

We agree with Battista et al. in general, but in this comment, we would like to point out that their interesting contribution about geometry underlines the necessity of considering a broader range of mathematical skills when looking at potential correlates of SNA. To date, the vast majority of studies have focused on the potential relation between SNA and arithmetic skill, which diverges from geometry skill (e.g., Lourenco, Bonny, Fernandez, & Rao, 2012). For an overview, Table 18.1 summarizes studies reporting a relation between the SNARC effect (perhaps the most prominent SNA) and measures of mathematics skill.

Inspection of Table 18.1 shows that mathematics skill was operationalized diversely across the different studies. In adult studies, it was either determined by field of study or by standardized arithmetic tasks. In studies with children, mathematics skill was operationalized according to scores in arithmetic tasks or in other tasks involving the use of numbers in Arabic notation. Three studies compared groups that were formed based on visuospatial skill level. One study considered school grades in mathematics. To sum up, one can say that these operationalizations differ considerably. Obviously, this can be a problem, especially if construct validities between different operationalizations of mathematics skills are mediocre or low. For instance, Lourenco et al. (2012) report that arithmetic performance in a task with and without time pressure correlated only moderately ($r = 0.57$). Another measure of mathematics skill—geometry performance only weakly correlated with timed arithmetic task performance ($r = 0.25$), and moderately with untimed one ($r = 0.56$). All these correlations account for no more than 33% of variance. In such cases, the choice of "mathematics" assessment itself contributes to the outcome of whether space and mathematics are found to be related.

Regarding adult studies, the field of study can be a proxy measure of general mathematics skill. However, it also may reflect a variety of other factors leading to the choice of a field of study, such as interests, attitudes, learning motivation, or anxiety (Maloney & Beilock, 2012). As regards using calculation tasks, its difficulty level covers skills that are formally required from all students at the level of secondary school (Cipora & Nuerk, 2013). Therefore, the differences in performance might originate from the use of inefficient strategies, working memory overload, mathematics anxiety, or decreased performance due to time pressure. Similar inconsistencies can be also found in children studies. Mathematics grade can be a kind of

Table 18.1 Arithmetic skill and directional implicit space-number associations in adults (top rows) and children (bottom rows), in chronological order

Study	Age	N	Impairments/ special skills	SNA task and effect	Mathematics skill assessment	Correlation	Group comparison	Direction	Bayesian or power testing (or just non-significance)
Adults									
Dehaene, Bossini, and Giraux (1993), Exp. 1	Adults	20	10 literary vs. 10 science students	Parity judgment: interaction term (SNARC effect)	Field of study	–	$ps < 0.10^{\dagger}$	No significant difference	–
Fischer and Rottmann (2005)	Adults	20	10 psychology vs. 10 mathematics students	Parity judgment b (SNARC effect) for positive and negative integers	Field of study	–	$p = 0.28$	No significant difference	–
Bonato, Fabbri, Umiltà, and Zorzi (2007), Exp. 1	Adults	20	10 psychology vs. 10 engineering students	Fraction magnitude comparison	Field of study	–	$p = 0.65$	No significant difference	–
Bull, Cleland, and Mitchell (2013), Exp. 2	Adults	40	–	Go-Nogo colour judgment: b (SNARC)	Numerical operations subtest (WIAT-II-UK)	$r = -0.07$	–	No significant correlation	–

(continued)

Table 18.1 (continued)

Study	Age	N	Impairments/ special skills	SNA task and effect	Mathematics skill assessment	Correlation	Group comparison	Direction	Bayesian or power testing (or just non-significance)
Cipora and Nuerk (2013)	Adults	71	Mathematics-related (18) and mathematics-unrelated studies (53)	Parity judgment: b (SNARC effect)	Equation verification	$r = 0.14$	$p = 0.58$	No significant correlation or difference	88% posterior probability of Null
Hoffmann, Mussolin, Martin, and Schiltz (2014)	Adults	95	38 literary-related vs. 38 science students vs. 19 students with mathematics difficulties (self-identified)	Parity judgment: b and ß (SNARC effect)	Arithmetic tasks (Arith and FastMath)	$r = 0.28$	$p < 0.05$	SNARC effect was increased with decreased mathematics proficiency	–
Göbel, Maier, and Shaki (2015)[a]	Adults	114	British and Arab adults	Parity judgment: ß (SNARC effect)	WRAT Arithmetics	$r = 0.02$	–	No significant correlation for either country	–
Cipora et al. (2016)	Adults	44	Doctoral students of mathematics (14), engineering (15), humanities (15), and social sciences (15)	Parity judgment: b and ß (SNARC effect)	Field of study	–	$p = 0.03$*	Smaller SNARC in mathematicians vs. humanities students	65–67% posterior probability of Null for the other group comparisons

Children									
Bachot, Gevers, Fias, and Roeyers (2005)	7–12-year olds	32	Visuospatial disability VSD (16), matched controls (16)	Magnitude comparison: b (SNARC effect)	Group comparison	–	$p < 0.05$*	No SNARC in VSD	–
Schneider, Grabner, and Paetsch (2009), Exp. 2	5–6 grade	110	–	Parity judgment: ß (SNARC effect)	Mathematics mark	$r = 0.08$	–	No significant correlation	–
Hoffmann, Hornung, Martin, and Schiltz (2013)	Kindergarten children	84	–	Magnitude judgment: b (SNARC effect)	Verbal counting strategies	$r = -0.30$*	–	SNARC emerges with transition to magnitude classification without counting	–
					Arabic digit writing	$r = -0.40$**		Stronger SNARC when Arabic digit writing was better	
Crollen, Vanderclausen, Allaire, Pollaris, and Noël (2015)	6–13-year-olds	30	Poor visuospatial ability NVLD (15), matched controls (15)	Magnitude comparison: b (SNARC effect) NLE, line bisection	Group comparison	–	$p < 0.05$*	No SNARC in LVLD in hands-crossed posture	–
Crollen and Noël (2015)	4 grade	70	Weak vs. high visuospatial groups (each 25, 35/65th percentile split)	Magnitude comparison: by category (SNARC effect)	Group comparison	–	$p = 0.20$	No significant differences between groups and postures	–

(continued)

Table 18.1 (continued)

Study	Age	N	Impairments/ special skills	SNA task and effect	Mathematics skill assessment	Correlation	Group comparison	Direction	Bayesian or power testing (or just non-significance)
Gibson and Maurer (2016)	6–8-year olds	60	–	Magnitude comparison: compatibility ratio (SNARC effect)	Test of Early Mathematics Ability (TEMA-3)	$r_p = 0.13$	–	No significant correlation	–
Georges, Hoffmann, and Schiltz (2017b)	3–4 grade	55	–	Parity judgment: b (SNARC effect)	Heidelberg Mathematics Test HRT 1–4	$r = -0.28$*	–	Better mathematics with stronger SNARC	–

Note. *b* indicates unstandardized regression coefficient, *β* standardized regression coefficient *by category* refers to comparison of dRTs between small and large magnitude numbers. *PAE* percentage of absolute error by category refers to comparison of dRTs between small and large magnitude numbers

*Significant correlation/comparison/interaction term at $p < 0.05$ or **$p < 0.01$ or [†]trend for significance at $p < 0.10$

[a]Correlation results from the study were reported at the Fifth Workshop on Numerical Cognition in Tübingen, Germany, 2017

proxy for general mathematics performance, but apart from the actual performance in different tasks it is affected by individuals' motivation, attitudes, anxiety, etc. Therefore, such measures seem to be largely heterogeneous and the construct validity of the dependent variable is at least questionable.

Keeping all that in mind, one might argue that despite the disadvantages listed above, (mixed) calculation scores are relatively valid measures of mathematics skill. Nevertheless, even within (speeded) arithmetic operations, there are considerable differences in cognitive processes needed to solve problems. The score might reflect different mathematics skills, but may also register the automaticity of processing. Multiplication, especially, has been attributed to fast fact retrieval (Grabner et al., 2009), which supposedly has different underlying neurocognitive components than other arithmetic operations. In case of multiplications of multi-digit numbers a considerable working memory load appears. On the other hand, addition and subtraction usually require more advanced cognitive processing than fact retrieval because different procedural steps have to be chosen and applied. Importantly, different operations seem to engage different brain regions both in adults (Arsalidou & Taylor, 2011; Prado et al., 2011), and in children (Prado, Mutreja, & Booth, 2014). The evidence on division is least conclusive, but it seems that easy problems are solved by means of inverse multiplication (Huber, Fischer, Moeller, & Nuerk, 2013), which is not necessarily the case in complex division.

Considering abovementioned dissociations, one can expect that the relation between SNAs and mathematics skill may be different within these operations, so that treating them as equivalent operationalizations of one construct called "mathematics" or "arithmetic" might in fact lead to confusing and contradictory results. One might speculate that SNAs relate more strongly to performance in tasks which do not solely depend on fact retrieval (like simple multiplications), but rather on magnitude manipulation (such as additions, subtractions, and possibly divisions). Nevertheless, this requires further investigation.

So far, we have largely referred to school mathematics like arithmetic operations. However, the picture gets even more complicated if one considers more complex mathematics reasoning, including algebra, conducting proofs, or calculus. In such occasions the processes required to successfully solve the problem comprise operating on symbols, transforming them according to certain rules, and finding complex relations between them. This set of operations approaches the definition criteria for fluid intelligence (see Cipora et al., 2016 for arguments), whereas the arithmetic calculations themselves play a secondary role. The multitude of mental operations required to succeed in mathematics gets ever broader if one considers geometry, to which we wish to draw reader's attention in this commentary.

Variation in Visuo-Spatial Skills Operationalization

In the literature one can find evidence that SNAs are related to visuospatial skills. Such results can be observed for different SNA categories, e.g., Number Line Estimation (Sella, Sader, Lolliot, & Cohen Kadosh, 2016). Unsurprisingly, individuals characterized with higher visuospatial skills perform the number line estimation task more accurately. There is also evidence indicating that the SNARC effect depends on individual differences in visuospatial skills. Namely, a stronger effect of mental rotation (i.e., larger increase in the time to decide on object identity depending on the degree to which it is rotated) relates to more pronounced SNARC (Viarouge, Hubbard, & McCandliss, 2014). In a more recent study it was shown that an individual's visualization profile moderates the relation between magnitude and parity SNARC (Georges, Hoffmann, & Schiltz, 2017a). Importantly, visuospatial skills themselves are also correlated with different types of mathematics skill at several stages of development (Mix et al., 2016). Like mathematics skills, visuospatial skills and processes must also be differentiated when their relation to mathematics is investigated (see Mix et al., 2016).

Keeping these developments in mind, one can postulate that the branch of mathematics, which quantifies physical space by means of mathematical concepts, that is, geometry, may be related to SNAs. Some spatial mapping of the magnitude of geometrical entities has already been demonstrated. Specifically, angle magnitude is spatially mapped (Fumarola et al., 2016): large angles are responded to faster with the left hand and small angles with the right hand. This effect was only present in engineering students, but not in psychology students. The authors' interpretation is that education and practice shape this kind of spatial association. All these findings together suggest that investigating the relation between SNAs and geometry skill will be an interesting new avenue of research.

Links Between Mathematics Skills and Visuospatial Skills

If one wishes to investigate the relation between mathematics skill and spatial skills, it is important to keep in mind that both these constructs are very heterogeneous. Therefore, the links between those skills, namely between space and number, are also rather heterogeneous. As we have already expressed in our contribution to this volume (Chapter 4), SNAs should not be treated as part of the same melting pot, but rather as a family of phenomena. They share some common traits (numbers are somehow related to space), but are also largely differentiated, and there is no strong common denominator of all SNA. They can rather be characterized by Wittgenstein's family resemblance (Wittgenstein, 1953), that is, their similarity comes from multiple and partly overlapping features and not from a core set of features shared by all SNA instances.

Conclusions

When investigating the relation between number and space, the heterogeneity on all sides, concerning mathematics skill, visuo-spatial skills, and their relations should be thoroughly considered. In particular, a multitude of different mathematics skills should be acknowledged as well as the fact that different members of the "SNA family" can particularly strongly relate to certain members of the "mathematics skill family". Such strong relations need to be investigated, and this can be far more instructive than attempts to capture the very blurred and inaccurate general picture. Finally, the relations between the space-number association, mathematics skills, and spatial skills need to be investigated in a differentiated way. As pointed out by Battista et al., there is compelling evidence for a relation between visuospatial skill and mathematics skill. This observation encourages us to investigate the relation between the "SNA family" and the relatively neglected member of the "mathematics skill family", that is the geometry skill.

Acknowledgments KC is supported by a DFG grant [NU 265/3-1] to HCN. KC and HCN are further supported by the LEAD Graduate School & Research Network [GSC1028], which is funded within the framework of the Excellence Initiative of the German federal and state governments. We thank Julianne Skinner for proofreading the manuscript.

References

Arsalidou, M., & Taylor, M. J. (2011). Is 2+2=4? Meta-analyses of brain areas needed for numbers and calculations. *NeuroImage, 54*(3), 2382–2393. https://doi.org/10.1016/j.neuroimage.2010.10.009

Bachot, J., Gevers, W., Fias, W., & Roeyers, H. (2005). Number sense in children with visuospatial disabilities: Orientation of the mental number line. *Psychology, 47*(1), 172–183. Retrieved from http://www.pabst-publishers.de/psychology-science/1-2005/ps_1_2005_172-183.pdf

Bonato, M., Fabbri, S., Umiltà, C., & Zorzi, M. (2007). The Mental Representation of Numerical Fractions: Real or Integer? *Journal of Experimental Psychology: Human Perception and Performance, 33*(6), 1410–1419. https://doi.org/10.1037/0096-1523.33.6.1410

Bull, R., Cleland, A. A., & Mitchell, T. (2013). Sex differences in the spatial representation of number. *Journal of Experimental Psychology: General, 142*(1), 181–192. https://doi.org/10.1037/a0028387

Cipora, K., Hohol, M., Nuerk, H.-C., Willmes, K., Brożek, B., Kucharzyk, B., & Nęcka, E. (2016). Professional mathematicians differ from controls in their spatial-numerical associations. *Psychological Research, 80*(4), 710–726. https://doi.org/10.1007/s00426-015-0677-6

Cipora, K., & Nuerk, H.-C. (2013). Is the SNARC effect related to the level of mathematics? No systematic relationship observed despite more power, more repetitions, and more direct assessment of arithmetic skill. *Quarterly Journal of Experimental Psychology, 66*(10), 1974–1991. https://doi.org/10.1080/17470218.2013.772215

Crollen, V., & Noël, M. P. (2015). Spatial and numerical processing in children with high and low visuospatial abilities. *Journal of Experimental Child Psychology, 132*, 84–98. https://doi.org/10.1016/j.jecp.2014.12.006

Crollen, V., Vanderclausen, C., Allaire, F., Pollaris, A., & Noël, M.-P. (2015). Spatial and numerical processing in children with non-verbal learning disabilities. *Research in Developmental Disabilities, 47*, 61–72. https://doi.org/10.1016/J.RIDD.2015.08.013

Dehaene, S., Bossini, S., & Giraux, P. (1993). The mental representation of parity and nummber magnitude. *Journal of Experimental Psychology: General, 122*(3), 371–396.

Fischer, M. H., & Rottmann, J. (2005). Do negative numbers have a place on the mental number line? *Psychology Science, 47*(1), 22–32.

Fumarola, A., Prpic, V., Fornasier, D., Sartoretto, F., Agostini, T., & Umiltà, C. (2016). The spatial representation of angles. *Perception, 45*(11), 1320–1330. https://doi.org/10.1177/0301006616661915

Georges, C., Hoffmann, D., & Schiltz, C. (2017a). How and why do number-space associations co-vary in implicit and explicit magnitude processing tasks? *Journal of Numerical Cognition, 3*(2), 182–211.

Georges, C., Hoffmann, D., & Schiltz, C. (2017b). Mathematical abilities in elementary school: Do they relate to number–space associations? *Journal of Experimental Child Psychology, 161*, 126–147. https://doi.org/10.1016/J.JECP.2017.04.011

Gibson, L. C., & Maurer, D. (2016). Development of SNARC and distance effects and their relation to mathematical and visuospatial abilities. *Journal of Experimental Child Psychology, 150*, 301–313. https://doi.org/10.1016/J.JECP.2016.05.009

Göbel, S. M., Maier, C. A., & Shaki, S. (2015). Which numbers do you have in mind? Number generation is influenced by reading direction. *Cognitive Processing, 16*(S1), 241–244. https://doi.org/10.1007/s10339-015-0715-8

Grabner, R. H., Ansari, D., Koschutnig, K., Reishofer, G., Ebner, F., & Neuper, C. (2009). To retrieve or to calculate? Left angular gyrus mediates the retrieval of arithmetic facts during problem solving. *Neuropsychologia, 47*(2), 604–608. https://doi.org/10.1016/J.NEUROPSYCHOLOGIA.2008.10.013

Hoffmann, D., Hornung, C., Martin, R., & Schiltz, C. (2013). Developing number-space associations: SNARC effects using a color discrimination task in 5-year-olds. *Journal of Experimental Child Psychology, 116*(4), 775–791. https://doi.org/10.1016/j.jecp.2013.07.013

Hoffmann, D., Mussolin, C., Martin, R., & Schiltz, C. (2014). The impact of mathematical proficiency on the number-space association. *PLoS One, 9*(1), e85048. https://doi.org/10.1371/journal.pone.0085048

Huber, S., Fischer, U., Moeller, K., & Nuerk, H.-C. (2013). On the interrelation of multiplication and division in secondary school children. *Frontiers in Psychology, 4*, 740. https://doi.org/10.3389/fpsyg.2013.00740

Lourenco, S. F., Bonny, J. W., Fernandez, E. P., & Rao, S. (2012). Nonsymbolic number and cumulative area representations contribute shared and unique variance to symbolic mathematics competence. *Proceedings of the National Academy of Sciences of the United States of America, 109*(46), 18737–18742. https://doi.org/10.1073/pnas.1207212109

Maloney, E. A., & Beilock, S. L. (2012). Math anxiety: Who has it, why it develops, and how to guard against it. *Trends in Cognitive Sciences, 16*(8), 404–406. https://doi.org/10.1016/j.tics.2012.06.008

Mix, K. S., Levine, S. C., Cheng, Y.-L., Young, C., Hambrick, D. Z., Ping, R., & Konstantopoulos, S. (2016). Separate but correlated: The latent structure of space and mathematics across development. *Journal of Experimental Psychology: General, 145*(9), 1206–1227. https://doi.org/10.1037/xge0000182

Prado, J., Mutreja, R., & Booth, J. R. (2014). Developmental dissociation in the neural responses to simple multiplication and subtraction problems. *Developmental Science, 17*(4), 537–552. https://doi.org/10.1111/desc.12140

Prado, J., Mutreja, R., Zhang, H., Mehta, R., Desroches, A. S., Minas, J. E., & Booth, J. R. (2011). Distinct representations of subtraction and multiplication in the neural systems for numerosity and language. *Human Brain Mapping, 32*(11), 1932–1947. https://doi.org/10.1002/hbm.21159

Schneider, M., Grabner, R. H., & Paetsch, J. (2009). Mental number line, number line estimation, and mathematical achievement: Their interrelations in grades 5 and 6. *Journal of Educational Psychology, 101*(2), 359–372. https://doi.org/10.1037/a0013840

Sella, F., Sader, E., Lolliot, S., & Cohen Kadosh, R. (2016). Basic and advanced numerical performances relate to mathematical expertise but are fully mediated by visuospatial skills. *Journal of Experimental Psychology: Learning Memory and Cognition, 42*(9), 1458–1472. https://doi.org/10.1037/xlm0000249

Viarouge, A., Hubbard, E. M., & McCandliss, B. D. (2014). The cognitive mechanisms of the SNARC effect: An individual differences approach. *PLoS One, 9*(4), e95756. https://doi.org/10.1371/journal.pone.0095756

Wittgenstein, L. (1953). *Philisophical Investigations*. New York: Macmillan.

Index

A

Absolute magnitude reasoning, 6–8
Advisory committees, 118
Analytical tool, 318
Angles, 189
Approximate number system (ANS), 17, 81, 91
Arabic notation, 362
Arithmetic sign and space, 88
Arithmetic skill, 363–366
Australian Curriculum Mathematics, 174
Australian school system, 342
Axiomatic system, 199

B

Battista hypothesizes, 187
Battista's Length Learning Progression, 152
Benchmark strategy, 11, 13, 14, 19
BI's reasoning, 218
Bidirectional analysis, 352
Boxspring, 288, 290, 291, 294, 295
Buildings tasks, 313–316, 318–322

C

Calculations, 285
Carol's students, 248
Challenging task, 311, 335
Children's natural propensity, 42
Classroom teachers in situ, 173
Classroom-based intervention research, 173
Cognition based assessment (CBA), 151–152
  AW's ruler reasoning, 158
  component, 152
  counting for length measurement, 160
  geometric measurement, 157
  non-measurement and measurement reasoning, 153, 154
  RC, 153
  TMM, 158
  unit-length iteration, 162
Cognitive complexity, 311
Cognitive demand
  doing mathematics, level of, 318
  procedures with connections, level of, 318
  procedures without connections, level of, 318
Cognitive psychology, 171, 173
Cognitive science, 118, 122, 124, 140
Collaborative visualization tasks, 313
Collaborative work, 333
Common Core State Standards for Mathematics (CCSS-M), 57, 67, 69, 121
Communication challenge, 288
Communication failures, 289
Communication tasks, 299
Compatibility effects, 84, 105
Conceptual Change in Childhood (1985), 347
Conceptual understanding of measurement, 32, 35, 39
Confirmatory factor analysis (CFA), 123, 128, 138
Congruence, 236–238, 243, 249
Contextual and mathematical information, 261
Conversion top, 292
Couch, 290, 296–298
Counting and cardinality, 121
Counting tasks, 88

K. S. S. Mix, M. T. Battista (eds.), *Visualizing Mathematics*, Research in Mathematics Education, https://doi.org/10.1007/978-3-319-98767-5

D
Decoding techniques, 265
Diagrammatical representations, 178
Diagramming, 229
Diagrams, 279
Didactical variables, 299–300
Digital technologies, 256
Dimensional decomposition, 234
Directional SNAs, 90
Disciplinary asymmetries, 347
Draw-A-Man Test, 231
Drawing
  Ancient Greece, 229
  and domain-specific knowledge, 231
  assessment, 232
  children, 230
  classroom teacher, 235
  congruence, 243
  dimensional deconstruction, 234
  fractions problems, 232
  geometric designs, 232
  geometrical structure, 247
  geometry education, 230
  gestures, 240, 244
  grid completion, 247
  high school students, 233
  interview protocol, 239
  interviewers, 243
  kindergarten children, 236
  kindergarten classroom, 235
  mathematical abstraction, 233
  mathematical concepts, 231
  mathematical ideas, 242
  1D and 2D geometric objects, 246
  procedure, 237
  psychological studies, 230
  qualitative study and analysis, 237
  structural awareness, 247, 248
  symmetry, 244
  2D shapes, 245
  two line segments, 245
  verbal/visual interplay, 234
Drawing behaviour, 231
Duval encourages educators, 245
Dynamic geometry learning system, 203
Dynamic spatial reasoning, 184

E
Education perspective, 173
Educational Studies in Mathematics, 171
Elementary arithmetic concepts, 26
Elementary school, 63, 356
Embodied learning approach, 97
Encoding and decoding graphics, 260–266
Encoding techniques, 260
Euclidean geometry, 304
Examination, 172
Experience-language-pictorial-symbolic-application (ELPSA)
  framework, 174
  learning framework, 174
  spatial reasoning, 174
Experiential world, 278
Experimental psychology
  attention, 80
  compatibility effect, 80
  domain-specific and domain-general processes, 81
Exploratory factor analysis (EFA), 123, 140
Extension SNAs, 90
Extrinsic-dynamic thinking, 184

F
Finger-tracking technologies, 95
Fraction card games
  alternate play–non-competitive, 59
  arrange the cards in order, 60
  materials, 59
  number line, 60
  solitaire, 59
  two-person practice game, 59
  two-person simpler practice game, 59
  WAR, 59
Fraction learning, 8, 9
Fraction magnitude representations, 58

G
Game, 303
Geogebra, 178
Geometric problem solving, 197
Geometric proofs, 356
Geometric reasoning, 197–198
Geometric structuring, 202
Geometry, 185, 196–198, 200, 234, 351
Geometry instruction, 281
Gestures, 233
Golf game, 215
Graphical languages, 258, 259
Graphing
  convergent series, 60–62
  successive sums, 61

H
Hatch-mark counting strategy, 39
Hegarty's spatial strategy, 205
High school geometry instruction, 278

I
Imagistic transformations, 120
Individualized Dynamic Geometry Instruction (*iDGi*) project, 203
Indoor staircase, 286
Intelligence, 118, 121, 135, 137
International Group for the Psychology of Mathematics Education (IGPME), 171
International Mathematics Fourth Grade Assessment, 35
Intervention programs, 173
Interview Protocol 1, 236
Interview Protocol 2, 237
iPad, 267
Isometries, 203, 205–210, 217

K
Kindergarten and grade 1 students, 187

L
Language conventions, 176
L-based right prism, 296
Learning progressions, 199, 203
Line bisection task, 86
Linear units of measure, 36, 39, 40
Logic-based reasoning, 351
L-strategy, 200

M
Macrospace, 280
Magnitude classification task, 87
Magnitude comparison task, 87
Math for Young Children (M4YC) project, 236
Mathematical ability, 118, 119
Mathematical array structure, 246
Mathematical diagrams, 233
Mathematical giftedness, 311
Mathematical modeling, 278, 302
Mathematics and spatial reasoning, 168
Mathematics assessment tools, 343
Mathematics education, 190, 233, 350
Mathematics education researchers, 352
Mathematics educators' perspectives
  arithmetic learning, 343
  geometry/measurement items, 344
  mathematics problems, 341
  numeric/algebra performance, 344
  numerical/algebraic problems, 343
  physical representations/graphics, 341
  problem solving strategies, 342
  problem-solving process, 342
  spatial problem solving, 342
  spatial reasoning, 341
  spatial representations progresses, 342
  spatial skills, 343
  spatial skills and numeric-based mathematics, 344
  visual-spatial tool, 341
Mathematics grade, 362
Mathematics learning, 356
Mathematics skills
  operationalization, 362
  and visuospatial skills, 368
Measurement functions, 210
Measurement of Intelligence by Drawings, 231
Measurement reasoning, 152
Mental models, 196, 198–203, 224, 225, 349
Mental number line (MNL), 79, 81
Mental rotation, 254, 255, 257, 265, 267, 268, 272, 357
  angular disparity, 133, 134
  complex and simple stimuli, 134
  fast flipping transformation, 134
  individual differences, 135
  mathematical reasoning, 134–136
  process view, 133
Mesospace objects
  activities, 282
  boxspring, 286
  couch, 296
  design features, 284
  devolution, 283
  didactical variables, 283, 284, 289
  education designers, 282
  geometric concepts, 281
  geometric figures, 284
  geometric knowledge, 281
  geometric modeling, 280
  geometry, 279, 283, 284
  macrospace, 281
  mathematical models, 278
  microspace, 280
  moving furniture task, 285
  moving objects, 285
  multimodal, 279
  multimodal modeling, 279

Mesospace objects (*cont.*)
- parallel and angle, 288
- pre-algebraic reasoning, 279
- scale, 297
- secondary school, 280
- shape and space, 283
- three-dimensional geometry, 279
- three-dimensional space, 284
- twin-sized boxspring, 290

Microspace, 280
Milieu, 283, 288, 299
Misconceptions, 31, 34, 37–42
Mnemonic techniques, 108
Modalities, 86
Modern education, 42
Movers, 285, 287–289
Moving furniture, 284
Multi-axis coordinate system, 78
Multi-digit number processing, 104–107
Multiplication, 367
Multi-symbol processing
- place identification, 105
- place-value activation, 105, 106
- place-value computation, 106, 107
- place-value system, 104

N

National Council of Teachers of Mathematics (NCTM), 4, 62, 118
National Research Council (NRC), 4
Networked analysis, 334
Networking strategy, 318
Networking theories, 312
Non-algorithmic thinking, 331
Nontraditional digital modes, 272
Nonverbal reasoning, 259
NSF Spatial Intelligence and Learning Center, 119
Number facility, 119
Number line estimation (NLE), 6, 7, 86, 93–95
Numerical cognition
- estimation, 81
- MNL, 81
- numerical distance effect, 81
- place-value processing, 82
- ratio effect, 81
- size effect, 82
- SNARC, 82
- SNAs, 82
- unit-decade compatibility effect, 82

Numerical competencies, 49
Numerical magnitude, 87
Numerical reasoning, 3
Numerically-based mathematics, 342

O

Operation sign spatial association (OSSA), 101
Operational Momentum, 101
Orthogonal projections, 225, 313, 320

P

Paper folding test, 267
Parallelograms, 203
Parity judgment task, 87
Pencil-and-paper modalities, 269
Perspective taking, 357
Physical comparison task, 87
Pictorial phase, 176
Pictorial reasoning, 176
Place identification, 105
Place-value activation, 105, 106
Place-value computation, 106, 107
Place-value integration, 93, 107
Place-value system, 104
Pointing gestures, 233, 240
Primary school leaving examination, 262
Probability task, 262
Procedural understanding of measurement, 33
Professional mathematics organizations, 118
Proportional reasoning, 5, 6, 8, 29, 30
Psychological literature, 231
PT's reasoning, 219

Q

Quadrant model, 184
Quantitative context, 186
Quantitative data, 334
Queen-sized boxspring, 286
Quick Draw program, 235

R

Reflections, 205
Relative magnitude reasoning
- children
  - analogical reasoning, 18
  - ANS, 17, 18
  - cognitive abilities, 19
  - geological scale, 18
  - hierarchical coding, 19
  - interventions, 17, 18

mental representations, 19
number line estimation, 17
cognitive abilities, 20
four-cell classification system, 5
fraction learning, 20
integers, 4
mathematical cognition, 5
mathematics education, 5
mathematics instruction, 4
mathematics learning
early fraction tasks, 6
fraction learning, 8, 9
number lines, 6, 7
proportional reasoning, 8
mental processing, 20
proportional reasoning, 5, 6
quantification types, 4, 5
referents, 4
spatial scaling, 4, 5
spatial thinking
Lego diagrams, 9
spatial representations (*see* Spatial representations)
spatial scaling, 9–11
Residential movers, 284

S
School mathematics, 367
Science, Technology, Engineering and Mathematics (STEM), 118
Shephard-Metzler block task, 357
Size congruity effect, 86, 90
SketchUp, 301
Solving mathematical tasks, 310
Spatial ability, 118, 119, 121, 140, 356
Spatial analytic reasoning, 197
Spatial and geometric reasoning, 196
Spatial and geometric structuring, 202
Spatial and mathematical thinking
children's conceptions of integers, 356
cognitive skills, 357
geometry, 356
space and mathematics, 356
Spatial arrangement, 86
Spatial biases, 87
Spatial cognition, 357
Spatial extent
encoding, 29
formal measurement and numerical systems, 28
infants, 28
interim conclusions, 31
intuitive proportional reasoning, 28
proportional reasoning, 29, 30
understanding of angles, 30, 31
Spatial grounding, 350
Spatial learning, 4, 6, 19, 20
Spatial orientation, 255, 267, 268, 272
Spatial play, 19
Spatial reasoning, 5, 184–186, 190, 196, 349, 350
assessment, 267
Australia, 271
childhood, 255
concepts, 254
decoding techniques, 265
dynamic transformations, 185
elementary curriculum, 266
in elementary school mathematics, 255
encoding techniques, 262, 264
extrinsic tasks, 184
geometry, 256, 257
graphical information, 264
graphics, 258
intervention and control groups, 270
intrinsic tasks, 184
large-scale spaces, 254
mathematical task information, 265
mathematics, 253
mathematics assessment, 270
mathematics curricula, 256
mathematics tasks, 271
multidimensionality, 254
nonrigid transformations, 186
pencil-and-paper test modes, 267
small-scale manipulation, 255
spatial training, 270
teaching practices, 266
temporal dimension, 183
temporal spatial reasoning, 184
temporal strategy, 188
temporalized conception, 186
testing regime, 266
TIMSS, 266
training, 270, 271
two- and three-dimensional objects, 256
visualization, 258
Spatial representation approach, 341
Spatial representations
adult strategy, 12
children's relational thinking, 12
coding distances and scaling, 12
landmark mapping and perceptual coding, 12
learning, 11
referent, 11
relative magnitude mathematics tasks

Spatial representations (*cont.*)
count-and-match method, 15
discrete *vs.* continuous proportional representations, 14
formulaic thinking, 16
inaccurate counting strategy, 15
manipulatives, 16
metaphors, 17
number line estimation, 13
part-whole diagrams, 15
proportional reasoning, 14
rules/procedures, 16
spatial scaling, 12
Spatial scaling, 4–6, 9–11, 166
Spatial sense, 48, 57, 69
Spatial skill and mathematics, 351, 352
Spatial skill and mathematics performance, 351
Spatial skills, 343, 357, 358
arithmetic strategy choices
addition and subtraction problems, 50
decomposition, 52–53
developmental changes, 55
early grades and later mathematics achievement, 51
gender differences, 53–55
mental imagery, 55
mental strategies, 51
retrieval and decomposition, 51
spatial reasoning skills, 52
visuospatial memory, 52
working memory, 52
cognitive processes, 133
components, 48
cross-loading tasks, 131
EFA/CFA, 129
factor analytic approach, 122
gender differences, 125
mathematical achievement, 47, 49, 50, 127, 128, 132, 133
mathematical concepts *vs.* procedures, 126
mathematical connections, 128
mathematical domain, 121
mathematical factors, 129
mathematical skills, 126, 129
mathematics content areas
forming and mental models, 56
fraction card games, 58–60
fraction magnitudes, mental visual arrays, 57, 58
graphing, 60–62
image generation/visualization, 56
physical instantiations/concrete models/ manipulatives, 57
spatial structuring, 56
translating verbal descriptions (*see* Visual representations)
mathematics curriculum, 138, 139
mental rotation (*see* Mental rotation)
meta-analysis and divisions, 124
multi-faceted cognitive science approach, 124
open questions, 131–132
principles, education, 139, 140
process models, 127, 132
spatial domain, 119
static/dynamic, 125, 126
thinking symbolically *vs.* non-symbolically, 127
visuospatial working memory (*see* Working memory)
Spatial structuring, 202
Spatial thinking, 4, 357
Spatial training, 270
Spatial training programs, 173
Spatial transformations, 235
Spatial visualization, 15, 20, 26, 119, 120, 125, 128, 132, 135, 196, 205, 230, 254–255, 257, 258, 264, 265, 267, 268, 272, 348–349
Spatializing the curriculum, 358
Spatial-Numerical Associations (SNAs), 82, 361
and arithmetic skill, 84, 85, 88, 362
behavioral phenomena, 78
denominator, 368
embodied influences, 83
geometry, 368
grounded influences, 83
implications, 88
modules/representations, 88
multi-digit numbers, 103
school mathematics achievement, 78
single melting pot, 80
situated influences, 83, 84
space and magnitude (*see* Multi-symbol processing)
space and numbers, 79
spatial-numerical trainings, 89
taxonomy (*see* Taxonomy)
Spatial-Numerical Associations of Response Codes (SNARC), 82, 98
Spatial-numerical linked structuring (SNLS)
activation, 211
angles, 211–214
cube building, 221
length, 214
NA and PE, 224

reasoning, 222
rectangular arrays, 217
Spatial-relational reasoning, 6, 9, 11, 17, 20, 173
Square, 234–238, 240
STEM-related subjects, 271
Street Map task, 269
Stretchy ruler, 188
Structural equation modeling, 343
Student KG Problem 1, 160
Student SA Problem 1, 161
Students' visualization behavior, 334
Students' visualization reasoning, 332
Symbolic reasoning, 178, 179
Symbolic thinking, 179
Symbolic understanding, 12
Symmetry, 175, 176, 180, 244
Symmetry task, 269

T
Table tennis conversion top, 290
Tangram pieces, 188
Taxonomy
implicit and explicit coding components, 89
MNL, 89
overview, 90
subcategory approximate
correlations with arithmetic skill, 91, 92
situatedness, 91
trainings, 92, 93
type and paradigms, 90, 91
subcategory cardinalities and operations, explicit coding, 103
subcategory cardinalities, implicit coding
correlations with arithmetic skills, 99
situatedness, 98, 99
trainings, 100
type and paradigms, 98
subcategory exact
correlations with arithmetic skill, 95, 96
situatedness, 95
trainings, 95–98
type and paradigms, 93–95
subcategory operations, implicit coding
correlations with arithmetic skill, 102
situatedness, 102
trainings, 102
type and paradigms, 101
subcategory ordinalities, explicit coding
correlations with arithmetic skill, 103
situatedness, 102
trainings, 103
type and paradigms, 102
subcategory ordinalities, implicit coding
correlations with arithmetic skill, 101
situatedness, 100, 101
trainings, 101
type and paradigms, 100
Teaching mathematics, 310
Test of Early Mathematics Achievement (TEMA-3), 18
The Geometer's Sketchpad and *Cabri-géomètre*, 185
Three-dimensional geometry, 301–302
Thumb-and-forefinger "pinching" gesture, 39
Tracing, 240
Training interventions
spatial visualization of linear measurement units, 39, 40
spatial visualization, angular measurement, 40, 41
teaching area measurement, 40
Trajectory of solution, 332
Transformational-saliency hypothesis, 185, 187
Transformations and measurement, 243
Transformation-saliency hypothesis, 190
Transition, conventional measurement
angular measurement, 37, 38
area measurement, 36
children to master, 31, 32
cognitive biases, 32, 33
higher order measurement skills, 35, 36
instructional needs, 34
interim conclusions, 38
international performance, 34, 35
proportions and fractions, 36, 37
traditional classroom instruction, 33, 34
Trends in International Mathematics and Science Study (TIMSS), 266
Triangles, 234, 243, 245, 246, 250
Two-digit number comparison, 88

U
Unaligned ruler measurement problems, 32
Units
concept, 26
instructional techniques, 26
intuitive spatial understanding, 27
learned numerical representations, 27
measurement tools, 27
ontogenetic development, 26
transitive inference, 27
young children and infants, 27, 31, 42

V

Vandenberg Mental 3D Rotation Test, 205
Virtual workshop, 311
Visual configuration, 202
Visual images and spatial mental models, 200
Visual representations
  2-d/3-d representation
    barrier games, 63, 64
    instructional activities, 67, 68
    mapping games, 65, 66
    mental images, 66, 67
    monster games, 64, 65
    orienteering, drawing and mapping routes, grid based, 65, 66
    surface, solids and cross-sections, 69, 70
  geometry learners, 62, 63
  image generation, 62
  pictorial representations, 62
  spatial skills and word problem solving, 62
Visualization
  abilities, 311
  and arithmetic skills, 167
  characteristics, 315
  characterization, 316
  cognitive demand, 312
  communication, 314
  definitions, 311
  educational psychologists and mathematics educators, 167
  educational psychology, 166–168
  educational psychology and mathematics education, 168
  elements, 315
  external representations, 311
  figure-ground perception, ability of, 321, 322, 327
  in school mathematics, 165
  learning numbers, 166
  learning processes, 167
  mathematics education, 167
  mathematization, 311
  mental images, 311
  middle primary school grades, 166
  perceptual constancy, ability of, 322, 327
  positions in space, ability of, 321, 323–327, 334, 335
  school mathematics, 165
  spatial relationships, ability of, 323, 327
  visual discrimination, ability of, 323, 327, 334–335
Visuospatial skills, 367
Volume, 222, 225–226

W

Weiming Chair Task, 262, 263
Wheatley's Quick Draw Program, 234
Word-based probability task, 261
Working memory
  attentional processes, 136
  discarding irrelevant information, 137
  mathematics, 137, 138
  modality specificity, 136
Working Memory Test Battery for Children, 52

CPSIA information can be obtained
at www.ICGtesting.com
Printed in the USA
BVHW011304171218
535800BV00001B/1/P

9 783319 987668